Nanoscale Science and Technology

Nanoscale Science and Technology

Edited by

Robert W. Kelsall
The University of Leeds, UK

Ian W. Hamley
The University of Leeds, UK

and

Mark Geoghegan
The University of Sheffield, UK

John Wiley & Sons, Ltd

Other Wiley Editorial Offices

John Wiley & Sons Inc., 111 River Street, Hoboken, NJ 07030, USA

Jossey-Bass, 989 Market Street, San Francisco, CA 94103-1741, USA

Wiley-VCH Verlag GmbH, Boschstr. 12, D-69469 Weinheim, Germany

John Wiley & Sons Australia Ltd, 33 Park Road, Milton, Queensland 4064, Australia

John Wiley & Sons (Asia) Pte Ltd, 2 Clementi Loop #02-01, Jin Xing Distripark, Singapore 129809

John Wiley & Sons Canada Ltd, 22 Worcester Road, Etobicoke, Ontario, Canada M9W 1L1

Library of Congress Cataloging in Publication Data

Nanoscale science and technology / edited by Robert W. Kelsall,
    Ian W. Hamley, Mark Geoghegan.
       p.   cm.
    ISBN 0-470-85086-8 (cloth : alk. paper)
    1. Nanotechnology.   2. Nanoscience.   3. Nanostructured materials—Magnetic properties.
    I. Kelsall, Robert W.   II. Hamley, Ian W.   III. Geoghegan, Mark.
    T174.7.N358 2005
    620′.5—dc22
                                                                                    2004016224

British Library Cataloguing in Publication Data

A catalogue record for this book is available from the British Library

ISBN 0-470-85086-8 (HB)

Typeset in 10/12pt Times by Integra Software Services Pvt. Ltd, Pondicherry, India
Printed and bound in Great Britain by Antony Rowe Ltd, Chippenham, Wiltshire
This book is printed on acid-free paper responsibly manufactured from sustainable forestry in which
at least two trees are planted for each one used for paper production.

Contents

List of contributors

EDITORS

Dr Robert W. Kelsall
Institute of Microwaves and Photonics
School of Electronic and Electrical
Engineering
University of Leeds
Leeds LS2 9JT
United Kingdom
r.w.kelsall@leeds.ac.uk

Dr Ian W. Hamley
Centre for Self Organising Molecular Systems
University of Leeds
Leeds LS2 9JT
United Kingdom
I.W.Hamley@chemistry.leeds.ac.uk

Dr Mark Geoghegan
Department of Physics and Astronomy
University of Sheffield
Sheffield S3 7RH
United Kingdom
mark.geoghegan@sheffield.ac.uk

AUTHORS

Dr Rik Brydson
Institute for Materials Research
School of Process, Environmental and
Materials Engineering
University of Leeds
Leeds LS2 9JT
United Kingdom
mtlrmdb@leeds.ac.uk

Prof. Mike R. J. Gibbs
Department of Engineering Materials
University of Sheffield
Sheffield S1 3JD
United Kingdom
M.R.Gibbs@Sheffield.ac.uk

Dr Martin Grell
Department of Physics and
Astronomy
University of Sheffield
Sheffield S3 7RH
United Kingdom
m.grell@sheffield.ac.uk

Dr Chris Hammond
Institute for Materials Research
School of Process, Environmental and
Materials Engineering
University of Leeds
Leeds LS2 9JT
United Kingdom
c.hammond@leeds.ac.uk

Prof. Richard Jones
Department of Physics and
Astronomy
Hicks Building
University of Sheffield
Sheffield S3 7HF
United Kingdom
r.a.l.jones@sheffield.ac.uk

Prof. Graham Leggett
Department of Chemistry
University of Sheffield
Sheffield S3 7HF
United Kingdom
graham.leggett@umist.ac.uk

Dr David Mowbray
Department of Physics and Astronomy
University of Sheffield
Sheffield S3 7RH
United Kingdom
d.mowbray@sheffield.ac.uk

Dr Iain Todd
Department of Engineering Materials
University of Sheffield
Sheffield S1 3JD
United Kingdom
i.todd@sheffield.ac.uk

Preface

In the two years since we first started planning this book, so much has been written about nanotechnology that the subject really needs no introduction. Nanotechnology has been one of the first major new technologies to develop in the internet age, and as such has been the topic of thousands of unregulated, unrefereed websites, discussion sites and the like. In other words, much has been written, but not all is necessarily true. The press has also made its own, unique contribution: 'nanotechnology will turn us all into grey goo' makes for a good story (in some newspapers at least), and then there's the 1960s image of nanotechnology, still present today, of Raquel Welch transported in a nanosubmarine through the bloodstream of an unsuspecting patient. This book isn't about *any* of that! One thing that the recent press coverage of nanotechnology has achieved is to draw attention to the possible hazards which accompany any new technology and to pose relevant questions about the likely impact of the various facets of nanotechnology on our society. Whilst we would certainly encourage investigation and discussion of such issues, they do not fall within the remit of this book.

Nanoscale Science and Technology has been designed as an educational text, aimed primarily at graduate students enrolled on masters or PhD programmes, or indeed, at final year undergraduate or diploma students studying nanotechnology modules or projects. We should also mention that the book has been designed for students of the physical sciences, rather than the life sciences. It is based largely on our own masters course, the Nanoscale Science and Technology MSc, which has been running since 2001 and was one of the first postgraduate taught courses in Europe in this subject area. The course is delivered jointly by the Universities of Leeds and Sheffield, and was designed primarily by several of the authors of this book. As in designing the course, so in designing the book have we sought to present the breadth of scientific topics and disciplines which contribute to nanotechnology. The scope of the text is bounded by two main criteria. Firstly, we saw no need to repeat the fine details of established principles and techniques which are adequately covered elsewhere, and secondly, as a textbook, *Nanoscale Science and Technology* is intended to be read, in its entirety, over a period of one year. In consideration of the first of these criteria, each chapter has a bibliography indicating where more details of particular topics can be found.

The expertise of the authors ranges from electronic engineering, physics and materials science to chemistry and biochemistry, which we believe has helped us achieve both breadth and balance. That said, this book is inevitably our take on nanotechnology, and any other group of authors would almost certainly have a different opinion on what should be included and what should be emphasised. Also, in such a rapidly developing field, our reporting is in danger of fast becoming out of date (one of our co-authors, who was the most efficient in composing his text, paid the rather undeserved penalty of having to make at least two sets of revisions simply to update facts and figures to reflect new progress in research). We should certainly be grateful to receive any information on errors or omissions.

Although most of the chapters have been written by different authors, we were keen that, to better fulfil its role as a textbook, this volume should read as one coherent whole rather than as a collection of individual monographs. To this end, not only have we as editors made numerous adjustments to improve consistency, and avoid duplication and omission, but in some places we have also made more substantial editorial changes. We should like to acknowledge the tolerance of our co-authors throughout this process. We are all still on speaking terms – just! It is not really necessary for us to tabulate in detail exactly who contributed what to each chapter in the final manuscript, except that we note that the nanostructured carbon section in Chapter 6 was provided by Rob Kelsall. Finally, we should like to acknowledge Terry Bambrook, who composed virtually all of the figures for chapters 1 and 2.

Robert W. Kelsall, Ian W. Hamley and Mark Geoghegan

Book cover acknowledgments

The nano images of silicon were taken by Dr Ejaz Huq and appear courtesy of the CCLRC Rutherford Appleton Laboratory Central Microstructure Facility; the images of carbon nanotubes appears courtesy of Z. Aslam, B. Rand and R. Brydson (University of Leeds); the image of a templated silica nanotube appears courtesy of J. Meegan, R. Ansell and R. Brydson (University of Leeds); the image of microwires is taken from E. Cooper, R. Wiggs, D. A. Hutt, L. Parker, G. J. Leggett and T. L. Parker, *J. Mater. Chem.* **7**, 435–441 (1997), reproduced by permission of the Royal Society of Chemistry, and the AFM images of block copolymers are adapted with permission from T. Mykhaylyk, O. O. Mykhaylyk, S. Collins and I. W. Hamley, *Macromolecules* **37**, 3369 (2004), copyright 2004 American Chemical Society.

Chapter authors

Chapter 1. Generic methodologies for nanotechnology: classification and fabrication
Rik M. Brydson and Chris Hammond

Chapter 2. Generic methodologies for nanotechnology: characterisation
Rik M. Brydson and Chris Hammond

Chapter 3. Inorganic semiconductor nanostructures
David Mowbray

Chapter 4. Nanomagnetic materials and devices
Mike R. J. Gibbs

Chapter 5. Processing and properties of inorganic nanomaterials
Iain Todd

Chapter 6. Electronic and electro-optic molecular materials and devices
Martin Grell

Chapter 7. Self-assembling nanostructured molecular materials and devices
Ian W. Hamley

Chapter 8. Macromolecules at interfaces and structured organic films
Mark Geoghegan and Richard A. L. Jones

Chapter 9. Bionanotechnology
Graham J. Leggett and Richard A. L. Jones

1

Generic methodologies for nanotechnology: classification and fabrication

1.1 INTRODUCTION AND CLASSIFICATION

1.1.1 What is nanotechnology?

Nanotechnology is the term used to cover the design, construction and utilization of functional structures with at least one characteristic dimension measured in nanometres. Such materials and systems can be designed to exhibit novel and significantly improved physical, chemical and biological properties, phenomena and processes as a result of the limited size of their constituent particles or molecules. The reason for such interesting and very useful behaviour is that when characteristic structural features are intermediate in extent between isolated atoms and bulk macroscopic materials; i.e., in the range of about 10^{-9} m to 10^{-7} m (1 to 100 nm), the objects may display physical attributes substantially different from those displayed by either atoms or bulk materials. Ultimately this can lead to new technological opportunities as well as new challenges.

1.1.2 Classification of nanostructures

As we have indicated above, a reduction in the spatial dimension, or confinement of particles or quasiparticles in a particular crystallographic direction within a structure generally leads to changes in physical properties of the system in that direction. Hence one classification of nanostructured materials and systems essentially depends on the number of dimensions which lie within the nanometre range, as shown in Figure 1.1: (a) systems confined in three dimensions, (b) systems confined in two dimensions, (c) systems confined in one dimension.

Nanoscale Science and Technology   Edited by R. W. Kelsall, I. W. Hamley and M. Geoghegan
© 2005 John Wiley & Sons, Ltd

(a)

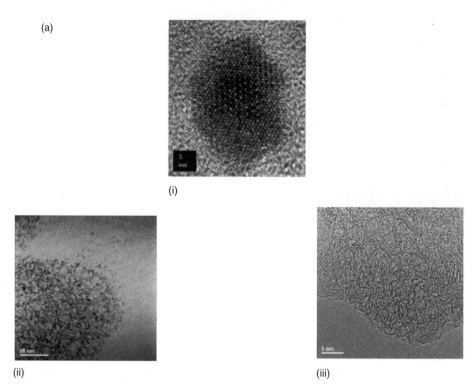

(i)

(ii)   (iii)

Figure 1.1   Classification of nanostructures. (a) Nanoparticles and nanopores (nanosized in three dimensions): (i) high-resolution TEM image of magnetic iron oxide nanoparticle, (ii) TEM image of ferritin nanoparticles in a liver biopsy specimen, and (iii) high-resolution TEM image of nanoporosity in an activated carbon). (b) Nanotubes and nanofilaments (nanosized in two dimensions): (i) TEM image of single-walled carbon nanotubes prepared by chemical vapour deposition, (ii) TEM image of ordered block copolymer film, and (iii) SEM image of silica nanotube formed via templating on a tartaric acid crystal. (c) Nanolayers and nanofilms (nanosized in one dimension): (i) TEM image of a ferroelectric thin film on an electrode, (ii) TEM image of cementite (carbide) layers in a carbon steel, and (iii) high-resolution TEM image of glassy grain boundary film in an alumina polycrystal. Images courtesy of Andy Brown, Zabeada Aslam, Sarah Pan, Manoch Naksata and John Harrington, IMR, Leeds

Nanoparticles and nanopores exhibit three-dimensional confinement (note that historically pores below about 100 nm in dimension are often sometimes confusingly referred to as micropores). In semiconductor terminology such systems are often called quasi-zero dimensional, as the structure does not permit free particle motion in any dimension.

Nanoparticles may have a random arrangement of the constituent atoms or molecules (e.g., an amorphous or glassy material) or the individual atomic or molecular units may be ordered into a regular, periodic crystalline structure which may not necessarily be the same as that which is observed in a much larger system (Section 1.3.1). If crystalline, each nanoparticle may be either a single crystal or itself composed of a number of different crystalline regions or grains of differing crystallographic orientations (i.e., polycrystalline) giving rise to the presence of associated grain boundaries within the nanoparticle.

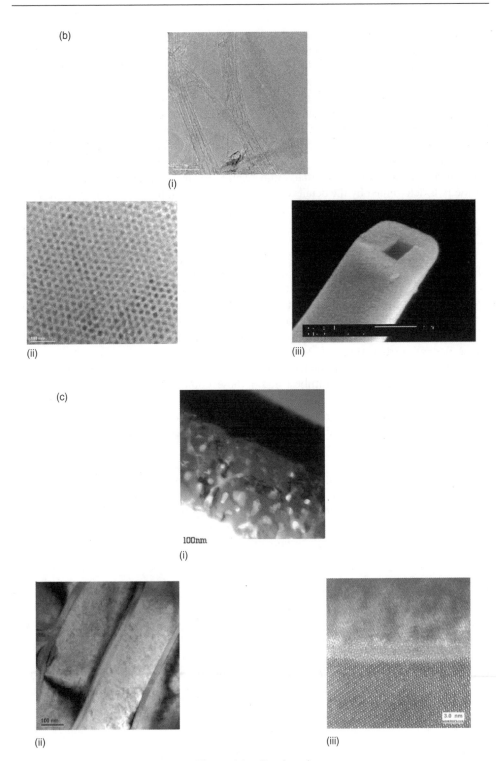

Figure 1.1 Continued

Nanoparticles may also be quasi-crystalline, the atoms being packed together in an icosahedral arrangement and showing non-crystalline symmetry characteristics. Such quasi-crystals are generally only stable at the nanometre or, at most, the micrometre scale.

Nanoparticles may be present within another medium, such as nanometre-sized precipitates in a surrounding matrix material. These nanoprecipitates will have a specific morphology (e.g., spherical, needle-shaped or plate-shaped) and may possess certain crystallographic orientation relationships with the atomic arrangement of the matrix depending on the nature (coherency) of the interface which may lead to coherency strains in the particle and the matrix. One such example is the case of self-assembled semiconductor quantum dots, which form due to lattice mismatch strain relative to the surrounding layers and whose geometry is determined by the details of the strain field (Chapter 3). Another feature which may be of importance for the overall transport properties of the composite system is the connectivity of such nanometre-sized regions or, in the case of a nanoporous material, nanopore connectivity.

In three dimensions we also have to consider collections of consolidated nanoparticles; e.g., a nanocrystalline solid consisting of nanometre-sized crystalline grains each in a specific crystallographic orientation. As the grain size d of the solid decreases the proportion of atoms located at or near grain boundaries, relative to those within the interior of a crystalline grain, scales as $1/d$. This has important implications for properties in ultrafine-grained materials which will be principally controlled by interfacial properties rather than those of the bulk.

Systems confined in two dimensions, or quasi-1D systems, include nanowires, nanorods, nanofilaments and nanotubes: again these could either be amorphous, single-crystalline or polycrystalline (with nanometre-sized grains). The term 'nanoropes' is often employed to describe bundles of nanowires or nanotubes.

Systems confined in one dimension, or quasi-2D systems, include discs or platelets, ultrathin films on a surface and multilayered materials; the films themselves could be amorphous, single-crystalline or nanocrystalline.

Table 1.1 gives examples of nanostructured systems which fall into each of the three categories described above. It can be argued that self-assembled monolayers and multi layered Langmuir–Blodgett films (Section 1.4.3.1) represent a special case in that they represent a quasi-2D system with a further nanodimensional scale within the surface film caused by the molecular self-organization.

1.1.3 Nanoscale architecture

Nanotechnology is the design, fabrication and use of nanostructured systems, and the growing, shaping or assembling of such systems either mechanically, chemically or biologically to form nanoscale architectures, systems and devices. The original vision of Richard Feynman[1] was of the 'bottom-up' approach of fabricating materials and devices at the atomic or molecular scale, possibly using methods of self-organization and self-assembly of the individual building blocks. An alternative 'top-down' approach is the

---

[1] R. Feynman, There's plenty of room at the bottom, *Eng. Sci.* **23**, 22 (1960) reprinted in *J. Micromech Systems* **1**, 60 (1992).

Table 1.1 Examples of reduced-dimensionality systems

3D confinement
Fullerenes
Colloidal particles
Nanoporous silicon
Activated carbons
Nitride and carbide precipitates in high-strength low-alloy steels
Semiconductor particles in a glass matrix for non-linear optical components
Semiconductor quantum dots (self-assembled and colloidal)
Quasi-crystals

2D confinement
Carbon nanotubes and nanofilaments
Metal and magnetic nanowires
Oxide and carbide nanorods
Semiconductor quantum wires

1D confinement
Nanolaminated or compositionally modulated materials
Grain boundary films
Clay platelets
Semiconductor quantum wells and superlattices
Magnetic multilayers and spin valve structures
Langmuir–Blodgett films
Silicon inversion layers in field effect transistors
Surface-engineered materials for increased wear resistance or corrosion resistance

---

ultraminiaturization or etching/milling of smaller structures from larger ones. These methods are reviewed in Section 1.4. Both approaches require a means of visualizing, measuring and manipulating the properties of nanostructures; computer-based simulations of the behaviour of materials at these length scales are also necessary. This chapter provides a general introduction to the preparation and properties of nanostructures, whilst the subsequent chapters give greater detail on specific topics.

1.2 SUMMARY OF THE ELECTRONIC PROPERTIES OF ATOMS AND SOLIDS

To understand the effects of dimensionality in nanosystems, it is useful to review certain topics associated with the constitution of matter, ranging from the structure of the isolated atom through to that of an extended solid.

1.2.1 The isolated atom

The structure of the atom arises as a direct result of the wave–particle duality of electrons, which is summarized in the de Broglie relationship, $\lambda = h/m_\mathrm{e}v$, where λ is the (electron) wavelength, m_e is the (electron) mass, v is the velocity and

$h = 6.63 \times 10^{-34}$ J s is the Planck constant. The wave–particle duality of the electron means that an electron behaves both as a wave (i.e., it is extended over space and has a wavelength and hence undergoes wave-like phenomena such as diffraction) and a particle (i.e., it is localized in space and has a position, a velocity and a kinetic energy). This is conveniently summarized in the idea of a wave packet a localized wave that is effectively the summation of a number of different waves of slightly differing wavelengths.

Using these ideas we come to our first model of the atom, the Rutherford–Bohr model. Here the small central nucleus of the atom consists of positively charged protons and (neutral) neutrons. Electrons orbit the nucleus in stable orbits. The allowed, stable orbits are those in which the electron wavelength, given by the de Broglie formula, is an integral multiple n of the circumference of the orbit r:

$$2\pi r = n\lambda = \frac{nh}{m_e v}. \tag{1.1}$$

This implies that

$$m_e vr = \frac{nh}{2\pi}, \tag{1.2}$$

in otherwords, the angular momentum $m_e vr$ is quantized in that it is an integral multiple of $h/2\pi$.

The Bohr model leads to the idea that only certain electron orbits or shells are allowed by this quantization of angular momentum (i.e., the value of n). The Bohr shells in an atom are labelled according to the quantum number, n, and are given the spectroscopic labels K, L, M, N, etc. (where $n = 1, 2, 3, 4, \ldots$). To understand the form of the periodic table of elements, it is necessary to assume that each Bohr shell can contain $2n^2$ electrons. For instance, a K shell ($n = 1$) can contain 2 electrons, whereas an L shell ($n = 2$) can accommodate 8 electrons. As well as having a distinct form and occupancy, each shell also has a corresponding well-defined energy. It is usual to define the zero of the energy scale (known as the vacuum level) as the potential energy of a free electron far from the atom. In order to correspond with atomic emission spectra measured experimentally, the energies of these levels E_n are then negative (i.e., the electrons are bound to the atom) and are proportional to $1/n^2$. Such a simplified picture of the structure of an isolated Mg atom and the associated energy level diagram are shown in Figure 1.2.

A much more sophisticated model of the atom considers the wave-like nature of the electrons from the very beginning. This uses wave mechanics or quantum mechanics.

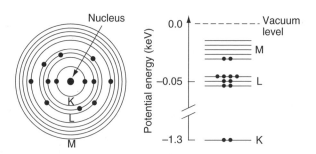

Figure 1.2   Bohr shell description of an Mg atom and the associated energy level diagram

Here each electron is described by a wavefunction ψ which is a function of spatial position (x,y,z) and, in general, of time. Physically $|\psi|^2$ represents the probability of finding the electron at any point. To work out the energy of each electron, we need to solve the Schrödinger equation which, in the time-independent case, takes the form

$$-\frac{\hbar^2}{2m_e}\nabla^2\psi + V(x,y,z)\psi = E\psi, \tag{1.3}$$

where $V(x,y,z)$ describes the potential energy function in the environment of the electron. Solution of the Schrödinger equation, under certain boundary conditions, leads to a set of solutions for the allowed wavefunctions ψ_n of the atomic electrons together with their associated energies E_n.

This equation can only be solved analytically for the case of the hydrogen atom, where there is only one electron moving in the potential of a single proton, the hydrogen nucleus. Only a certain set of electronic wavefunctions and associated energy levels fulfil this Schrödinger equation. The wavefunctions may be expressed as a radial part, governing the spatial extent of the wavefunction, multiplied by a spherical harmonic function which determines the shape. The allowed wavefunctions form the electron orbitals, which we term 1s, 2s, 2p, 3s, 3p, 3d, etc. (here $1, 2, 3, \ldots$ are alternative labels for $K, L, M, \ldots$). These allowed wavefunctions now depend on not just one quantum number but four: n, l, m and s. These numbers may be summarized as follows:

- n is the principal quantum number; it is like the quantum number used for the case of Bohr shells ($n = 1, 2, 3, \ldots$).

- l is the angular momentum quantum number; it can vary from $l = 0, 1, 2, \ldots, (n-1)$. The value of l governs the orbital shape of the subshell: $l = 0$ is an s orbital, which is spherical; $l = 1$ is a p orbital, which has a dumbbell shape; while $l = 2$ is a d orbital, which has a more complex shape such as a double dumbbell.

- m is the magnetic quantum number; it can vary from $m = 0, \pm1, \ldots, \pm l$. The value of m governs the spatial orientation of the different orbitals within a subshell; i.e., there are three p orbitals ($l = 1$) p_x, p_y, and p_z corresponding to the three values of m which are 0, $+1$ and -1. In the absence of a magnetic field, all these orbitals within a particular subshell will have the same energy.

- s is the spin quantum number which, for an electron, can take the values $\pm1/2$. Each (n, l, m) orbital can contain two electrons of opposite spin due to the Pauli exclusion principle, which states that no two electrons can have the same four quantum numbers.

Using this identification in terms of the quantum numbers, each electron orbital in an atom therefore has a distinct combination of energy, shape and direction (x, y, z) and can contain a maximum of two electrons of opposite spin.

In an isolated atom, these localized electronic states are known as Rydberg states and may be described in terms of simple Bohr shells or as combinations of the three quantum numbers n, l and m known as electron orbitals. The Bohr shells (designated $K, L, M, \ldots$) correspond to the principal quantum numbers n equal to 1, 2, 3, etc. Within each of these shells, the electrons may exist in $(n-1)$ subshells (i.e., s, p, d, or f subshells, for which the angular momentum quantum number l equals 0, 1, 2, 3, respectively).

The occupation of the electronic energy levels depends on the total number of electrons in the atom. In the hydrogen atom, which contains only one electron, the set of Rydberg states is almost entirely empty except for the lowest-energy 1s level which is half full. As we go to higher energies, the energy spacing between these states becomes smaller and smaller and eventually converges to a value known as the vacuum level ($n = \infty$), which corresponds to the ionization of the inner-shell electron. Above this energy the electron is free of the atom and this is represented by a continuum of empty electronic states. In fact, the critical energy to ionize a single isolated hydrogen atom is equal to 13.61 eV and this quantity is the Rydberg constant.

This description is strictly only true for hydrogen; however, other heavier atoms are found to have similar wavefunction (hydrogenic-like) solutions, which ultimately leads to the concept of the periodic table of elements, as each atom has more and more electrons which progressively fill the allowed energy levels. This is shown for a magnesium atom in Figure 1.2. The chemical properties of each atom are then principally determined by the number of valence electrons in the outermost electron shell which are relatively loosely bound and available for chemical reaction with other atomic species.

1.2.2 Bonding between atoms

One way to picture the bonding between atoms is to use the concept of Molecular Orbital (MO) Theory. MO theory considers the electron wavefunctions of the individual atoms combining to form molecular wavefunctions (or molecular orbitals as they are known). These orbitals, which are now delocalized over the whole molecule, are then occupied by all the available electrons from all the constituent atoms in the molecule. Molecular orbitals are really only formed by the wavefunctions of the electrons in the outermost shells (the valence electrons); i.e., those which significantly overlap in space as atoms become progressively closer together; the inner electrons remain in what are essentially atomic orbitals bound to the individual atoms.

A simple one-electron molecule is the H_2^+ ion, where we have to consider the interactions (both attractive and repulsive) between the single electron and two nuclei. The Born–Oppenheimer approximation regards the nuclei as fixed and this simplifies the Hamiltonian used in the Schrödinger equation for the molecular system. For a one-electron molecule, the equation can be solved mathematically, leading to a set of molecular wavefunctions ψ which describe molecular orbitals and depend on a quantum number λ which specifies the angular momentum about the internuclear axis. Analogous to the classification of atomic orbitals (AOs) in terms of angular momentum l as s, p, d, etc., the MOs may be classified as σ, π, δ depending on the value of λ ($\lambda = 0, 1, 2$, respectively). Very simply a σ MO is formed from the overlap (actually a linear combination) of AOs parallel to the bond axis, whereas a π MO results from the overlap of AOs perpendicular to the bond axis. For the H_2^+ ion, the two lowest-energy solutions are known as $1s\sigma_g$ and $1s\sigma_u$. Here 1s refers to the original atomic orbitals; the subscripts g and u refer to whether the MO is either symmetrical or non-symmetrical with respect to inversion about a line drawn between the nuclei (viz. an even or odd mathematical function). This is shown in figure 1.3.

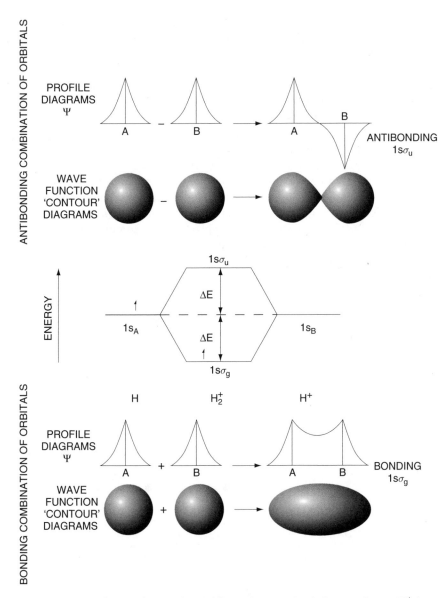

Figure 1.3 Molecular orbital description and energy level diagram for an H_2^+ ion

As can be seen the electron density is concentrated between the nuclei for the $1s\sigma_g$ MO, which is known as a bonding orbital since the energy of the molecular wavefunction is lower (i.e., more stable) than the corresponding isolated atomic wavefunctions. Conversely, the electron density is diminished between the nuclei for $1s\sigma_u$, which is known as an antibonding orbital since the energy of the molecular wavefunction is higher (i.e., less stable) than the corresponding isolated atomic wavefunction.

More generally, it is necessary to be able to solve the Schrödinger equation for molecules containing more than one electron. One way to do this is to use approximate solutions similar to those obtained for the hydrogen atom, since when an electron is near a particular nucleus it will have a hydrogen-like form. Using this approach we can then construct a set of molecular orbitals from a linear combination of atomic orbitals (LCAO). For instance, as shown in Figure 1.4, the $1s\sigma_g$ bonding MO is formed from the in-phase overlap (i.e., addition) of two 1s atomic orbitals, whereas the $1s\sigma_u$ antibonding MO is formed from the out of-phase overlap (i.e., subtraction) of two 1s atomic orbitals. Similar considerations apply to overlap of p orbitals, although now these may form both σ and π bonding and antibonding MOs.

The stability of simple diatomic molecules such as H_2, H_2^- and He_2 depends on the relative filling of the bonding and antibonding MOs; e.g., H_2^- contains three electrons, two of which fill the bonding MO ($1s\sigma_g$ level) while the third enters the antibonding MO ($1s\sigma_u$ level); consequently, the overall bond strength is approximately half that in H_2. Meanwhile He_2 is unstable as there are an equal number of electrons in bonding MOs as in antibonding MOs. The same principles apply to more complicated diatomic molecules. However, if the atoms are different then the energy levels of the electrons associated with the constituent atoms will also be different and this will lead to an asymmetry in the MO energy level diagram.

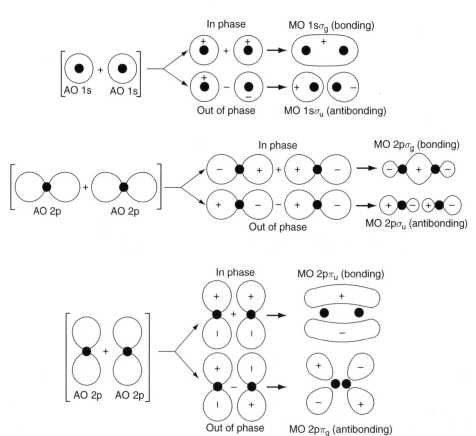

Figure 1.4 Formation of molecular orbitals from a linear combination of atomic orbitals; the + and − signs indicate the signs (phases) of the wavefunctions

For polyatomic molecules such as BF_3 a greater variety of molecular orbitals can be formed. MO theory emphasizes the delocalized nature of the electron distribution, so in general these MOs are extended over not just two, but all the constituent atoms. The total number of MOs (bonding, antibonding or non-bonding) is equal to the number of valence atomic orbitals used to construct them.

1.2.3 Giant molecular solids

When atoms come into close proximity with other atoms in a solid, most of the electrons remain localized and may be considered to remain associated with a particular atom. However, some outer electrons will become involved in bonding with neighbouring atoms. Upon bonding the atomic energy level diagram is modified. Briefly, the well-defined outer electron states of the atom overlap with those on neighbouring atoms and become broadened into energy bands. One convenient way of picturing this is to envisage the solid as a large molecule. Figure 1.5 shows the effect of increasing the number of atoms on the electronic energy levels of a one-dimensional solid (a linear chain of atoms).

For a simple diatomic molecule, as discussed previously, the two outermost atomic orbitals (AOs) overlap to produce two molecular orbitals (MOs) which can be viewed as a linear combination of the two constituent atomic orbitals. As before, the bonding MO is formed from the in-phase overlap of the AOs and is lower in energy than the corresponding AOs, whereas the other MO, formed from the out-of-phase overlap, is higher in energy than the corresponding AOs and is termed an antibonding MO. Progressively increasing the length of the molecular chain increases the total number of MOs, and gradually these overlap to form bands of allowed energy levels which are separated by forbidden energy regions (band gaps). These band gaps may be thought of as arising from the original energy gaps between the various atomic orbitals of the isolated atoms.

Note that the broadening of atomic orbitals into energy bands as the atoms are brought closer together to form a giant molecular solid can sometimes result in the overlapping of energy bands to give bands of mixed (atomic) character. The degree to which the orbitals are concentrated at a particular energy is reflected in a quantity known as the density of states (DOS) $N(E)$, where $N(E)\,dE$ is the number of allowed

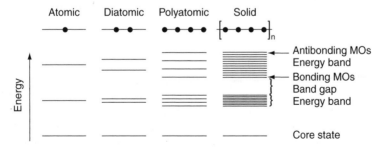

Figure 1.5   Electron energy level diagram for a progressively larger linear chain of atoms showing the broadening of molecular orbitals into energy bands for a one-dimensional solid

energy levels per unit volume of the solid in the energy range between E and $E + dE$. As in a simple molecule, each MO energy level in the energy band can accommodate two electrons of opposite spin. The total number of electrons from all the interacting atomic orbitals in the large molecule fill this set of MOs, the highest occupied energy level being known as the Fermi level E_F. The sum of the energies of all the individual electrons in the large molecule gives the total energy of the system, which gives a measure of the stability of the atomic arrangement in terms of the system free energy.

1.2.4 The free electron model and energy bands

An alternative view of the electronic band structure of solids is to consider the electron waves in a periodic crystalline potential. The starting point for this approach is the Drude–Lorentz free electron model for metals. In this model a metallic solid is considered as consisting of a close packed lattice of positive cations surrounded by an electron sea or cloud formed from the ionization of the outer shell (valence) electrons. We can then treat the valence electrons as if they were a gas inside a container and use classical kinetic gas theory. This works best for the electropositive metals of Groups I and II as well as aluminium (the so-called free electron metals) and can explain many of the fundamental properties of metals such as high electrical and thermal conductivities, optical opacity, reflectivity, ductility and alloying properties.

However, a more realistic approach is to treat the free electrons in metals quantum mechanically and consider their wave-like properties. Here the free valence electrons are assumed to be constrained within a potential well which essentially stops them from leaving the metal (the 'particle-in-a-box' model). The box boundary conditions require the wavefunctions to vanish at the edges of the crystal (or 'box'). The allowed wavefunctions given by the Schrödinger equation then correspond to certain wavelengths as shown in Figure 1.6. For a one-dimensional box of length L, the permitted wavelengths are $\lambda_n = 2L/n$, where $n = 1, 2, 3 \ldots$ is the quantum number of the state; the permitted wavevectors $k_n = 2\pi/\lambda$ are given by $k_n = n\pi/L$.

This simple particle-in-a-box model results in a set of wavefunctions given by

$$\psi_n = (2/L)^{1/2} \sin(n\pi x/L), \tag{1.4}$$

where $n = 1, 2, 3 \ldots$, and for each n the corresponding energy of the electronic level is

$$E_n = \frac{n^2 h^2}{8mL^2}. \tag{1.5}$$

E_n represents solely kinetic energy since the potential energy is assumed to be zero within the box. Thus there is a parabolic relationship between E_n and n, and therefore between E_n and k since k depends directly on n as described above. The permitted energy levels on this parabola are discrete (i.e., quantized): however in principle the size of L for most metal crystals (ranging from microns to millimetres or even centimetres) means that the separation between levels is very small compared with the thermal energy $k_B T$,

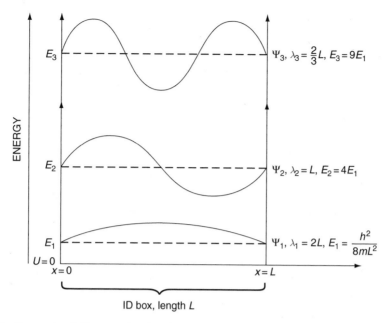

Figure 1.6   Energy level diagram also showing the form of some of the allowed wavefunctions for an electron confined to a one-dimensional potential well

and we can regard the energy distribution as almost continuous (quasi-continuous) so that the levels form a band of allowed energies as shown in figure 1.7.

Note that as the electron becomes more localized (i.e., L decreases), the energy of a particular electron state (and more importantly the spacing between energy states) increases; this has important implications for bonding and also for reduced-dimensionality or quantum-confined systems which are discussed later.

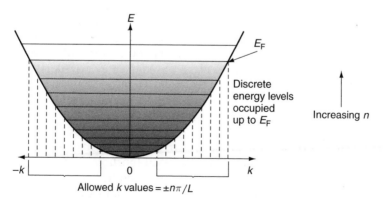

Figure 1.7   Schematic version of the parabolic relationship between the allowed electron wave vectors and the their energy for electrons confined to a one-dimensional potential well. Shaded energy regions represent those occupied with electrons

1.2.5 Crystalline solids

The above arguments may be extended from one to three dimensions to consider the electronic properties of bulk crystalline solids. For a perfectly ordered three dimensional crystal, the periodic repetition of atoms (or molecules) along the one dimensional linear chain considered in Section 1.2.2 is replaced by the periodic repetition of a *unit cell* in all three dimensions. The unit cell contains atoms arranged in the characteristic configuration of the crystal, such that contiguous replication of the unit cell throughout all space is sufficient to generate the entire crystal structure. In otherwords, the crystal has translational symmetry, and the crystal structure may be generated by translations of the unit cell in all three dimensions. Translation symmetry in a periodic structure is a so-called discrete symmetry, because only certain translations – those corresponding to integer multiples of the *lattice translation vectors* derived from the unit cell – lead to symmetry-equivalent points. (This may be contrasted with the case of empty space, which displays a continuous translation symmetry because *any* translation leads to a symmetry-equivalent point.) Common unit cells are simple cubic, face centred cubic, body centred cubic, and the diamond structure, which comprises two interlocking face-centred cubic lattices. However, in general, the lattice spacing may be different along the different principal axes, giving rise to the orthorhombic and tetragonal unit cells, and sides of the unit cell may not necessarily be orthogonal, such as in the hexagonal unit cell (refer to the Bibliography for further reading on this topic).

Generally, symmetries generate conservation laws; this is known as *Noether's theorem*. The continuous translation symmetry of empty space generates the law of momentum conservation; the weaker discrete translation symmetry in crystals leads to a weaker quasi-conservation law for quasi- or crystal momentum. An important consequence of discrete translation symmetry for the electronic properties of crystals is *Bloch's theorem*, which is described below.

1.2.6 Periodicity of crystal lattices

The three dimensional periodicity of the atomic arrangement in a crystal gives rise to a corresponding periodicity in the internal electric potential due to the ionic cores. Incorporating this periodic potential into the Schrödinger equation results in allowed wavefunctions that are modulated by the lattice periodicity. Bloch's theorem states that these wavefunctions take the form of a plane wave (given by $\exp(i\mathbf{k}.\mathbf{r})$) multiplied by a function which has the same periodicity as the lattice; i.e.,

$$\psi(\mathbf{r}) = u_k(\mathbf{r})\exp(i\mathbf{k}.\mathbf{r}), \qquad (1.6)$$

where the function $u_k(\mathbf{r})$ has the property $u_k(\mathbf{r}+\mathbf{T}) = u_k(\mathbf{r})$, for any lattice translation vector $\mathbf{T}$. Such wavefunctions are known as Bloch functions, and represent travelling waves passing through the crystal, but with a form modified periodically by the crystal potential due to each atomic site. For a one dimensional lattice of interatomic spacing a, these relationships reduce to

$$\psi(x) = u_k(x)\exp(ikx) \qquad (1.7)$$

with $u_k(x+na)=u_k(x)$ for any integer n. Now, if we impose periodic boundary conditions at the ends of the chain of atoms of length $L=Na$:

$$\psi(Na) = \psi(0),\qquad(1.8)$$

we find that

$$u_k(Na)\exp(ikNa) = u_k(0),\qquad(1.9)$$

from which $\exp(ikNa)=1$, which has the solutions

$$k = \pm 2n\pi/Na = \pm 2n\pi/L\qquad(1.10)$$

for integer n. This result has two important consequences: firstly, it tells us that the difference between consecutive k values is always $2\pi/L$, which can be interpreted as representing the volume (or properly, in this simplified one dimensional case, the *length*) of k-space occupied by each wavevector state. Applying this argument to each dimension in turn gives, for a 3D crystal, a k-space volume of $8\pi^3/V$ occupied by each wavevector state, where $V=L^3$ is the crystal volume. Secondly, once the upper limit on n is determined, equation (1.10) will also tell us how many wavevector states are contained within each energy band. This point is examined below.

The lattice periodicity also gives rise to diffraction effects. Diffraction of X-rays in crystals is discussed in detail in Chapter 2, as an important structural characterisation technique. However, since electrons exhibit wave-like properties, the free electrons present in the crystal also experience the same diffraction phenomena, and this has a crucial effect on the spectrum of allowed electron energies. If we consider an electron wave travelling along a one dimensional chain of atoms of spacing a, then each atom will cause reflection of the wave. These reflections will all be constructive provided that $m\lambda = 2a$, for integer m, where λ is the electron de Broglie wavelength (this is a special case of the Bragg Law of diffraction introduced in Section 2.1.2.5). When this condition is satisfied, both forward and backward travelling waves exist in the lattice, and the superposition of these creates standing waves. The standing waves correspond to electron density distributions $|\psi(x)|^2$ which have either all nodes, or all antinodes, at the lattice sites $x=a, 2a, 3a, \ldots$, and these two solutions, although having the same wavevector value, have quite different associated energies, due to the different interaction energies between the electrons and the positively charged ions. Consequently a *band gap* forms in the electron dispersion curve at the corresponding values of wavevector: $k = \pm m\pi/a$ (see figure 1.8). The fact that the electron waves are standing waves means that the electron group velocity

$$v_g = \frac{\partial\omega}{\partial k} = \frac{1}{\hbar}\frac{\partial E}{\partial k}\qquad(1.11)$$

tends to zero at these points. This represents a fundamental difference between the behaviour of electrons in crystalline solids and that in free space, where the dispersion relationship remains purely parabolic ($E \propto k^2$) for all values of k.

The region of k space which lies between any two diffraction conditions is know as a Brillouin zone: thus, in a one-dimensional crystal, the first Brillouin zone lies between $k = -\pi/a$ and $k = +\pi/a$. However, any value of k which lies outside the first Brillouin zone corresponds, mathematically, to an electron wave of wavelength $\lambda < 2a$. Such

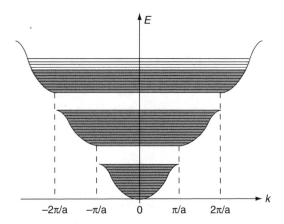

Figure 1.8 Schematic version of the parabolic relationship between the allowed electron wavevectors and the their energy for electrons confined to a one-dimensional potential well containing a periodically varying potential of period a. Shaded energy regions represent those occupied with electrons

a wave has too high a spatial frequency to be uniquely defined by a set of wave amplitudes which are only specified at lattice sites: an equivalent wave of wavelength $\lambda > 2a$ can always be identified. In **k**-space, this transformation is represented by the fact that any value of k lying outside the first Brillouin zone is equivalent to some point lying inside the first Brillouin zone, where the equivalent point is found from the relation

$$k' = k \pm 2m\pi/a \tag{1.12}$$

and the set of values $2m\pi/a$ are known as the reciprocal lattice vectors for the crystal.

In a three dimensional crystal, the location of energy gaps in the electron dispersion is still determined from electron diffraction by the lattice planes, but the Brillouin zones are no longer simple ranges of k, as in 1D: rather, they are described by complex surfaces in 3D **k**-space, the geometry of which depends on the unit cell and atomic structure. When the energy–wavevector relationship for such a crystal extending over multiple Brillouin zones is mapped entirely into the first Brillouin zone, as described above, this results in a large number of different energy bands and consequently the density of energy states takes on a very complex form. An example of the multiple energy bands and corresponding density of states in a real crystal is shown for the case of silicon in figure 1.9.

1.2.7 Electronic conduction

We may now observe that the series of allowed k values in equation 1.12 extends up to the edges of the Brillouin zone, at $k = \pm\pi/a$. Since one of these endpoints may be mapped onto the other by a reciprocal lattice vector translation, the total number of allowed k values is precisely N. Recalling that each k state may be occupied by both a spin up and a spin down electron, the total number of states available is $2N$ per energy band. In three dimensions, this result is generalised to $2N_u$ states per band, where N_u is the number of unit cells in the crystal. Now, the total number of valence electrons in the

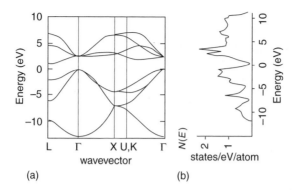

(a)   (b)

Figure 1.9   Electron energy band structure diagram and density of states for crystalline silicon. The symbols Γ, L, X, U and K on the horizontal (wavevector) axis of the band structure plot represent different symmetry points in 3D k-space. Γ corresponds to $k = 0$, the origin of the Brillouin zone; the range Γ–X represents a path through the Brillouin zone from centre to edge along the 100 direction; Γ–L and Γ–K represent middle to edge paths along the $\langle 111 \rangle$ and $\langle 110 \rangle$ directions, respectively, and X–U represents a path along the Brillouin zone boundary starting from the zone edge on the $\langle 100 \rangle$ axis and moving in a direction parallel to $\langle 101 \rangle$

crystal is zN_u, where z is the number of valence electrons per unit cell. This leads to two very different electronic configurations in a solid. If z is even, then one energy band is completely filled, with the next band being completely empty. The highest filled band is the valence band, and the next, empty band, is the conduction band. The electrons in the valence band cannot participate in electrical conduction, because there are no available states for them to move into consistent with the small increase in energy required by motion in response to an externally applied voltage: hence this configuration results in an insulator or, if the band gap is sufficiently small, a semiconductor. Alternatively, if z is odd, then the highest occupied energy band is only half full. In such a material, there are many vacant states immediately adjacent in energy to the highest occupied states, therefore electrical conduction occurs very efficiently and the material is a metal. Figure 1.10 shows schematic energy diagrams for insulators, metals and semiconductors respectively. There is one further, special case which gives rise to metallic behaviour: namely, when the valence band is completely full (z is even), but the valence and

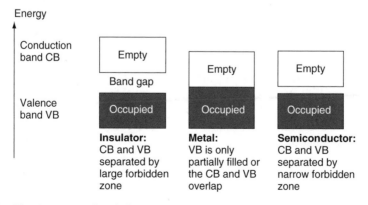

Figure 1.10   Electron energy band diagram for an insulator, a conductor and a semiconductor

conduction bands overlap in energy, such that there are vacant states immediately adjacent to the top of the valence band, just as in the case of a half-filled band. Such a material is called a semi-metal.

In the same way as was defined in the molecular orbital theory of section 1.2.3, the uppermost occupied energy level in a solid is the Fermi level E_F, and the corresponding Fermi wavevector is given by $E_F = \hbar k_F^2/2m_e$. As mentioned above, the volume of **k**-space per state is $8\pi^3/V$. Therefore, the volume of **k** space filled by N electrons is $4N\pi^3/V$, accounting for the fact that 2 electrons of opposite spins can occupy each wavevector state. If we equate this volume to the volume of a sphere in **k** space, of radius k_F (the Fermi sphere), we obtain the result

$$k_F = (3\pi^2 n_e)^{1/3} \tag{1.13}$$

where $n_e = N/V$ is the electron density, and hence

$$E_F = \hbar(3\pi^2 n_e)^{2/3}/2m_e. \tag{1.14}$$

If the Fermi sphere extends beyond the first Brillouin zone, as occurs in many metals, then the appropriate mapping back into the zone results in a Fermi surface of complex topology.

The density of states $N(E)dE$ is defined such that $N_s = \int N(E)dE$ gives the total number of states per unit crystal volume in an energy band. Now, from the above argument, the number of wavevector states per unit volume of **k** space is $V/8\pi^3$. Thus, the total number of states per band may be calculated from

$$N_s = 2 \times V/8\pi^3 \int d\mathbf{k} \tag{1.15}$$

where the factor of 2 accounts for the 2 spin states per **k** value. In three dimensions, $d\mathbf{k} = 4\pi k^2 dk$ and thus we may write

$$N_s = \frac{V}{4\pi^3} \int 4\pi k^2 \frac{dk}{dE} dE. \tag{1.16}$$

For parabolic bands, $E = \hbar^2 k^2/(2m_e)$ and hence $dk/dE = m_e/(\hbar^2 k)$, from which

$$N(E) = \frac{4\pi(2m_e)^{3/2} E^{1/2}}{h^3}. \tag{1.17}$$

The dependence of the density of states on $E^{1/2} (\propto k)$ is simply a consequence of the increased volume of phase space available at larger values of energy. The actual population of electrons as a function of energy is given by the product of the density of states and the occupation probability $f(E)$ which, for electrons or holes, is given by the Fermi Dirac function

$$f(E) = \frac{1}{\exp((E - E_F)/k_B T) + 1}, \tag{1.18}$$

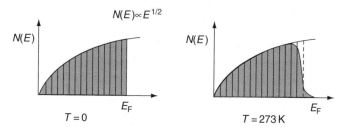

Figure 1.11   The density of electron states for free electrons and the occupation of electron energy levels (shaded region) at zero and room temperature

from which we may observe that the Fermi energy corresponds to the energy at which the occupation probability is exactly one half.

In the zero temperature limit, $f(E) = 1$ for all $E < E_F$, and $f(E) = 0$ for all $E > E_F$: in otherwords, all electron states below the Fermi energy are filled, and all those above E_F are empty, as previously described; the electron population at any energy $E < E_F$ is then given just by $N(E)$. At non-zero temperatures $f(E)$ describes the fact that some electrons are thermally excited from states just below E_F to states just above E_F, and the sharpness of the cut-off of $N(E)$ at E_F decreases with increasing temperature. Both zero temperature and non-zero temperature cases are shown in Figure 1.11.

In addition to the total DOS, which has already been mentioned, it is possible to project the DOS onto a particular atomic site in the unit cell and determine the so-called local DOS; this is the contribution of that particular atomic site to the overall electronic structure. If a unit cell contains a particular type of atom in two distinct crystallographic environments, then the local DOS will be correspondingly different. Similar projections may be performed in terms of the angular momentum symmetry (i.e., the s, p, d or f atomic character of the DOS).

Until now we have been concerned with crystalline systems. However, it is also possible to consider the DOS of an amorphous material; here the DOS is primarily determined by the short-range order in the material; i.e., the nearest neighbours. An alternative approach is to represent the amorphous solid by a very large unit cell with a large number of slightly different atomic environments.

1.3 EFFECTS OF THE NANOMETRE LENGTH SCALE

The small length scales present in nanoscale systems directly influence the energy band structure and can lead indirectly to changes in the associated atomic structure. Such effects are generally termed quantum confinement. The specific effects of quantum confinement in one, two and three dimensions on the density of states are discussed in detail in the Chapter 3 for the case of semiconductor nanostructures; however, initially we outline two general descriptions that can account for such size-dependent effects in nanoscale systems.

1.3.1 Changes to the system total energy

In the free electron model, it is clear that the energies of the electronic states depend on $1/L^2$ where L is the dimension of the system in that particular direction; the spacing between successive energy levels also varies as $1/L^2$. This behaviour is also clear from the description of a solid as a giant molecule: as the number of atoms in the molecule increases, the MOs gradually move closer together. Thus if the number of atoms in a system, hence the length scale, is substantially different to that in a normal bulk material, the energies and energy separations of the individual electronic states will be very different. Although in principle the Fermi level (Section 1.2.5) would not be expected to change since the free electron density N/V should remain constant, there may be associated modifications in structure (see below) which will change this quantity. Furthermore, as the system size decreases, the allowed energy bands become substantially narrower than in an infinite solid. The normal collective (i.e., delocalized) electronic properties of a solid become severely distorted and the electrons in a reduced-dimensional system tend to behave more like the 'particle in a box' description (Section 1.2.5); this is the phenomenon of quantum confinement. In otherwords, the electronic states are more like those found in localized molecular bonds rather than those in a macroscopic solid.

The main effect of these alterations to the bulk electronic structure is to change the total energy and hence, ignoring entropy considerations, the thermodynamic stability of the reduced length scale system relative to that of a normal bulk crystal. This can have a number of important implications. It may change the most energetically stable form of a particular material; for example, small nanoparticles or nanodimensional layers may adopt a different crystal structure from that of the normal bulk material. For example, some metals which normally adopt a hexagonal close-packed atomic arrangement have been reported to adopt a face-centred cubic structure in confined systems such as metallic multilayers. If a different crystallographic structure is adopted below some particular critical length scale, then this arises from the corresponding change in the electronic density of states which often results in a reduced total energy for the system.

Reduction of system size may change the chemical reactivity, which will be a function of the structure and occupation of the outermost electronic energy levels. Correspondingly, physical properties such as electrical, thermal, optical and magnetic characteristics, which also depend on the arrangement of the outermost electronic energy levels, may be changed. For example, metallic systems can undergo metal–insulator transitions as the system size decreases, resulting from the formation of a forbidden energy band gap. Other properties such as mechanical strength which, to a first approximation, depends on the change in electronic structure as a function of applied stress and hence interatomic spacing, may also be affected. Transport properties may also change in that they may now exhibit a quantized rather than continuous behaviour, owing to the changing nature and separation of the electron energy levels.

1.3.2 Changes to the system structure

A related viewpoint for understanding the changes observed in systems of reduced dimension is to consider the proportion of atoms which are in contact with either a free surface, as in the case of an isolated nanoparticle, or an internal interface, such as a grain boundary in a nanocrystalline solid. Both the surface area to volume ratio (S/V) and the specific surface area (m^2g^{-1}) of a system are inversely proportional to particle size and both increase drastically for particles less than 100 nm in diameter. For isolated spherical particles of radius r and density ρ, the surface area per unit mass of material is equal to $4\pi r^2/(4/3\pi r^3\rho) = 3/r\rho$. For 2 nm diameter spherical particles of typical densities, the specific surface area (SSA) can approach 500 $m^2\,g^{-1}$. However, for particles in contact this value will be reduced by up to approximately half. This large surface area term will have important implications for the total energy of the system. As discussed above this may lead to the stabilization of metastable structures in nanometre-sized systems, which are different from the normal bulk structure or, alternatively, may induce a simple relaxation (expansion or contraction) of the normal crystalline lattice which could in turn alter other material properties.

If an atom is located at a surface then it is clear that the number of nearest-neighbour atoms are reduced, giving rise to differences in bonding (leading to the well-known phenomenon of surface tension or surface energy) and electronic structure. In a small isolated nanoparticle, a large proportion of the total number of atoms will be present either at or near the free surface. For instance, in a 5 nm particle approximately 30–50% of the atoms are influenced by the surface, compared with approximately a few percent for a 100 nm particle. Similar arguments apply to nanocrystalline materials, where a large proportion of atoms will be either at or near grain boundaries. Such structural differences in reduced-dimensional systems would be expected to lead to very different properties from the bulk.

1.3.2.1 Vacancies in nanocrystals

Another important consideration for nanostructures concerns the number of atomic vacancies n_v which exist in thermal equilibrium in a nanostructure. Vacancies are point defects in the crystalline structure of a solid and may control many physical properties in materials such as conductivity and reactivity. In microcrystalline solids at temperatures above 0 K, vacancies invariably exist in thermal equilibrium. In the simple case of metals with one type of vacancy, the number of vacancies in a crystal consisting of N atom sites is approximated by an Arrhenius-type expression

$$n_v = N\exp(-Q_f/RT), \tag{1.19}$$

where T is the absolute temperature, R is the gas constant and Q_f is the energy required to form one mole of vacancies. Q_f is given by the relationship $Q_f = N_A q_f$, where N_A is the Avogadro number and q_f is the activation energy for the formation of one vacancy. However, the value of q_f is not well defined but is generally estimated to be the energy required to remove an atom from the bulk interior of a crystal to its surface. As a rough approximation, a surface atom is bonded to half the number of atoms compared with an interior atom, so q_f represents half the bonding energy per atom. Since the melting temperature T_m of a metal is also a measure of the bond energy, then q_f is expected to be a near linear function of T_m.

From a continuum model, Q_f may be estimated from the latent heat of vaporisation, since on leaving the surface an atom breaks the remaining (half) bonds. In practice it is found that the latent heat of vaporisation is considerably higher than Q_f. Alternatively, Q_f may be estimated from the surface energy per unit area. Given that one atom occupies an area b^2, the number of atoms per unit area is equal to $1/b^2$ and the surface energy σ is therefore q_f/b^2. Surface energies depend on melting temperature and vary within the range 1.1 J m^{-2} (for aluminium) to 2.8 J m^{-2} (for tungsten). Taking an average value of σ as 2.2 J m^{-2} and $b = 2.5 \times 10^{-10}$ m, we may calculate $Q_f = N_A \sigma b^2$ as 83×10^3 J mol^{-1}, which is close to the accepted value of 90 kJ mol^{-1}.

Furthermore, the value of Q_f may be modified for nanoparticles through the influence of the surface energy term, σ, which is related to the internal pressure, P, by the simple relationship $P = 4\sigma/d$, where d is the diameter of the nanoparticle. The effect of P is to require an additional energy term, q_n, for the formation of a vacancy, which is approximately given by Pb^3. Again taking σ as 2.2 J m^{-2}, we may calculate this additional energy per mole $Q_n = N_A q_n$ as 8.3×10^3 J mol^{-1} for a 10 nm diameter nanoparticle. This term is only approximately 10% of Q_f and rapidly decreases for larger particle sizes. Thus we may conclude that the effect of the surface energy (internal pressure) factor on the vacancy concentration will be small. Additionally, the internal pressure P results in an elastic, compressive volume strain, and hence linear strain, ε, given approximately as

$$\varepsilon = P/3E = 4\sigma/3dE \tag{1.20}$$

where E is the Young's modulus. This expression suggests that the linear strain will be inversely proportional to particle size and that there will be a decrease in lattice parameter or interatomic spacing for small nanoparticles. This prediction correlates reasonably well with the data in Figure 1.12.

Finally, substituting a value of $Q_f = 90 \times 10^3$ J mol^{-1} into the Arrhenius expression (1.19) for the vacancy concentration, we obtain values for the ratio n_v/N of 2.4×10^{-16} (at 300 K), 6.5×10^{-7} (at 600 K) and 4.8×10^{-4} (at 1000 K), illustrating the exponential

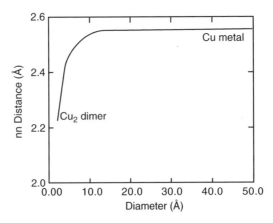

Figure 1.12 Schematic diagram of the change in nearest-neighbour (nn) distance as a function of cluster size or particle size for copper

increase in vacancy concentration with temperature. Now consider a spherical nanoparticle, say 50 nm in diameter, and in which each atom occupies a volume b^3. Again taking $b = 0.25$ nm, there are a total of 4.2×10^6 atoms in the particle, which implies that $n_v \ll 1$, except at very high temperatures. Therefore nanocrystals are predicted to be essentially vacancy-free; their small size precludes any significant vacancy concentration. This simple result also has important consequences for all thermomechanical properties and processes (such as creep and precipitation) which are based on the presence and migration of vacancies in the lattice.

1.3.2.2 Dislocations in nanocrystals

Planar defects, such as dislocations, in the crystalline structure of a solid are extremely important in determining the mechanical properties of a material. It is expected that dislocations would have a less dominant role to play in the description of the properties of nanocrystals than in the description of the properties of microcrystals, owing to the dominance of crystal surfaces and interfaces. The free energy of a dislocation is made up of a number of terms: (i) the core energy (within a radius of about three lattice planes from the dislocation core); (ii) the elastic strain energy outside the core and extending to the boundaries of the crystal, and (iii) the free energy arising from the entropy contributions. In microcrystals the first and second terms increase the free energy and are by far the most dominant terms. Hence dislocations, unlike vacancies, do not exist in thermal equilibrium.

The core energy is expected to be independent of grain size. Estimates are close to 1 eV per lattice plane which, for an interplanar spacing b of 0.25 nm, translates to a value of about 6.5×10^{-10} J m^{-1}. The elastic strain energy per unit length for an edge dislocation is given by

$$E = \frac{Gb^2}{4\pi(1-\nu)} \times \ln\left(\frac{r_1}{r_0}\right) \qquad (1.21)$$

where G is the bulk modulus, r_0 is the core radius, r_1 is the crystal radius and ν is Poisson's ratio. ν is typically around $1/3$ for a crystalline sample. The expression for a screw dislocation omits the $(1 - \nu)$ term, giving an energy about $2/3$ that of an edge dislocation. For $G = 40 \times 10^9$ Pa, the constant term $Gb^2/4\pi(1 - \nu)$ has a value of 3×10^{-10} J m^{-1}. The grain size dependence is given in the $\ln(r_1/r_0)$ term, which for grain size $(2r_1)$ values of 10, 50, 1000 and 10 000 nm increases as 3, 4.6, 7.6 and 9.9 respectively. Hence it can be seen that the elastic strain energy of dislocations in nanoparticles and nanograined materials is about one-third of that in microcrystals and that, for a 10 nm grain size the core energy is comparable with the elastic strain energy. In comparison, the core energy is about one-tenth of the elastic strain energy for a microcrystal.

This reduction in the elastic strain energy of dislocations in nanocrystals has important consequences. The forces on dislocations due to externally applied stresses are reduced by a factor of about three and the interactive forces between dislocations are reduced by a factor of about 10. Hence recovery rates and the annealing out of dislocations to free surfaces are expected to be reduced. For a dislocation near the surface of a semi-infinite solid, the stress towards the surface is given by the interaction of the stress field of an image dislocation at an equal distance on the opposite side. Since nanocrystals do not approximate to semi-infinite solids, such image stresses will operate across all surfaces and the net effect, together with the reduced elastic strain energy, results in dislocations that are relatively immobile.

Finally, we estimate the entropy contributions to the free energy. These arise as a result of (i) configurational entropy (i.e., the dislocation can be arranged in a variety of ways), (ii) a further contribution if the dislocation is assumed to be perfectly flexible, and (iii) the effect of the dislocation on the thermal vibrations of the crystal. Factors (ii) and (iii) are independent of crystal size and their values can be estimated to be $2k_B T$ and $3k_B T$, respectively, per atomic plane. Assuming a temperature of 300 K, these values correspond to about 3×10^{-11} J m^{-1}, considerably less than the core and elastic strain energy terms. The configurational entropy contribution to the free energy is given by

$$E = \frac{bk_B T}{L} \ln\left(\frac{L^2}{b^2}\right) \tag{1.22}$$

per atom plane, where L is the length of the dislocation. At 300 K this gives values of $3.0 \times 10^{-12}, 5.7 \times 10^{-14}$ and 7.6×10^{-15} J m^{-1} for $L = 10$, 1000 and 10 000 nm, respectively. These values are again much smaller than the core and elastic strain energy terms and hence it may be concluded that dislocations in nanocrystals, as with microcrystals, do not exist as thermodynamically stable lattice defects.

1.3.3 How nanoscale dimensions affect properties

Many properties are continuously modified as a function of system size. Often these are extrinsic properties, such as resistance, which depend on the exact size and shape of the specimen. Other properties depend critically on the microstructure of the material; for example, the Hall–Petch equation for yield strength, σ, of a material as a function of average grain size $\langle d \rangle$ is given by

$$\sigma = k\langle d \rangle^{-1/2} + \sigma_0 \tag{1.23}$$

where k and σ_0 are constants. Intrinsic materials properties, such as resistivity, should be independent of specimen size, however, even many of the intrinsic properties of matter at the nanoscale are not necessarily predictable from those observed at larger scales. As discussed above this is because totally new phenomena can emerge, such as: quantum size confinement leading to changes in electronic structure; the presence of wave-like transport processes, and the predominance of interfacial effects.

1.3.3.1 Structural properties

The increase in surface area and surface free energy with decreasing particle size leads to changes in interatomic spacings. For Cu metallic clusters the interatomic spacing is observed to decrease with decreasing cluster size, as shown in Figure 1.12. This effect can be explained by the compressive strain induced by the internal pressure arising from the small radius of curvature in the nanoparticle (Section 1.3.2.2). Conversely, for semiconductors and metal oxides there is evidence that interatomic spacings increase with decreasing particle size.

A further effect, previously mentioned, is the apparent stability of metastable structures in small nanoparticles and clusters, such that all traces of the usual bulk atomic arrangement become lost. Metallic nanoparticles, such as gold, are known to adopt polyhedral shapes such as cube–octahedra, multiply twinned icosahedra and multiply twinned decahedra (Figure 1.13). These nanoparticles may be regarded as multiply twinned crystalline particles (MTPs) in which the shapes can be understood in terms of the surface energies of various crystallographic planes, the growth rates along various crystallographic directions and the energy required for the formation of defects such as twin boundaries. However, there is compelling evidence that such particles are not crystals but are quasiperiodic crystals or crystalloids. These icosahedral and decahedral quasicrystals form the basis for further growth of the nanocluster, up until a size where they will switch into more regular crystalline packing arrangements.

Crystalline solids are distinct from amorphous solids in that they possess long-range periodic order and the patterns and symmetries which occur correspond to those of the 230 space groups. Quasiperiodic crystals do not possess such long-range periodic order and are distinct in that they exhibit fivefold symmetry, which is forbidden in the 230 space groups. In the cubic close-packed and hexagonal close-packed structures, exhibited by many metals, each atom is coordinated by 12 neighbouring atoms. All of the coordinating atoms are in contact, although not evenly distributed around the central atom. However, there is an alternative arrangement in which each coordinating atom is situated at the apex of an icosahedron and in contact only with the central atom. If however we relax this 'rigid atomic sphere' model and allow the central atom to reduce in diameter by 10%, the coordinating atoms come into contact and the body now has the shape and symmetry of a regular icosahedron with point group symmetry 235, indicating the presence of 30 twofold, 20 threefold and 12 fivefold axes of symmetry. This geometry represents the nucleus of a quasiperiodic crystal which may grow in the forms of icosahedra or pentagonal dodecahedra. These are dual solids with identical symmetry, the apices of one being replaced by the faces of the other. Such quasiperiodic crystals are known to exist in an increasing number of aluminium-based alloys and may be stable up to microcrystalline sizes. It should be noted that their symmetry is precisely the same as that of the fullerenes C_{20} (dodecahedrene with 12 pentagonal faces of a pentagonal dodecahedron, but unstable) and C_{60} (the well-known buckyball with 12 pentagonal faces and 20 hexagonal faces of a truncated icosahedron). Hence, like the fullerenes, quasiperiodic crystals are expected to have an important role to play in nanostructures.

Cubo-octahedron   Decahedron   Icosahedron

Figure 1.13 Geometrical shapes of cubo-octahedral particles and multiply twinned decahedral and icosahedral particles

The size-related instability characteristics of quasiperiodic crystals are not well understood. A frequently observed process appears to be that of multiple twinning, such crystals being distinguished from quasiperiodic crystals by their electron diffraction patterns. Here the five triangular faces of the fivefold symmetric icosahedron can be mimicked by five twin-related tetrahedra (with a close-packed crystalline structure) through relatively small atomic movements.

1.3.3.2 Thermal properties

The large increase in surface energy and the change in interatomic spacing as a function of nanoparticle size mentioned above have a marked effect on material properties. For instance, the melting point of gold particles, which is really a bulk thermodynamic characteristic, has been observed to decrease rapidly[2] for particle sizes less than 10 nm, as shown in Figure 1.14. There is evidence that for metallic nanocrystals embedded in a continuous matrix the opposite behaviour is true; i.e., smaller particles have higher melting points.[3]

1.3.3.3 Chemical properties

The change in structure as a function of particle size is intrinsically linked to the changes in electronic properties. The ionization potential (the energy required to remove an electron) is generally higher for small atomic clusters than for the corresponding bulk material. Furthermore, the ionization potential exhibits marked fluctuations as a function of cluster size. Such effects appear to be linked to chemical reactivity, such as the reaction of Fe_n clusters with hydrogen gas (Figure 1.15).

Nanoscale structures such as nanoparticles and nanolayers have very high surface area to volume ratios and potentially different crystallographic structures which

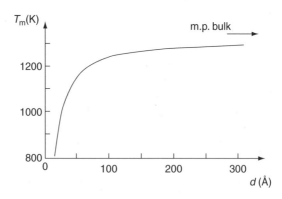

Figure 1.14 Schematic diagram of the variation in melting point of gold nanoparticles as a function of particle size

[2] *Nanomaterials: Synthesis, Properties and Applications*, ed. A. S. Edelstein and R. C. Cammarata (Institute of Physics 1996) and references therein.
[3] U. Dahmen *et al.*, *Inst.Phys. Conf. Ser.* **168**, 1 (IOP Publishing 2001).

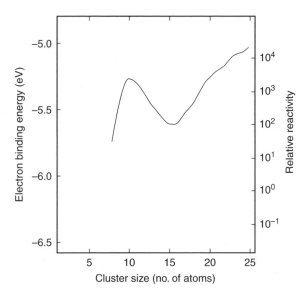

Figure 1.15 Schematic diagram of the dependence of the electron binding energy and relative chemical reactivity of iron clusters to hydrogen gas as a function of cluster size

may lead to a radical alteration in chemical reactivity. Catalysis using finely divided nanoscale systems can increase the rate, selectivity and efficiency of chemical reactions such as combustion or synthesis whilst simultaneously significantly reducing waste and pollution. Gold nanoparticles smaller than about 5 nm in diameter are known to adopt icosahedral structures rather than the normal face centred cubic arrangement. This structural change is accompanied by an extraordinary increase in catalytic activity. Furthermore, nanoscale catalytic supports with controlled pore sizes can select the products and reactants of chemical reactions based on their physical size and thus ease of transport to and from internal reaction sites within the nanoporous structure. Additionally, nanoparticles often exhibit new chemistries as distinct from their larger particulate counterparts; for example, many new medicines are insoluble in water when in the form of micron-sized particles but will dissolve easily when in a nanostructured form.

1.3.3.4 Mechanical properties

Many mechanical properties, such as toughness, are highly dependent on the ease of formation or the presence of defects within a material. As the system size decreases, the ability to support such defects becomes increasingly more difficult and mechanical properties will be altered accordingly. Novel nanostructures, which are very different from bulk structures in terms of the atomic structural arrangement, will obviously show very different mechanical properties. For example, single- and multi-walled carbon nanotubes show high mechanical strengths and high elastic limits that lead to considerable mechanical flexibility and reversible deformation.

As the structural scale reduces to the nanometre range, for example, in nano-ayered composites, a different scale dependence from the usual Hall–Petch relationship

(Equation 1.23) for yield strength often becomes apparent with large increases in strength reported. In addition, the high interface to volume ratio of consolidated nanostructured materials appears to enhance interface-driven processes such as plasticity, ductility and strain to failure. Many nanostructured metals and ceramics are observed to be superplastic, in that they are able to undergo extensive deformation without necking or fracture. This is presumed to arise from grain boundary diffusion and sliding, which becomes increasingly significant in a fine-grained material. Overall these effects extend the current strength–ductility limit of conventional materials, where usually a gain in strength is offset by a corresponding loss in ductility.

1.3.3.5 Magnetic properties

Magnetic nanoparticles are used in a range of applications, including ferrofluids, colour imaging, bioprocessing, refrigeration as well as high storage density magnetic memory media. The large surface area to volume ratio results in a substantial proportion of atoms (those at the surface which have a different local environment) having a different magnetic coupling with neighbouring atoms, leading to differing magnetic properties. Figure 1.16 shows the magnetic moments of nickel nanoparticles as a function of cluster size.

Whilst bulk ferromagnetic materials usually form multiple magnetic domains, small magnetic nanoparticles often consist of only one domain and exhibit a phenomenon known as superparamagnetism. In this case the overall magnetic coercivity (Section 4.1) is then lowered: the magnetizations of the various particles are randomly distributed due to thermal fluctuations and only become aligned in the presence of an applied magnetic field.

Giant magnetoresistance (GMR) is a phenomenon observed in nanoscale multi-layers consisting of a strong ferromagnet (e.g., Fe, Co) and a weaker magnetic or non-magnetic buffer (e.g., Cr, Cu); it is usually employed in data storage and sensing. In the absence of a magnetic field the spins in alternating layers are oppositely aligned through antiferromagnetic coupling, which gives maximum scattering from the interlayer interface and hence a high resistance parallel to the layers. In an

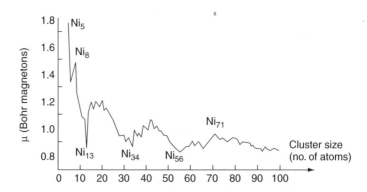

Figure 1.16 Schematic diagram of the variation in magnetic moments of clusters as a function of cluster size. The Bohr magneton is the classical magnetic moment associated with an electron orbiting a nucleus which has a single positive charge

oriented external magnetic field the spins align with each other and this decreases scattering at the interface and hence resistance of the device. Further details are given in Chapter 4.

1.3.3.6 Optical properties

In small nanoclusters the effect of reduced dimensionality on electronic structure has the most profound effect on the energies of the highest occupied molecular orbital (HOMO), essentially the valence band, and the lowest unoccupied molecular orbital (LUMO), essentially the conduction band. Optical emission and absorption depend on transitions between these states; semiconductors and metals, in particular, show large changes in optical properties, such as colour, as a function of particle size. Colloidal solutions of gold nanoparticles have a deep red colour which becomes progressively more yellow as the particle size increases; indeed gold colloids have been used as a pigment for stained glass since the seventeenth century. Figure 1.17 shows optical absorption spectra for colloidal gold nanoparticles of varying sizes. Semiconductor nanocrystals in the form of quantum dots show similar size-dependent behaviour in the frequency and intensity of light emission as well as modified non-linear optical properties and enhanced gain for certain emission energies or wavelengths. Other properties which may be affected by reduced dimensionality include photocatalysis, photoconductivity, photoemission and electroluminescence.

1.3.3.7 Electronic properties

The changes which occur in electronic properties as the system length scale is reduced are related mainly to the increasing influence of the wave-like property of the electrons (quantum mechanical effects) and the scarcity of scattering centres. As the size of the

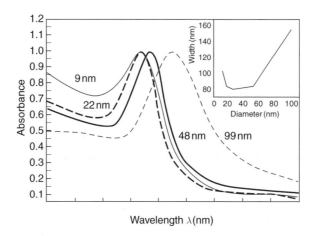

Figure 1.17 Size dependence of the optical absorption wavelength for gold nanoparticles and (inset) the corresponding value of the full width at half maximum (FWHM) of the absorption peak

system becomes comparable with the de Broglie wavelength of the electrons, the discrete nature of the energy states becomes apparent once again, although a fully discrete energy spectrum is only observed in systems that are confined in all three dimensions. In certain cases, conducting materials become insulators below a critical length scale, as the energy bands cease to overlap. Owing to their intrinsic wave-like nature, electrons can tunnel quantum mechanically between two closely adjacent nanostructures, and if a voltage is applied between two nanostructures which aligns the discrete energy levels in the DOS, *resonant tunnelling* occurs, which abruptly increases the tunnelling current.

In macroscopic systems, electronic transport is determined primarily by scattering with phonons, impurities or other carriers or by scattering at rough interfaces. The path of each electron resembles a random walk, and transport is said to be *diffusive*. When the system dimensions are smaller than the electron mean free path for inelastic scattering, electrons can travel through the system without randomization of the phase of their wavefunctions. This gives rise to additional localization phenomena which are specifically related to phase interference. If the system is sufficiently small so that all scattering centres can be eliminated completely, and if the sample boundaries are smooth so that boundary reflections are purely specular, then electron transport becomes purely *ballistic*, with the sample acting as a waveguide for the electron wavefunction.

Conduction in highly confined structures, such as quantum dots, is very sensitive to the presence of other charge carriers and hence the charge state of the dot. These *Coulomb blockade* effects result in conduction processes involving single electrons and as a result they require only a small amount of energy to operate a switch, transistor or memory element.

All these phenomena can be utilised to produce radically different types of components for electronic, optoelectronic and information processing applications, such as resonant tunnelling transistors and single-electron transistors. Further details of these concepts and their applications are given in Chapter 3.

1.3.3.8 Biological systems

Biological systems contain many examples of nanophase materials and nanoscale systems. Biomineralization of nanocrystallites in a protein matrix is highly important for the formation of bone and teeth, and is also used for chemical storage and transport mechanisms within organs. Biomineralization involves the operation of delicate biological control mechanisms to produce materials with well-defined characteristics such as particle size, crystallographic structure, morphology and architecture. Generally complex biological molecules such as DNA have the ability to undergo highly controlled and hierarchical self-assembly, which makes them ideal for the assembling of nanosized building blocks. Self-assembly is discussed in Section 1.4.3.1 and in detail in Chapters 7 and 8. Methods for altering and controlling these interactions and building nanoscale building blocks and assembling nanoscale architectures are also discussed.

Biological cells have dimensions within the range 1–10μm and contain many examples of extremely complex nano-assemblies, including molecular motors, which are complexes embedded within membranes that are powered by natural biochemical processes (Figure 1.18).

Generally, naturally occurring biological nanomaterials have been refined by evolutionary processes over a long timescale and are therefore highly optimized. We can often use biological systems as a guide to producing synthetic nanomaterials or nanosystems, a process known as *biomimicry*.

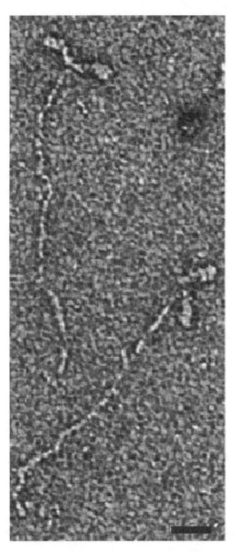

Figure 1.18   A transmission electron microscope image of two myosin molecules, each of which has two heads that can bind an actin filament and split a biochemical compound called ATP to provide chemical energy for motion. A large head shape change while attached to actin causes muscle contraction and a wide variation in cell motility. A 20 nm scale bar is shown in the image which reproduced from The Journal of Cell Biology, permission of the Rockefeller University Press and courtesy of Professor John Trinick, University of Leeds

1.4 FABRICATION METHODS

Nanostructures can be made in numerous ways. As highlighted in Section 1.1.3, a broad classification divides methods into either those which build from the bottom up, atom by atom, or those which construct from the top down using processes that involve the removal or reformation of atoms to create the desired structure. The two approaches are schematically represented in Figure 1.19.

In the bottom-up approach, atoms, molecules and even nanoparticles themselves can be used as the building blocks for the creation of complex nanostructures; the useful size of the building blocks depends on the properties to be engineered. By altering the size of the building blocks, controlling their surface and internal chemistry, and then controlling their organization and assembly, it is possible to engineer properties and functionalities of the overall nanostructured solid or system. These processes are essentially highly controlled, complex chemical syntheses. On the other hand, top-down approaches are inherently simpler and rely either on the removal or division of bulk material, or on the miniaturization of bulk fabrication processes to produce the desired structure with the appropriate properties. When controlled, both top-down and bottom-up methods may be viewed as essentially different forms of microstructural engineering. As shown in Figure 1.19, in terms of scale, biological processes are essentially intermediate between top-down and bottom-up processes, however in reality they usually constitute complex bottom-up processes. A brief overview of some of the more common fabrication methods for nanostructures is given below; further details are discussed in subsequent chapters.

1.4.1 Top-down processes

Top-down processes are effectively examples of solid-state processing of materials.

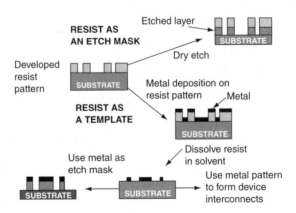

Figure 1.19 Schematic representation of the top-down and bottom-up processes and their relationship to biological processes and structures

1.4.1.1 Milling

One nanofabrication process of major industrial importance is high-energy ball milling, also known as mechanical attrition or mechanical alloying. As shown in Figure 1.20, coarse-grained materials (usually metals but also more recently ceramics and polymers) in the form of powders are crushed mechanically in rotating drums by hard steel or tungsten carbide balls, usually under controlled atmospheric conditions to prevent unwanted reactions such as oxidation. This repeated deformation can cause large reductions in grain size via the formation and organization of grain boundaries within the powder particles. Different components can be mechanically alloyed together by cold welding to produce nanostructured alloys. A nanometre dispersion of one phase in another can also be achieved. Microstructures and phases produced in this way can often be thermodynamically metastable. The technique can be operated at a large scale, hence the industrial interest.

Generally any form of mechanical deformation under shear conditions and high strain rates can lead to the formation of nanostructures, since energy is being continuously pumped into crystalline structures to create lattice defects. The severe plastic deformation that occurs during machining, cold rolling, drawing, cyclic deformation or sliding wear has also been reported to form nanostructured material.

1.4.1.2 Lithographic processes

Conventional lithographic processes are akin to the emulsion-based photographic process and can be used to create nanostructures by the formation of a pattern on a substrate via the creation of a resist on the substrate surface. One lithographic method uses either visible or ultraviolet (UV) light, X-rays, electrons or ions to project an image containing the desired pattern onto a surface coated with a photoresist material; this method requires the prior fabrication of an absorbing mask through which the parallel radiation passes before shadowing onto the photoresist. Alternatively, primary patterning (or direct writing) of the resist is possible using a focused electron, ion or possibly X-ray beam; here either the focused beam or the resist itself is scanned according to the desired pattern design. These different techniques are generally termed photolithography, X-ray lithography, electron beam lithography or ion beam lithography depending on the radiation employed.

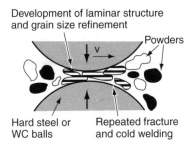

Figure 1.20 Schematic representation of the mechanical alloying process

The resist material, typically a polymer, metal halide or metal oxide, is chemically changed during irradiation, often altering the solubility or composition of the exposed resist. In a positive resist, irradiated areas are dissolved as they are more soluble than unexposed material, whilst in a negative resist irradiated areas are insoluble. As shown in Figure 1.21, the resist can then be used as a template, or as an etch-mask, for subsequent deposition onto, or etching of, the underlying substrate. Resists based on self-assembled monolayers on substrate surfaces are currently being investigated.

The pattern transfer processes which utilize the patterned resist may be divided into solution-based wet chemical etching procedures, dry etching in a reactive plasma, doping using ion implantation techniques, or thin film deposition. Dry etching is the collective term for a range of techniques such as reactive ion etching (RIE) and chemically assisted ion beam etching (CAIBE), which are used extensively for high-resolution pattern transfer; both methods produce, either directly or indirectly, reactive ion species which combine with the elements in the substrate material to form volatile reaction products which are removed into a vacuum system.

Lithographic techniques are very heavily used in the semiconductor processing industry for the fabrication of integrated circuits, optoelectronic components, displays and electronic packaging. The important considerations here are not only the uniformity and reproducibility of the fabrication process, but also the time required to pattern a given area of a planar device, which is summarized in terms of the areal throughput (in microns per hour). A further consideration for nanostructures, particularly for the production of integrated circuits and microelectromechanical systems (MEMS), is the ultimate resolution of the lithographic technique. Fundamentally, the wavelength of the radiation used in the lithographic process determines the detail in the resist and hence the final planar nanostructure; additional considerations may involve the limitations of the projection optics and the nature of the interaction of the radiation with the resist material. Typically the resolution ranges from a few hundred nanometres for optical techniques to tens of nanometres for electron beam techniques. Phenomenologically, throughput and resolution of lithographic techniques broadly follow a power-law dependence; the resolution is approximately equal to $23A^{0.2}$, where A is the areal throughput. Optical, UV and X-ray lithography are fast, parallel exposure techniques capable of micron and submicron resolution, whereas electron and ion beam methods provide nanometre resolution, but

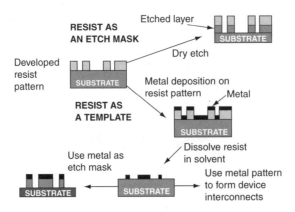

Figure 1.21   Schematic representation of various types of photolithographic process

are, at present, considerably slower as they involve serial exposure using scanned focused probes controlled by computer-aided design (CAD) software. Parallel projection electron beam systems (such as the Scalpel method[4]) and X-ray proximity printing methods are currently being developed for high-resolution, large-throughput mask fabrication. A further area which may be developed in the future is the use of massive arrays of individually controllable atomic force microscope (AFM) or scanning tunnelling microscope (STM) tips for parallel atom manipulation on planar substrates; these are essentially MEMS devices integrated with microprocessor control for the direct patterning of resists (e.g., self-assembled monolayers) or even substrates at reasonable throughputs.

Soft lithography techniques (Figure 1.22) pattern a resist by physically deforming (or embossing) the resist shape with a mould or stamp, rather than by modifying the resist chemical structures with radiation as in conventional lithography. Additionally the stamp may be coated with a chemical that reacts with the resist solely at the edges of the stamp. These methods circumvent many of the resolution limitations inherent in conventional lithographic processes that arise due to the diffraction limit of the radiation, the projection or scanning optics, the scattering process and the chemistry within the resist material. Ultimately, nanoimprinting should represent a cheaper process for mass production. Currently, these soft lithography techniques can produce patterned structures in the range 10 nm and above. One of the main limitations on resolution arises from plastic flow of the polymeric materials involved. Master moulds may be fabricated using either conventional lithographic techniques, micromachining or naturally occurring surface relief on the substrate materials. Once the master mould is formed, a low molecular weight prepolymer is poured and then cured to yield a polydimethylsiloxane (PDMS) elastomeric 'stamp' which may simply be peeled off the master. The relief features that result on the bottom of the PDMS stamp are an inverted replica of those on the master. They may be inked with a molecule of interest (typically by wiping a Q-tip, dampened with the molecule of interest, across their surfaces) and simply placed against a substrate, transferring molecules in a pattern that reflects the features of the stamp. Elastomeric PDMS stamps have been used

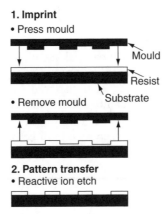

1. Imprint
• Press mould

Mould

Resist

• Remove mould   Substrate

2. Pattern transfer
• Reactive ion etch

Figure 1.22 Schematic representation of the soft lithographic process which patterns a resist using a stamp followed, if required, by subsequent etching

[4] J. A. Liddle *et al.*, *J Vac. Sci. Technol. B* **19**, 476 (2001).

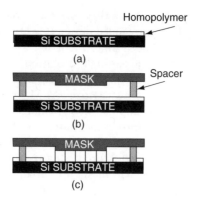

Figure 1.23 Schematic representation of the self-assembly process of a surface polymer film lithographically induced by a mask

to impressive effect as conformal contact masks for phase-shifting photolithography, facilitating the fabrication of structures as small as 50 nm in photoresist. Channels formed in PDMS stamps have also been used to mould three-dimensional structures, a process known as micromoulding in capillaries (MIMIC).

Another variation of this technique is lithographically induced self-assembly (Figure 1.23), in which a mask is used to induce and control, via electrostatic and hydrodynamic instabilities, the formation of periodic supramolecular pillar arrays on a thin polymer melt initially deposited flat onto a substrate. Alternatively, microcontact printing of self-assembled monolayers (SAMs) can be achieved by transfer of a SAM 'ink' onto a substrate surface from a patterned 'stamp' possessing elastomeric properties so as to achieve intimate contact with the surface. A related, and rapidly emerging technique is direct inkjet printing of (currently) micron-scale structures onto surfaces.

1.4.1.3 Machining

Lithographic techniques, whilst being a parallel batch processes, essentially consist of a two-dimensional chemical or mechanical patterning of the surface of a material. More intricate three-dimensional patterning of a material can be achieved by techniques analogous to more conventional machining. Currently the resolution limits of conventional machining are of the order of 5 μm, however, in recent years focused ion beams (FIBs) and high-intensity lasers have been used to directly pattern or shape materials at micron and submicron levels. Figure 1.24 shows a submicron structure produced in a FIB. The length scale achievable in such direct sculpture of materials is not only determined by the spot size and power (or current) density of the radiation employed but also by the nature of the material itself and the removal process.

Modern FIB machines have very accurate sample manipulation stages and are not only capable of material removal, via sputtering of surface substrate atoms by the computer-controlled scanning of focused ion beams produced by liquid metal gallium ion guns, but also of material deposition using the interaction of the ion beam with reactive gases from localized microinjectors. Modern dual-beam FIBs can

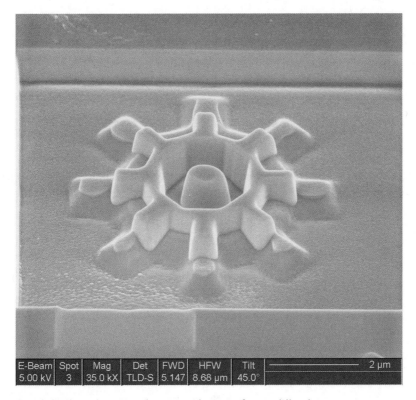

E-Beam	Spot	Mag	Det	FWD	HFW	Tilt		2 μm
5.00 kV	3	35.0 kX	TLD-S	5.147	8.68 μm	45.0°		

Figure 1.24 Scanning electron microscope image of a multilevel gear structure created by focused ion beam sputtering of silicon. Image courtesy of FEI

simultaneously view the etching or deposition procedures using a separate scanning electron microscope column. As well as inorganic materials, there is increasing interest in FIB patterning of biological structures. Thin sample preparation of specimens for transmission electron microscopy (Chapter 2) is routinely undertaken in a FIB.

1.4.2 Bottom-up processes

Bottom-up processes effectively encompass chemical synthesis and/or the highly controlled deposition and growth of materials. Chemical synthesis may be carried out in either the solid, liquid or gaseous state. Solid-state synthesis usually involves an iterative procedure of bringing solid reaction precursors into intimate contact by mixing and grinding and then promoting atomic diffusion processes via heat treatment at high temperatures to form a reaction product. Such elevated temperatures often lead to rapid grain growth and ultimately a final product with a relatively large grain size unless grain growth inhibitors are present. Consequently, true nanoscale systems are difficult to obtain via solid-state routes, apart from perhaps the controlled growth of second phase nanoscale precipitates within a primary matrix such as in precipitation-strengthened steels, aluminium alloys or other nanocomposite systems. These methods are discussed in Chapter 5. Here we focus primarily on liquid and gas phase fabrication routes. Diffusion of matter in the liquid and gas phases is typically many orders of magnitude greater than in the solid phase; therefore these synthetic methods can be implemented at much lower temperatures, thus inhibiting unwanted grain growth and so resulting in true nanoscale systems.

1.4.2.1 Vapour phase deposition methods

Vapour phase deposition can be used to fabricate thin films, multilayers, nanotubes, nanofilaments or nanometre-sized particles. The general techniques can be classified broadly as either physical vapour deposition (PVD) or chemical vapour deposition (CVD).

PVD involves the conversion of solid material into a gaseous phase by physical processes; this material is then cooled and redeposited on a substrate with perhaps some modification, such as reaction with a gas. Examples of PVD conversion processes include thermal evaporation (such as resistive or electron beam heating or even flame synthesis), laser ablation or pulsed laser deposition (where a short nanosecond pulse from a laser is focused onto the surface of a bulk target), spark erosion and sputtering (the removal of a target material by bombardment with atoms or ions).

One example of a PVD technique is vapour phase expansion, which relies on the expansion of a high-pressure vapour phase through a jet into a low-pressure ambient background to produce supersaturation of the vapour. Flow rates can approach supersonic speeds and the process can lead to the nucleation of extremely small clusters, ranging from a few atoms upwards, which can be analysed using mass spectrometry. Although the quantities of such clusters are low, these can be of great use for studying the physics of small, usually metallic, clusters, which often are composed of magic numbers of atoms due to their very stable geometric and electronic configurations. A schematic diagram of such a system is shown in Figure 1.25.

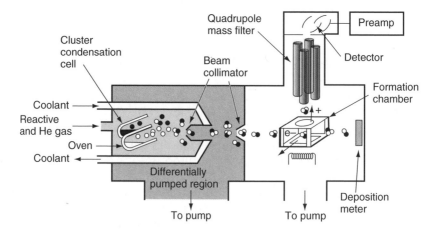

Figure 1.25   Schematic diagram of cluster formation via vapour phase expansion from an oven source

Another PVD process is direct gas phase condensation. Here a material, often a metal, is evaporated from a temperature-controlled crucible into a low-pressure, inert gas environment – ultra-high vacuum (UHV) inert gas condensation. The metal vapour cools through collisions with the inert gas species, becomes supersaturated and then nucleates homogeneously; the particle size is usually in the range 1–100 nm and can be controlled by varying the inert gas pressure. Particles can be collected on a cold finger cooled by liquid nitrogen, scraped off and compacted to produce a dense nanomaterial. Figure 1.26 shows the experimental apparatus. Further details are given in Section 5.3.3.

CVD involves the reaction or thermal decomposition of gas phase species at elevated temperatures (typically 500–1000 °C) and subsequent deposition onto a substrate. A simple example is aerosol spray pyrolysis involving aqueous metal salts which are sprayed as a fine mist, dried and then passed into a hot flow tube where pyrolysis converts the salts to the final products. Since materials are mixed in solution, homogeneous mixing at the atomic level is possible and pyrolysis at relatively low temperatures provides particles in the size range 5–500 nm.

Several CVD processes employ catalysts to enhance the rates of certain chemical reactions. When the catalyst is in the form of a nanometre-sized dispersion of particles, this can often provide a templating effect; for example, in the prodution of carbon nanotubes using the decompostion of ethyne or ethene with hydrogen, Fe-, Co- or Ni-based

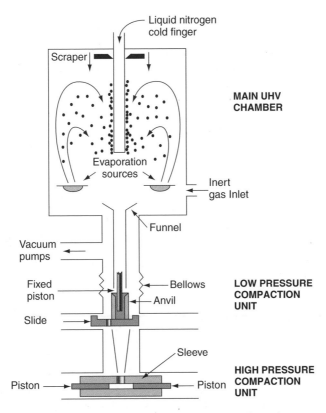

Figure 1.26 Schematic diagram of an inert gas condensation apparatus

catalysts are employed. The size and size distribution of the catalyst particles may determine the internal diameter of the nanotube.

1.4.2.2 Plasma-assisted deposition processes

The use of plasmas (i.e., ionized gases) during vapour deposition allows access to substantially different chemical and physical processes and also higher-purity final materials relative to the conventional PVD and CVD processes described above. There are several different types of plasma deposition reactor for plasma-assisted PVD.

DC glow discharge

DC glow discharge involves the ionization of gas atoms by electrons emitted from a heated filament. The gas ions in the plasma are then accelerated to produce a directed ion beam. If the gas is a reactive precursor gas, this ion beam is used to deposit directly onto a substrate; alternatively an inert gas may be used and the ion beam allowed to strike a target material which sputters neutral atoms onto a neighbouring substrate (Figure 1.27).

Magnetron sputtering

Magnetron sputtering involves the creation of a plasma by the application of a large DC potential between two parallel plates (Figure 1.28). A static magnetic field is applied near a sputtering target and confines the plasma to the vicinity of the target. Ions from the high-density plasma sputter material, predominantly in the form of neutral atoms, from the target onto a substrate. A further benefit of the magnetic field is that it prevents secondary electrons produced by the target from impinging on the substrate and causing heating or damage. The deposition rates produced by magnetrons are high enough ($\sim 1\,\mu m/min$) to be industrially viable; multiple targets can be rotated so as to produce a multilayered coating on the substrate.

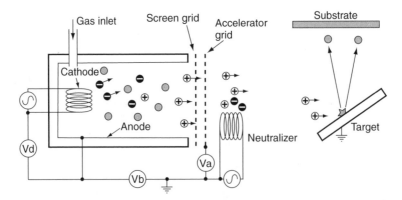

Figure 1.27 Schematic diagram of a DC glow discharge apparatus in which gas atoms are ionized by an electron filament and either deposit on a substrate or cause sputtering of a target

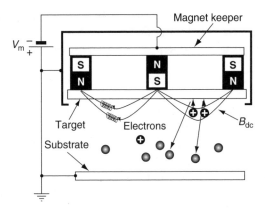

Figure 1.28 Schematic diagram of a magnetron sputtering apparatus that uses a magnetic field to contain ionized gas molecules close to a sputtering target, giving large deposition rates

Vacuum arc deposition

Vacuum arc deposition involves the initiation of an arc by contacting a cathode, constructed from the target material, with an igniter attached to an anode; this generates a low-voltage, high-current arc, which is self-sustaining. The arc ejects predominantly ions and large, micrometre-sized droplets from a small area on the cathode. The ions in the arc are accelerated towards a substrate and deflected using a magnetic field if desired; any large particles being filtered out before deposition (Figure 1.29). A vacuum cathodic arc can operate without a background gas under high-vacuum conditions, however, reactive deposition can be achieved by introducing a background gas such as nitrogen. The high ion energy produces dense films even at low substrate temperatures and consequently arc technology is commonly used for the deposition of hard coatings. As well as DC sources, plasmas can also be produced using radio (MHz) frequency and microwave (GHz) frequency power; these methods have the advantage of being able to provide higher current densities and higher deposition rates.

In the plasma-assisted PVD processes described above, all vapour phase species originate from a solid target. Instead, plasma-enhanced CVD (PECVD) employs gas phase precursors that are dissociated to form molecular fragments which condense to form thin

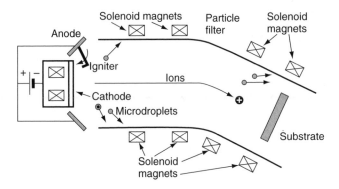

Figure 1.29 Schematic diagram of a vacuum arc deposition apparatus

films or nanoparticles. The dissociation temperatures required for PECVD tend to be much lower than for conventional CVD processes due to the high energy of the plasma, and this may be of importance for deposition on sensitive substrates such as semiconductors and polymers.

1.4.2.3 MBE and MOVPE

A molecular beam epitaxy (MBE) machine is essentially an ultra-high-precision, ultra-clean evaporator, combined with a set of in-situ tools, such as RHEED and Auger spectroscopy (see Chapter 2), for characterization of the deposited layers during growth. The reactor consists of an ultra-high-vacuum chamber (typically better than $\sim 5 \times 10^{-14}$ atmospheric pressure) of approximately 1.5 m diameter (Figure 1.30).

MBE is a growth technique in which epitaxial, single atomic layers ($\sim 0.2-0.3$ nm) are grown on a heated substrate under UHV conditions, using either atomic or molecular beams evaporated from effusion sources with openings directed towards a heated substrate usually consisting of a thin (~ 0.5 mm) wafer cut from a bulk single crystal. The sources can be either solid or gaseous and an MBE machine will typically have an array of multiple sources, which can be shuttered to allow layered, alternating heterostructures to be produced. Semiconductor quantum wells, superlattices and quantum wires (Chapter 3) and metallic or magnetic multilayers for spin valve structures (Chapter 4) are deposited using this technique.

Standard MBE uses elements in a very pure form as solid sources contained within a number of Knudsen cells. In operation the cells are heated to the temperature at which the elements evaporate, producing beams of atoms which leave the cells. The beams intersect at the substrate and deposit the appropriate semiconductor, atomic layer by atomic layer. The substrate is rotated to ensure even growth over its surface. By operating mechanical

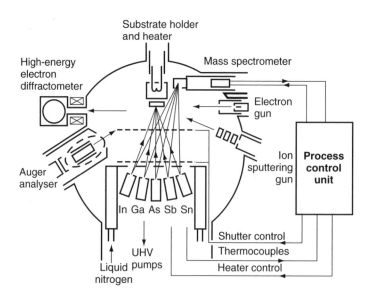

Figure 1.30   Schematic diagram of a molecular beam epitaxy thin film deposition system

shutters in front of the cells, it is possible to control which semiconductor or metal is deposited. For example, opening the Ga and As cell shutters results in the growth of GaAs. Shutting the Ga cell and opening the Al cell switches the growth to AlAs. As the shutters can be switched rapidly, in comparison to the rate at which material is deposited, it is possible to grow very thin layers exhibiting very sharp interfaces. Other effusion cells contain elements required for doping, and it is possible to monitor the growth by observing the electron diffraction pattern produced by the surface (Chapter 2). MBE can also be performed using gaseous sources, and this is often termed chemical beam epitaxy (CBE). When the sources are metallorganic compounds, the process is known as metallorganic MBE (MOMBE).

The second main epitaxial technique is metallorganic vapour phase epitaxy (MOVPE), also known as metallorganic CVD (MOCVD). In this technique the required elements are transported as components of gaseous compounds such as metal alkyls and non-metal hydrides (e.g., trimethylgallium $(CH_3)_3Ga$, arsine AsH_3) to a suitable chamber where they flow over the surface of a heated substrate. These compounds break down and react so as to deposit the relevant semiconductor, with the remaining waste gases being removed from the chamber. MOVPE is used extensively for the production of compound semiconductor (e.g., III–V) thin films, such as the production of GaAs and AlGaAs layers using trimethyl gallium (TMG):

$$Ga(CH_3)_3 + AsH_3(g) \rightarrow GaAs(s) + 3CH_4(g), \tag{1.24a}$$

$$xAl(CH_3)_3(g) + (1-x)Ga(CH_3)_3(g) + AsH_3(g)$$
$$\rightarrow Al_xGa_{1-x}As(s) + 3CH_4(g). \tag{1.24b}$$

Valves in the gas lines allow gas switching, thereby enabling layered structures to be deposited. However, since rapid switching of a gas is more difficult than for an atomic beam, and because of residual gas adsorption on the walls of the feed pipework, the interfaces between adjacent layers grown by the gas-source epitaxy methods are not as abrupt as those attainable with solid-source MBE. For the same reason, solid-source MBE is capable of growing thinner layers, and the slow growth rates involved (of order 1 μm per hour) generally give rise to higher precision in layer thicknesses compared to MOVPE and MOMBE. On the other hand, the high growth rate of MOVPE is better suited for commercial production, especially for semiconductor lasers where thick cladding or waveguiding layers must be grown in a reasonable timescale, in addition to the thin quantum well layers. MOVPE is also easier to scale up to larger systems, permitting simultaneous growth on multiple substrates. Gas source MBE, including MOMBE, generally operates with speeds and precisions which are intermediate between those of solid-source MBE and MOVPE. All the metallorganic epitaxy techniques have significant safety implications as the gases used are highly toxic.

By definition, the aim of molecular beam epitaxy is to produce a perfect single-crystal structure whose morphology and lattice spacing exactly match those of the substrate material. This requires the equilibrium lattice constant of all the different epitaxial layers to be closely matched to that of the substrate, otherwise dislocations will form. (The epitaxial growth of strained semiconductors is discussed in Chapter 3.) For some combinations of layers, a suitable substrate cannot be identified, in which case an initial buffer layer is deposited to serve as a 'virtual' substrate. In some material systems the deposited layers are sometimes annealed to produce an atomically smooth surface prior to further deposition of another material.

MBE growth is generally far from equilibrium and various growth modes can be identified as shown in Figure 1.31. The prevalence of each mode depends on the exact substrate temperature and deposition rate.

- 1D step propagation growth occurs at high temperatures and/or low growth rates. It occurs on vicinal, stepped surfaces when the surface diffusion of the growing species is pronounced. Deposited material diffuses to a step and is incorporated into the laterally growing terrace. This growth mode produces very flat and smooth interfaces.

- 2D nucleation and growth occurs at intermediate temperatures and/or growth rates and involves the initial nucleation of islands, which grow and coalesce to form a network of interconnected islands separated by holes. These holes are finally filled and a flat surface is formed. The process then repeats in a cyclical fashion.

- 3D rough growth generally occurs at low temperatures and high deposition rates. Islands nucleate on existing islands and extensive faceting occurs, producing very rough interfaces.

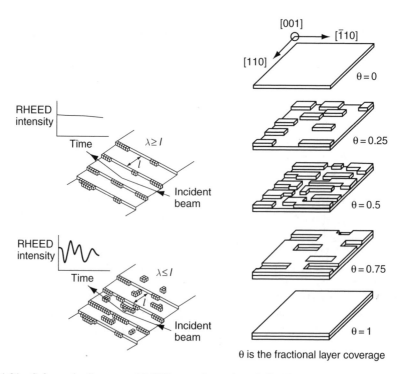

Figure 1.31 Schematic diagram of MBE growth modes: (left) 1D step propagation and (right) 2D nucleation and growth

Because the dopants in both MBE and MOVPE systems exist as additional switchable sources, they may be added in known amounts and distributed over known depth concentration profiles such an abrupt delta-doped layers or extended and/or graded distributions. These processes are discussed in more detail in Chapter 3.

Both MBE and MOVPE are very successful in growing quantum well structures with a sensitive control of layer thicknesses. However, it is much more difficult to achieve fine control of lateral dimensions in devices for the production of either quantum wires or quantum dots. One technique for achieving lateral control is by the use of patterning. Patterning can be carried out before or after epitaxial growth (or both) using lithography or etching. In addition, under certain growth conditions, quantum wires or dots may form spontaneously during the epitaxial growth of strained materials. This *self-assembly* growth technique is described in Chapter 3.

1.4.2.4 Liquid phase methods

A variant of many of the PVD processes described above are thermal spraying techniques in which a spray of molten or semi-molten solid particles generated by either an electrical thermal source (e.g., plasma spraying) or by chemical combustion (e.g., flame spraying or high-velocity oxygen fuel spraying) are deposited onto a substrate and undergo rapid solidification. This is extensively used to produce nanocrystalline coatings from nanocrystalline powder, wire or rod feedstocks previously fabricated by the mechanical milling or precipitation routes discussed above.

More generally, liquid phase chemical synthesis involves the reaction of a solution of precursor chemicals in either an aqueous or non-aqueous solvent; these precursors react and/or naturally self-assemble to form a solution supersaturated with the product. Thermodynamically this is an unstable situation and ultimately results in nucleation of the product either homogeneously (in solution) or heterogeneously (on external species such as vessel walls or impurities). Initial nuclei then grow into nanometre-sized particles or architectures according to both thermodynamic and kinetic factors. The nature, size and morphological shape of the precipitated structures can often be controlled by parameters such as temperature, pH, reactant concentration and time. A multicomponent product may require careful control of co-precipitation conditions in order to achieve a chemically homogeneous final product.

1.4.2.5 Colloidal methods

Colloidal methods rely on the precipitation of nanometre-sized particles within a continuous fluid solvent matrix to form a colloidal sol. Generally a finely dispersed system is in a high free energy state, as work has essentially been done to break up the solid, which this is equivalent to the free energy required to produce the increased surface area. The colloidal material will therefore tend to aggregate due to attractive van der Waals forces and lower its energy unless a substantial energy barrier to this process exists. The presence and magnitude of an energy barrier to agglomeration will depend on the balance of attractive and repulsive forces between the particles, as shown in Figure 1.32. The energy of a colloidal solution which is available for aggregation arises from Brownian motion and is typically of order $k_B T$. Agglomeration (the formation of strong compact aggregates of nanoparticles) and also flocculation (the formation of a loose network of particles) can be prevented by increasing the repulsive energy term, which is normally short range. This can be achieved by electrostatic or steric stablization, both of which lead to a repulsive contribution to the potential energy.

Surface charges on colloid particles can be easily produced by ionization of basic or acidic groups as the pH is varied. For example, $TiCl_4$ hydrolyses in the presence of a base and forms a colloidal solution of TiO_2; here surface OH^- groups on TiO_2 clusters act as an electrostatic colloid stabilizer. Stabilizing steric effects can be produced by the attachment of a capping layer to the particle surfaces. Additional chemicals are added to the colloidal solution which bind to the cluster surface and block vacant coordination sites, thus preventing further growth. These additives can be polymeric surfactants or stabilizers that attach electrostatically to the surface, or anionic capping agents which covalently bind to the cluster. A further possibility is to precipitate a material within a volume of space defined by a micelle or membrane which acts a barrier to further growth.

Subsequent processing of colloids can involve a number of approaches, including additional colloidal precipitation on particle surfaces to produce a core–shell nanoparticle structure, deposition on substrates to produce quantum dots, self-assembly on substrates as ordered 2D and even 3D arrays, and finally embedding in other media to form a nanocomposite. Colloidal methods are relatively simple and inexpensive and have been extensively used for the production of metal and semiconductor nanocrystals using optimized reactions and reaction conditions. One problem inherent in many colloidal methods is that colloid solutions can often age; that is, the particles can increase their size as a function of time.

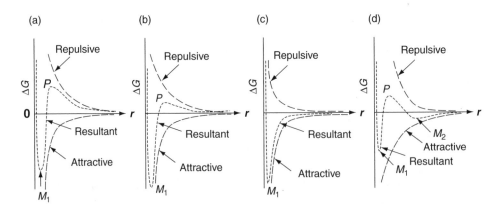

Figure 1.32 Schematic diagram of the possible energy balances between attractive and repulsive interparticulate forces as a function of interparticle separation. In (a) there exists a large energy barrier (P) to strong aggregation of particles (minimum M_1); this is smaller in (b) and absent in (c) owing to reduced long-range repulsive forces. Long-range attractive forces in (d) can lead to weak flocculation of particles (minimum M_2)

1.4.2.6 Sol–gel methods

Sol–gel methods involve a set of chemical reactions which *irreversibly* convert a homogeneous solution of molecular reactant precursors (a sol) into an infinite molecular weight three-dimensional polymer (a gel) forming an elastic solid filling the same volume as the solution. Typically this involves a hydrolysis reaction followed by condensation polymerization, for example:

$$\text{tetraethylorthosilicate (TEOS)}(Si(OC_2H_5)_4)$$
$$+ \text{ethanol}(C_2H_5OH) \xrightarrow{\text{react and dry}} SiO_2 + \text{other products} \qquad (1.25)$$

Mixtures of precursors can also be used to produce binary or ternary systems, each molecular precursor having its own reaction rate that is dependent on the conditions (e.g., pH, concentration, solvent and temperature). The polymer gel so formed is a 3D skeleton surrounding interconnected pores, and this can be dried and shrunk to form a rigid solid form. In a single precursor component system, the final material can be designed to have interconnected nanoscale porosity and hence a high surface area. For example, nanoporous silica can have a variable percentage porosity that depends on the precursor and the solvent employed.

As well as true molecular precursors, precursors containing nanometre-sized particles can be used, such as aqueous colloidal sols. The transformation to a gel is most often achieved by changing the pH or the concentration of solution; this causes aggregation of the colloidal particles in the sol and results in a skeleton composed of interconnected nanometre particles surrounding pores that are generally bigger than in the corresponding material derived from molecular precursor.

The main benefits of sol–gel processing are the high purity and uniform nanostructure achievable at low temperatures. The method is often used to produce metal-oxide nanomaterials. Further processing usually involves forming the gel using a number of techniques (see below) followed by gentle drying to remove the solvent (this often leads to shrinkage so care is needed to prevent cracks forming during this process). Gels can be cast and moulded to form a microporous preform and dried to produce a monolithic bulk material (e.g., a xerogel or an aerogel) that can be used to form filters and membranes. They can also be spin coated or dipped to produce thin (typically 50–500 nm) films on substrates. These films are used for electronic thin film devices, for wear, chemical or oxidation protection, as well as for their optical properties (e.g., anti-reflection). Alternately, fibres can be drawn from the gel; e.g., silica fibres for light transmission.

The interconnected nanoscale porosity in the dried gel can be filled via incorporation of a second material using techniques such as liquid infiltration or chemical reaction. These materials may then be classed as nanocomposites. Diphasic gels use the initial gel host for the precipitation of a second phase by sol–gel routes. In another variant, organic material can be incorporated as a monomer within an inorganic gel host; the monomers can be subsequently polymerized to form a hybrid material. If a dense rather than a nanoporous material is desired, drying is followed by sintering at higher temperatures. The high surface area leads to rapid densification, which can be accompanied by significant grain growth if temperatures are too high.

1.4.2.7 Electrodeposition

Electrodeposition involves inducing chemical reactions in an aqueous electrolyte solution via use of an applied voltage: methods may be classified as either anodic or cathodic processes. In addition to the simple deposition of metallic thin films, typical examples relevant to nanostructured materials include the deposition of metal oxides and chalcogenides. The electrodeposition process obviously requires an electrically conducting substrate.

In an *anodic* process, a metal anode is electrochemically oxidized in the presence of other ions in solution, which then react together and deposit on the anode. For example:

$$Cd \rightarrow Cd^{2+} + 2e^-,$$
$$Cd^{2+} + Te^{2-} \rightarrow CdTe. \tag{1.26}$$

Meanwhile in a *cathodic* process, components are deposited onto the cathode from solution precursors. In the case of metals this process is known as electroplating:

$$Cd^{2+} + 2e^- \rightarrow Cd,$$
$$HTeO_2^+ + 4e^- + 3H^+ \rightarrow Te + 2H_2O, \tag{1.27}$$
$$Cd^{2+} + Te^{2-} \rightarrow CdTe.$$

If a metal is oxidized and then undergoes hydrolysis or the local pH at the electrode is changed by liberating hydrogen gas electrochemically and producing OH^- ions, then metal oxide materials may be deposited.

Electrodeposition is relatively cheap and can be performed at low temperatures which will minimize interdiffusion if, say, a multilayered thin film material is being prepared. The film thickness can be controlled by monitoring the amount of charge delivered, whereas the deposition rate can be followed by the variation of the current with time. The composition and defect chemistry can be controlled by the magnitude of the applied potential, which can be used to deposit non-equilibrium phases. Pulsing or cycling the applied current or potential in a solution containing a mixture of precursors allows the production of a multilayered material. The potential during the pulse will determine the species deposited whilst the thickness of individual layers is determined by the charge passed. Alternatively, the substrate can be transferred periodically from one electrolytic cell to another. The final films can range in thickness from a few nanometres to tens of microns and can be deposited onto large specimen areas of complex shape, making the process highly suitable for industrial use.

Electrodeposition can also be performed within a nanoporous membrane which serves to act as a template for growth; for example, anodized aluminium has cylindrical nanopores of uniform dimensions (see Figure 1.36) and electrodeposition within this membrane can produce nanocylinders. Deposition on planar substrates can also limit nanocrystal growth and produce ordered arrays; if the growth is epitaxial then any strain due to lattice mismatch between the nanocrystal and the substrate can be growth-limiting. Furthermore, it is possible to modify the surface of substrates (e.g., by STM or AFM) to produce arrays of defects which can act as nucleation sites for the electrodeposition of nanocrystals.

Electroless deposition processes electrons by chemical reactions rather than involve the generation of using an external current as in electroplating and anodization. Unlike electrodeposition, substrates are not required to be electrical conductors, hence biological materials may be coated. One method employs deposition from solutions containing reducing agents where the deposited metal acts as a catalyst for the reduction/deposition process.

1.4.3 Methods for templating the growth of nanomaterials

As discussed in relation to lithographic methods (Section 1.4.1.2), resists can be used as templates for subsequent deposition or etching procedures. Here we consider templating methods utilized principally for wet chemical, bottom-up fabrication methods. One simple method of templating is to precipitate or deposit a material on the outside of another crystal and control the growth. If the template is organic then subsequently it may be washed or burnt out, leaving a porous centre if desired, such as the sol–gel derived silica nanotube templated on tartaric acid shown in Figure 1.1(b). Alternatively a material may be precipitated inside a host material that regulates its growth, as described in detail below.

Small semiconductor nanocrystals, such as CdS_xSe_{1-x}, can be embedded in a glass by the addition of cadmium, sulphur and selenium to a silicate glass melt. The glass is cast and annealed below the melting point, causing nanocrystals to form in the dense glass matrix. However, control of precipitate size can be difficult using this method. Alternatively a porous glass, obtained from sol–gel routes, can be infiltrated at low temperatures with a precursor solution which is then allowed to undergo in-situ precipitation; the pore size of the glass controls the nanoparticle size. Such materials can be used as colour filters and employed in optical devices. Precipitation of general nanocrystals within polycrystalline matrices, such as ferrite (α-Fe) or a ceramic, can be complicated by the crystallization of the matrix; variation in crystallization rates and segregation effects at grain boundaries can lead to an inhomogeneous distribution of precipitates within the microstructure and an inhomogeneous size distribution.

Crystalline hosts such as zeolites can be used as reaction vessels for nanoparticle formation. Zeolites are a range of aluminosilicates, based on building blocks such as sodalite cages, which contain well-ordered, well-defined, interconnected cavities in three dimensions within the structure, having dimensions of the order of nanometres (Figure 1.33).

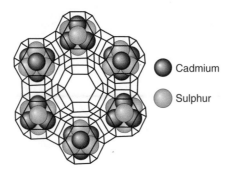

Cadmium

Sulphur

Figure 1.33 The structure of zeolite Y: note isolated Cd_4S_4 units are shown contained within the constituent sodalite cages

The cavities are often charged and are accessed by well-defined windows, which provide a means of transporting reactants to the internal cages. The physical size of the cavities limits the growth of nanoparticles.

Metallic or magnetic nanowires can be formed using a mica or plastic template in which straight, cylindrical tracks have been formed by a high-energy particle beam. Alternatively, metal nanowires can be produced by electrodeposition onto the cleaved edge of an MBE-grown semiconductor heterostructure. The metal preferentially deposits onto the edge of one specific semiconductor layer, yielding a nanowire of a few nanometres in thickness. This approach is an example of the templated growth of a quasi-1D system on a quasi-2D system.

Nanoparticle growth can be regulated within self-organized, biological and synthetic organic membrane assemblies; e.g., micelles, microemulsions, liposomes and vesicles (Figure 1.34). The molecules of these assemblies have a polar head group and a non-polar hydrocarbon tail, which self-assemble into membrane structures in an aqueous environment. Aqueous and reverse micelles have diameters in the range 3–6 nm, whereas microemulsions possess diameters of 5–100 nm. Liposomes and vesicles are closed bilayer aggregates formed from either phospholipids (liposomes) or surfactants (vesicles). Single bilayer vesicles are 30–60 nm in diameter, but multilamellar vesicles can be as large as 100–800 nm. The membrane structures described above can serve as reaction cages to control the nucleation and growth of particles and also to prevent agglomeration.

Infiltration of cations into the structure of a polymer by ion-exchange methods followed by solution or gas chemical treatment to produce in situ precipitation results in polymer composites with a dispersed nanophase. Alternatively a polymer or monomer solution may be simply mixed with a stable colloid and polymerized and dried to produce a composite material. These techniques are used extensively for semiconductor and electroceramic polymer composites, which may then be spin coated onto a substrate to produce a doped polymer thin film.

The macromolecular structure of organic polymers is defined in terms of four variables: the molecular size of the polymer chains, their composition, the sequence of the respective monomer units in the polymer, and their stereochemistry. Increased

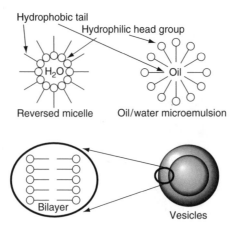

Figure 1.34 Schematic diagram of different self-assembled membrane structures. Image courtesy of IBM Zurich Research Laboratory

control over these variables during polymerization significantly enhances the properties of polymers for industrial application, such as crystallinity and mechanical strength. Usually polymers are synthesized in the laboratory using the iterative (stepwise) addition of selectively activated monomer units to a growing polymer chain, however this becomes increasingly difficult with increasing length and complexity of the polymer molecule. Template-directed polymerization overcomes these problems by using a master template to direct polymer growth in a parallel process from the component monomers; this also provides a verification and proof-reading mechanism for the final fabricated polymer structure. Prime examples of template-directed growth exist in biology; for example, DNA serves as a template for the synthesis of polypeptides through a messenger RNA sequence. Well-defined polypeptides, unlike most synthetic polymers, undergo self-assembly to form ordered three-dimensional structures in solution and in the solid state, as described in the following section.

1.4.4 Ordering of nanosystems

As outlined in Section 1.3, individual nanocrystals or nanoparticles often show very different properties from those of the corresponding bulk material. A further consideration, governing the overall properties of complete nanosystems, is the interaction and coupling between the individual nanocrystalline building blocks. The overall electric, optical, magnetic and transport properties of the nanostructure will then depend not only on the individual nanocrystalline units but also on the coupling and interaction between the nanocrystals, which may be arranged with long-range, translational and even orientational order. A prime example of this phenomenon is a photonic crystal – a periodic array of dielectric particles having separations of the order of the wavelength of light.

There are several ways of producing ordered nanoarchitectures: multicomponent fabrication of multilayered systems by MBE growth of metal or semiconductor heterostructures (Section 1.4.2.3), by processing of particles with core–shell structures, or by fabrication followed by further processing such as etching or focused ion beam milling. Manual or robotic manipulation of nanoparticles on a surface by scanning probe microscopy tips to produce complex structures such as quantum rows and quantum corrals or a molecular abacus (Figure 1.35) with well-defined electronic properties. Initial imaging using AFM or STM is followed by using the tip to push each particle along a desired trajectory, and although overall throughput is slow, parallel arrays of tips in a MEMS device could, in principle, assist larger-scale production. Chemical, physical or geometrical self-assembly or self-organization of atoms, molecules or nanoparticles to form highly ordered nanostructures, as described below.

1.4.4.1 Self-assembly and self-organization

Self-assembly and self-organization of nanostructures is an important area which often bridges the divide between organic and inorganic systems. Many self-assembly processes rely on the self-assembling nature of organic molecules, including complex species such as DNA; these methods are termed chemical or molecular self-assembly. Generally molecular self-assembly is the spontaneous organization of relatively rigid molecules

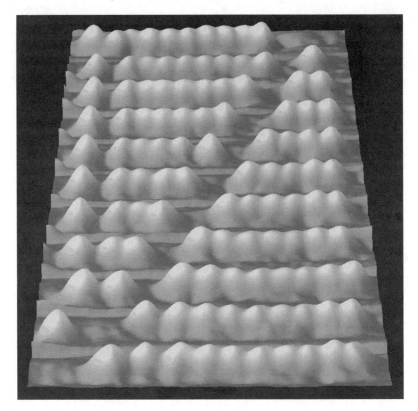

Figure 1.35 A 'molecular abacus'. Image courtesy of IBM Zurich Research Laboratory

into structurally well-defined aggregates via weak, reversible interactions such as hydrogen bonds, ionic bonds and van der Waals bonds. The aggregated structure represents a minimum energy structure or equilbrium phase. Self-assembly is also found throughout biological systems, micelles and liquid crystals and is being increasingly used in synthetic supramolecular chemistry.

Other simpler methods rely on geometric self-organization, in which hard spheres or hard rods will arrange themselves into two- and three-dimensional structures based on packing considerations. For example, solutions of colloidal metal particles can spontaneously order themselves into 2D hexagonally close-packed sheets on substrates. The individual nanoparticles are often encapsulated in a protective organic coating to provide stabilization and a degree of control over the final self-organized structure; for example, gold nanoparticles stabilized with an organic surfactant will self-organize into ordered arrays in which the interparticle spacing depends on the length of the surfactant molecule. Molecular systems, such as rod-like and disc-like liquid crystals, also exhibit geometric self-organization properties.

A variation on geometric self-organization is templated self-organization, in which an ordered nanostructure is formed by deposition of a material around a previously self-organized template. This approach can be used to produce porous metallic structures via electrodeposition on geometrically self-organized polystyrene spheres of submicron dimensions. The spheres are subsequently dissolved to leave a highly porous, ordered structure. A similar process has been used to produce 3D ordered mesoporous ceramics such as MCM41, which contains an ordered hexagonal pore structure (2–15 nm in size) and has extensive applications in catalysis. Figure 1.36 shows an AFM image of an ordered hexagonal array of pores in an aluminium oxide thin film grown on a pure aluminium substrate by anodization (electrochemically induced oxidation).

More complex self-assembly processes involve the use of self-assembled monolayers (SAMs). SAMs comprise organic molecules whose functionality can be modified by chemical treatment or radiation (e.g., lithography) so that subsequent layers can be selectively attached and used to direct oriented crystal growth. The ends of the molecules are usually terminated with a thiol group to provide good adhesion to a gold substrate. The molecules will order on the surface under given conditions of concentration, pH and temperature. An extensive discussion of molecular self-assembly is given in Chapter 7.

Another important category of self-assembly processes is the self-assembled growth of semiconductor quantum dots. This is achieved via use of a three-dimensional MBE growth mode (the Stranski–Krastonov mode) on a lattice-mismatched substrate (e.g., InAs on GaAs). The strain fields resulting from the lattice mismatch give rise to island growth with a well-defined geometry (e.g., pyramids, cones or lenses) depending on the precise material combination and growth conditions. Unlike colloidal quantum dots, the Stranski–Krastonov dots cannot exist in isolation, and are encapsulated by a subsequently grown semiconductor layer having a sufficient band-gap difference to provide electronic confinement. The quantum dots have no in-plane ordering, but if

Figure 1.36 A 1 micron square AFM image of an ordered hexagonal array of pores in an anodic alumina film. Image courtesy of Joe Boote and Professor Steve Evans, Department of Physics and Astronomy, University of Leeds

subsequent InAs layers are deposited, the second layer of dots will preferentially nucleate above the first layer. In this way, long columns of quantum dots can be built up with almost perfect vertical ordering. Further details are given in Chapter 3.

1.5 PREPARATION, SAFETY AND STORAGE ISSUES

The preparation of nanoscale systems generally requires a well-regulated laboratory environment since the systems are so small that contamination, even at low levels, becomes a highly important issue. In addition to the use of ultra-high purity reagents, this means that there should also be good control of temperature and humidity, which necessitates the provision of a good air-conditioning system, as well as a well-defined, clean water or solvent supply. Generally the environment should be dust-free and often vibration- and draught-free; it should possess adequate ventilation in the form of fume cupboards, and it should even have facilities for fabrication under a controlled atmosphere environment through use of gloveboxes. Such laboratories are termed clean rooms and will have a standard ISO designation based on the degree of environmental control; this is usually based on the number of submicron particles in a given volume of air. Overall the room is under positive pressure to prevent dust ingress, and incoming air is filtered and recirculated and protective clothing is worn to minimize sources of contamination. The air in a typical room would contain around 100 000 particles per cubic foot, and a fairly clean area may contain around 10 000 particles per cubic foot, formerly known as Class 10 000. Class 10 000 is now more commonly known as ISO Class 7; note that this new standard measures the number of particles per cubic metre and also includes particles smaller than 0.5 microns. ISO Class 7 is used for measurement and packaging areas in integrated circuit fabrication, whereas ISO Class 3 is used during fabrication. Class 1 implies minimal human contact.

Health and safety issues are paramount during any industrial revolution. From the field of environmental particulates there is increasing evidence that nanoscale materials are often highly biologically active, in that they can easily penetrate through biological matter and, owing to their high surface area, will have a large associated chemical reactivity which may result in extensive tissue damage if inhaled or ingested. This means that nanoscale particles and fibres may be biotoxic and, if in doubt, care should be taken when handling or breathing them.

Besides the problem of ageing in nanoparticle solutions, many nanoparticle solutions and more particularly powders can be very reactive (even explosive) due to their high surface area to volume ratio (such as metals in air) and they may need to be stored under an inert atmosphere and at low temperatures. In consolidated nanostructured materials, reactivity and diffusion processes (for example, grain growth) may also be a problem, particularly as the temperature is raised.

BIBLIOGRAPHY

R. R. H. Coombs and D. W. Robinson (Eds), *Nanotechnology in Medicine and Biosciences* (Gordon and Breach, New York, 1996)
A. S. Edelstein and R. C. Cammarata (Eds), *Nanomaterials: Synthesis, Properties and Applications* (Institute of Physics, 1998)

A. N. Goldstein (Ed.), *Handbook of Nanophase Materials* (Marcel Dekker, New York, 1997)

C. Hammond, *The Basics of Crystallography and Diffraction* (Oxford University Press, 1997)

C. C. Koch, *Nanostructured Materials: Processing, Properties and Applications* (Institute of Physics, 2002)

M. Kohler and W. Fritzsche, *Nanotechnology: An Introduction to Nanostructuring Techniques* (Wiley-VCH, Weinheim, 2004)

K. W. Kolasinski, *Surface Science: Foundations of Catalysis and Nanoscience* (Wiley, Chichester, 2002)

C. P. Poole and F. J. Owens, *Introduction to Nanotechnology* (Wiley, New Jersey, 2003)

M. C. Roco, R. S. Williams and P. Alivisatos, *Nanotechnology Research Directions* (Kluwer, Dordecht, 2000)

C. Suryanarayana, J. Singh and F. H. Froes (Eds), *Processing and Properties of Nanocrystalline Materials* (TMS, Warrendale, PA, 1996)

G. Timp (Ed.), *Nanotechnology* (Springer-Verlag, New York, 1999)

Z. L. Wang (Ed.) *Characterisation of Nanophase Materials* (Wiley-VCH, Weinheim, 2000)

2

Generic methodologies for nanotechnology: characterization

2.1 GENERAL CLASSIFICATION OF CHARACTERIZATION METHODS

In this section we will confine ourselves to a general description of characterization methods for both imaging (*microscopy*) and analysis (*spectroscopy*); in subsequent sections we discuss in more detail some of the common techniques employed for the investigation of nanostructures. Considering the large range of possible physical analysis techniques, very few microscopical analytical techniques are currently in widespread use. Generally there exist around ten possible primary probes (e.g., electrons, X-rays, ions, atoms, light (visible, ultraviolet, infrared), neutrons, sound) which may be used to excite secondary effects (electrons, X-rays, ions, light, neutrons, sound, heat, etc.) from the region of interest in the sample. The chosen secondary effects may then be monitored as a function of one or more of at least seven variables (e.g., energy, temperature, mass, intensity, time, angle, phase). This implies a theoretical maximum of around 700 single-signal characterization techniques (although some permutations of probe and effect are physically impractical) as well as the technically more difficult multi-signal techniques. To date, around 100 different techniques have been attempted, most of which use ions, electrons, neutrons or photons as the primary probes.

When ions are used as probes, generally ions and/or photons are emitted from the sample and analysed; examples include secondary ion mass spectrometry (SIMS), Rutherford backscattering spectrometry (RBS) and proton-induced X-ray emission

---

Nanoscale Science and Technology   Edited by R. W. Kelsall, I. W. Hamley and M. Geoghegan
© 2005 John Wiley & Sons, Ltd

(PIXE). When electrons are used as probes, generally electrons and/or photons are emitted and analysed; examples include scanning electron microscopy (SEM), electron probe microanalysis (EPMA), analytical transmission electron microscopy (TEM) including energy-dispersive X-ray (EDX) analysis and electron energy loss spectroscopy (EELS), Auger electron spectroscopy (AES), low-energy electron diffraction (LEED) and reflection high-energy electron diffraction (RHEED). When photons are used as probes, generally electrons and/or photons are emitted and analysed; examples include light microscopy, X-ray diffraction (XRD), X-ray fluorescence (XRF), X-ray absorption spectroscopy (XAS), infrared spectroscopy (IR), Raman spectroscopy and X-ray photoelectron spectroscopy (XPS).

In recent years there has been the development of proximal probe techniques such as atomic force microscopy (AFM), scanning tunnelling microscopy (STM) and scanning tunnelling spectroscopy (STS) which monitor the interaction between a localized probe and a sample surface. In other cases, such as position-sensitive atom probe (POSAP) spectroscopy, the sample itself is effectively the probe.

2.1.1 Analytical and imaging techniques

Before using a particular characterization technique it is essential to consider what information is required about a sample and at what resolution. Simplistically we may think of this information as being divided into:

- morphology (the microstructural or nanostructural architecture);

- crystal structure (the detailed atomic arrangement in the chemical phases contained within the microstructure);

- chemistry (the elements and possibly molecular groupings present);

- electronic structure (the nature of the bonding between atoms).

In addition, information on the mechanical, thermal or electronic properties of the sample may also be required. However, many of these properties will arise as a direct consequence of the structural factors outlined above and, ultimately, we need to establish a correlation between nanostructure and properties and determine how both these factors depend on the processing conditions employed to fabricate the material or the device. Eventually this information may be employed as input to computer-based modelling procedures, which may then allow the prediction of nanostructure–property–processing relationships.

Having decided on the type of information required, the next question to address is from what sample volume or area should this information be obtained? The resolution of a particular technique may be conveniently divided into lateral or spatial resolution (i.e., from what area on a sample surface or from what volume within a sample a particular signal originates?) and depth resolution (i.e., from how far below the sample surface a signal originates). Very simplistically a large number of techniques may be classified as either surface or bulk (volume) analytical techniques. Different types of nanostructure (e.g., films, dots, particles) require different combinations of lateral and

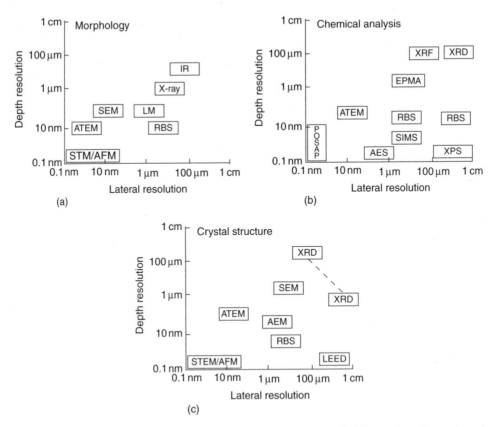

Figure 2.1 Schematic classification of the information content of different imaging and analytical techniques in terms of their lateral and depth resolutions

depth resolutions. Figure 2.1 presents a schematic classification of some of the more common techniques in terms of these two resolution parameters.

2.1.2 Some scattering physics

Electromagnetic radiation (e.g., X-rays, light) can be described in terms of a stream of photons, whilst correspondingly, particles (e.g., neutrons, ions and electrons) can be described in terms of an associated wavelength. This wave–particle duality of incident radiation is summarized by the de Broglie relationship, $\lambda = h/mv$, where λ and v are the wavelength and velocity, h is Planck's constant and m is the mass. When a beam of incident radiation interacts with matter it will be scattered. For each type of radiation there are a variety of possible scattering processes that may occur, each of which will have a particular scattering cross section σ which represents an area the scattering particle (e.g., electron, nucleus or atom in the material) appears to present to the incident particle or radiation. Alternatively the mean free path Λ (which is proportional to the inverse of σ) represents the average distance travelled by the incident particle between scattering events of a particular type. The frequency of occurrence of different scattering events will depend on the thickness of the specimen and the probabilities will follow Poisson statistics. When many scattering events occur, averaging approaches such as computer-based Monte Carlo simulations are commonly used to model the overall process.

During scattering, the incident wave may undergo changes in amplitude and/or phase, and scattering processes may generally be divided into two types:

- elastic scattering involves no energy transfer during the scattering process (hence no wavelength change of the scattered radiation), although there may be changes in the direction of the incident wave following scattering;

- inelastic scattering involves energy transfer during scattering (hence a gain or loss in the energy of the associated particles or quasiparticles) and changes in the direction of the incident wave.

Another distinction is made between (spatially) coherent scattering, where the phase relationship between scattered waves from neighbouring scattering centres is preserved, and incoherent scattering, where any phase relationships between scattered waves are lost. Generally small-angle elastic scattering is coherent, while inelastic scattering is incoherent; large-angle elastic scattering becomes rapidly more incoherent with increasing scattering angle.

2.1.2.1 X-rays and their interaction with matter

X-rays are produced via the electron bombardment of a metal target or anode. The emission spectrum consists of (a) white radiation or *bremsstrahlung* (from the German for 'braking radiation'), the wavelengths of which are related to the electron energy losses (ΔE) as they are decelerated through Planck's equation, $\Delta E = hc/\lambda$, and (b) characteristic radiations, of discrete wavelengths, which arise from ionization of inner shell electron levels followed by electron relaxation. The emitted X-rays are generally collimated and made monochromatic using filters or crystal monochromators which may be flat or curved to provide a focusing action. Alternatively extremely high intensity X-ray beams are produced by accelerating electrons around the storage ring of a synchrotron and undulating their trajectories ('wiggling') to release electromagnetic radiation of varying wavelength.

When X-rays interact with matter, elastic scattering occurs when X-rays are scattered by the electrons in the material and involves the interaction between the negatively charged electron cloud and the electromagnetic field of the incident X-rays. Electrons respond to the applied field, oscillate and emit an electromagnetic wave (X-ray) identical in wavelength and phase to the incident X-ray. Inelastic scattering occurs when the incident X-ray photons give up all or part of their energy to individual electrons associated with atoms. These electrons are excited to higher energy levels or ionized and escape from the solid as photoelectrons (Section 2.8.1). The ionized atoms in the solid undergo de-excitation to produce a variety of secondary signals. Alternatively, X-rays can lose part of their energy to an electron in a high-energy collision known as

Compton scattering. Overall, X-rays tend to undergo mainly elastic scattering with heavy elements (i.e., those with a large atomic number Z), while for light elements inelastic scattering predominates.

2.1.2.2 Electrons and their interaction with matter

Electrons are produced either by thermionic (heat) or by (electric) field emission from sharp metallic tips. Low-angle (1–10°), coherent elastic scattering of electrons occurs via the interaction of the electrons with the electron cloud associated with atoms in a solid. High-angle, incoherent elastic (back)scattering (10–180°) occurs via interaction of the negatively charged electrons with the nuclei of atoms. The cross section for elastic scattering of electrons varies as the square of the atomic number of the element.

Inelastic scattering of electrons occurs through smaller angles than elastic scattering and the cross section varies linearly with atomic number. Inelastic scattering of electrons occurs predominantly via the four major mechanisms outlined below.

Phonon scattering

The incident electrons excite phonons (atomic vibrations) in the sample, typically the energy loss is <1 eV, the scattering angle is quite large (~10°) and the mean free path Λ is ~1 μm for carbon. This is the basis for heating the specimen by an electron beam.

Plasmon scattering

The incident electrons excite collective, resonant oscillations (plasmons) of the valence electrons of a solid. The energy loss from the incident beam is 5–30 eV and Λ is ~100 nm, causing this to be the dominant scattering process in electron–solid interactions.

Single-electron excitation

The incident electron transfers energy to single electrons, resulting in ionization of atoms. The mean free path for this event is of the order of micrometres. Lightly bound valence electrons may be ejected from atoms, and if they escape from the specimen surface, they may be used to form secondary electron images in the SEM. Energy losses for such excitations typically range up to 50 eV. If inner shell electrons are removed, the energy loss can be up to keV. For example, the energy loss required to ionize carbon 1 s (K shell) electrons is 284 eV. The energy loss of the incident beam can be used in EELS analysis and the secondary emissions (e.g., X-ray or Auger electron production) produced when the ionized atom relaxes can also be used in analysis (energy-dispersive or wavelength-dispersive X-ray spectroscopy) as discussed in Section 2.7.3.

Direct radiation losses

The principal direct radiation losses are the bremsstrahlung X-ray emission caused by the deceleration of electrons, as described above. The energy losses can approach the total incident beam energy in the limit of full deceleration.

2.1.2.3 Neutrons and their interaction with matter

Low-energy thermal neutrons produced in nuclear reactors are most commonly used in materials analysis. Neutrons are neutral particles but have a relatively large mass and can interact with atomic nuclei through nuclear and magnetic forces. They can undergo elastic scattering and also inelastic scattering where the low-energy (slow) thermal neutrons induce atomic vibrations (phonons). High-energy (fast) neutrons can displace and even ionize atoms.

2.1.2.4 Ions and their interaction with matter

In comparison with electrons, ions are relatively heavy, negatively or positively charged particles and various effects occur upon their interaction with matter. These include ion backscattering (Rutherford backscattering spectrometry), the excitation of electrons and photons, X-ray bremsstrahlung, the displacement of atoms and sputtering, as well as the possible implantation of ions within the surface of the material. The latter is extensively used for doping semiconductors.

2.1.2.5 Elastic scattering and diffraction

The atomic scattering amplitude (factor) $f(\theta)$ describes the amplitude of X-rays, electrons or neutrons that are elastically scattered by individual atoms. For X-rays and electrons, this function decreases with increasing angle. In the case of X-rays, it is primarily the electrons (rather than the nucleons) which give rise to elastic scattering and $f_x(\theta)$ is defined as the ratio of the amplitude scattered by the atom to that of a single electron. It is a dimensionless number, equal to Z at low scattering angles but decreases below Z with increasing angles because of interference arising from path differences between the scattered rays from all Z electrons. In the case of electrons, it is primarily the protons in the atomic nuclei which give rise to elastic scattering. However, because of the shielding effect of the electrons, which is proportional to f_x, the electron scattering amplitude $f_e(\theta)$ is proportional to $(Z-f_x)$. It is normally expressed as a scattering length, hence electron intensities (the square of the amplitude) are expressed as areas or atomic scattering cross sections. On average, f_e is proportional to $Z^{3/2}$. Similarly, neutrons are scattered by the nucleons in atoms, but since they also possess spin, they interact and are scattered by unpaired electrons in magnetic materials. The neutron scattering amplitude (again expressed as a scattering length) varies in an irregular way with Z and also between different isotopes of the same element; in particular, hydrogen and oxygen scatter strongly in comparison with heavy atoms. Neutron scattering therefore complements X-ray and electron scattering in providing information about the positions of light atoms and atoms of similar Z. The important practical difference between X-ray, electron and neutron elastic scattering is that the magnitude of the scattering amplitude for electrons is roughly 10^4 times greater than the magnitude for X-rays, which are in turn greater than the magnitude for neutrons.

Since the elastically scattered X-rays, neutrons and electrons are coherent, they interfere and give rise to diffraction patterns; the diffracted beams arise from strong constructive interference from all the atoms in the material. The nature of the diffraction pattern depends on the nature of the material. For perfect crystals, with long-range periodic order, the diffraction pattern is sharply defined. As crystals become smaller and also when they become increasingly imperfect (due to the presence of defects and internal stresses), the diffraction pattern is less sharply defined. In amorphous materials (such as glasses and liquids) the diffraction pattern shows diffuse maxima which are related to the average interatomic spacings between the atoms; a radial profile through the diffraction pattern essentially gives a radial distribution function of the atoms surrounding a particular scattering centre in the amorphous solid. In short, in all cases diffraction patterns reveal the nature (i.e., the structure and the imperfections) of the material under investigation.

In crystalline solids, atoms are arranged periodically in regular, three-dimensional repeat motifs known as unit cells and this also leads to a corresponding periodic distribution of electrons. Waves coherently scattered from such periodically arranged centres therefore undergo interference leading to intense diffracted waves in specific directions which will depend on the exact three-dimensional periodicity (the crystallography). For radiation of wavelength λ incident on a crystal whose lattice planes are regularly spaced by a distance d, the allowed angles θ for diffraction are given by the Bragg equation

$$m\lambda = 2d\sin\theta \qquad (2.1)$$

for integer m. Summing the atomic scattering amplitudes of atoms in a particular plane of Miller indices (hkl), taking into account the path differences between scattered waves, allows the calculation of a structure factor F_{hkl}, the scattering amplitude for diffraction from a particular set of (hkl) atomic planes. The analysis of the angular distribution and the intensities of these diffracted waves (usually for a given wavelength of incident radiation) therefore provides information on the atomic identities and arrangements in the material. An introduction to the relationship between crystallography and diffraction is provided in Section 2.7. Incoherent high-angle elastic scattering is reasonably unaffected by this periodicity and depends only on the nature of the atoms not their relative positions and generally follows the decreasing $f(\theta)$ function at high angles.

A summary of the analytical applications of a variety of interactions between radiation and matter is presented in Table 2.1. It also provides a basis for the classification of many of the imaging and analytical techniques that are discussed in the following sections.

2.2 MICROSCOPY TECHNIQUES

Radiation–matter interactions provide a wide range of analytical and spectroscopic techniques, and although it is possible to obtain an analytical signal from a single nanostructured system or an averaged signal from a collection of nanostructured systems, images are a very powerful means of presenting information. Imaging the morphology, as well as the structural and chemical arrangements in nanostructures, necessarily requires microscopical techniques capable of high spatial and/or depth resolution.

Table 2.1 Summary of radiation–matter interactions and their analytical applications for characterization of nanomaterials (see Section 2.1 for a list of acronyms)

Primary radiation	Modification of matter	Modification of radiation	Secondary radiation (signal)	Applications
Photons (e.g., X-rays)	Elastic vibration	Elastic and quasi-elastic scattering		IR, Raman, light microscopy, XRD
	Recoil electron	Inelastic scattering		Compton spectrometry
	Bond breaking	Inelastic scattering: absorption		Photolithography
	Ionization			Optical absorption, XAS
			Photoelectrons	XPS
	De-excitation		Photons	Luminescence, XRF
			Auger electrons	Auger, XPS
Electrons	Thermal vibration	Elastic and quasi-elastic scattering		Backscattered electron imaging in SEM/TEM imaging and electron diffraction
	Plasmon	Inelastic scattering: energy loss		EELS
Electrons	Bond breaking	Inelastic scattering: absorption		Electron beam lithography
	Atom displacement Sputtering			
	Ionization		Bremsstrahlung Secondary electrons	EDX or EPMA SEM
		Inelastic scattering: energy loss		EELS or energy-filtered TEM
	De-excitation		X-rays	EDX or EPMA
			Auger electrons	AES
Neutrons	Thermal vibration	Elastic and quasi-elastic scattering		Neutron diffraction and reflectometry
	Bond breaking	Inelastic scattering		
	Atom displacement Sputtering			
Ions	Thermal vibration	Elastic and quasi-elastic scattering		RBS
	Bond breaking	Inelastic scattering: absorption	Emission of ions and radiation (e.g., gamma rays)	NRA
	Atom displacement Sputtering		Secondary ions	SIMS FReS Ion beam lithography
	Ion implantation			Implantation or doping
	Ionization		Bremsstrahlung	
	De-excitation		X-rays	PIXE

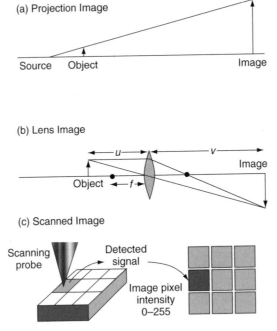

Figure 2.2 Formation of (a) a projection image, (b) a magnified image by a lens and (c) a scanned image

2.2.1 General considerations for imaging

The general purpose of microscopy is to resolve finer details than are discernible with the naked eye, which is 0.1–0.2 mm at the near point of vision, 250 mm. For nanostructured materials we need to resolve details down to a level of 100 nm and below. In transforming an object to an image, there are essentially three methods available, as shown in Figure 2.2:

- A projection image formed in parallel; e.g., field ion microscopy (FIM).

- A lens image formed in parallel; e.g., TEM and light microscopy. Here a lens of focal length f will form an image of an object at a magnification v/u, where u and v are the object and image distances, respectively. The relationship between these parameters is given approximately by the thin lens formula

$$\frac{1}{f} = \frac{1}{v} + \frac{1}{u} \tag{2.2}$$

from which the magnification v/u can be obtained.

- A scanned image formed in serial; e.g., scanning electron microscopy (SEM), scanning transmission electron microscopy (STEM), confocal laser scanning microscopy (CLSM), scanning tunnelling microscopy (STM) and scanning force microscopy (SFM).

Signals used to form an image can be reflected (e.g., reflected light microscopy or backscattered electrons in SEM), transmitted (e.g., transmitted or polarized light microscopy, or electrons in TEM and scanning TEM) or stimulated (e.g., secondary electrons, cathodoluminesence, X-ray emission or electron beam induced current (EBIC) in SEM, tunnelling current in STM or force in SFM). The concept of picture elements (pixels) is applicable to all images collected in either serial or parallel. Each pixel in the image is defined by x and y coordinates and may be assigned an intensity value; e.g., in terms of grey levels a number between 0 and 255, allowing images to be stored digitally on a computer. Imaging techniques capable of a 3D representation of the object (e.g., tomographic techniques) involve 3D volume elements (voxels). Digital images may be processed to reduce noise (e.g., by use of a framestore), alter contrast (e.g., by use of a non-linear intensity scaling) or analysed and quantified mathematically.

2.2.2 Image magnification and resolution

The magnification is defined as the ratio of the image size to the object size. This can be altered by changing the focal length of the lenses (Equation 2.2). However, making an image bigger does not necessarily increase the resolution. The (spatial) resolution is the smallest separation of two object points that are discernible as separate entities in the image.

When illumination from a point object passes through an aperture – either a physical aperture or a lens – diffraction (scattering and interference) occurs at the aperture giving rise to a series of cones (Airy's rings) with a characteristic intensity distribution. As shown in Figure 2.3, the diameter of the central Airy disc D_d (which contains most of the intensity) is inversely proportional to aperture diameter and is given by $1.22\lambda/n_r \sin \alpha$, where λ is the wavelength of the radiation, n_r is the refractive index of the medium between object and lens, and α is the semi-angle subtended by the aperture. Note that $n_r \sin \alpha$ is often called the numerical aperture (NA). Hence, due to diffraction, each point in the object becomes a small Airy disc in the image, known as a disc of confusion. The *Rayleigh criterion* states that two points in the object may be just distinguished when the maximum intensity of one Airy disc in the image

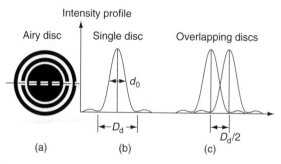

Figure 2.3 Schematic diagram of the Airy diffraction pattern in the image of a point, the intensity variation along the diameter of the Airy pattern and the overlap of the images of two points at the resolution limit

corresponds to the minimum of the second. Hence the limit of resolution r set by diffraction is given by

$$r = \frac{D_d}{2} = \frac{0.61\lambda}{n_r \sin \alpha}. \tag{2.3}$$

In light microscopy the limit of resolution may be decreased by decreasing λ and using green, blue ($\lambda = 0.4\,\mu m$) or UV light ($\lambda = 0.2\,\mu m$), or increasing n_r and using an oil immersion objective lens. Generally the achievable resolution of light microscopy is $0.2\,\mu m$, which is is not generally sufficient for the characterization of nanostructures. However, some specialized light microscopy techniques are applicable that are not diffraction-limited and that allow much lower limits of resolution, such as scanning near-field optical microscopy (SNOM) as discussed in Section 2.2.4.1. In electron microscopy the accelerated electron wavelength is $\sim 10^5$ times smaller than that of light, which leads to increased resolution. For electrons accelerated through a potential V, the corresponding de Broglie wavelength is given by

$$\lambda = \frac{h}{(2m_e eV)^{1/2}}. \tag{2.4}$$

ignoring relativistic effects, where m_e and e are the electron mass and charge, respectively. For an accelerating potential of $100\,kV$, this gives an electron wavelength of $0.0037\,nm$. For electrons in a vacuum $n_r = 1$, and electron lenses employ very small values of α ($\sim 10^{-3}$ radians), thus the diameter of the disc of confusion due to Airy diffraction simplifies to $D_d = 1.22\lambda/\alpha$, giving a limit of resolution of $\sim 0.2\,nm$. Electron microscopy techniques are discussed in Section 2.3. Finally, for scanned imaging techniques, the resolution is determined primarily by the size of the scanned probe: the smaller the probe, the higher the resolution. However, many scanning techniques such as SEM and STEM necessarily employ condenser lenses to demagnify the probe, which may limit the attainable resolution.

The depth of field z is the range of positions of the object, on either side of the object plane, which give no loss in resolution. Since the resolution is limited by diffraction, it is relatively simple to show that

$$z = \frac{0.61\lambda}{n_r \sin \alpha \tan \alpha}. \tag{2.5}$$

Electron microscopes operate with small values of α, thus the depth of field is relatively large compared to light microscopy, where α is much bigger. Typical values for a magnification of $\times 1000$ are $z = 40\,\mu m$ (SEM) and $1\,\mu m$ (light microscopy). A large depth of field is particularly useful for imaging three-dimensional (3D) structures and devices.

The depth resolution of a signal and hence an image depends on either the penetration depth of incident radiation or the escape depth of the monitored signal. This can vary from a few atom layers (e.g., XPS, Auger, SIMS) to micrometres (e.g., X-ray emission in SEM).

A large number of imaging techniques (for example, TEM) essentially provide a two-dimensional (2D) projection of what is in reality a 3D object; essentially there is no depth resolution and information is averaged through thickness. Tomographic techniques essentially reconstruct the 3D image, and hence the object, from a set of 2D projections measured along different directions of the incident radiation. An example of tomographic reconstruction using scanning force microscopy is given in Section 2.5.2. Alternatively, confocal techniques are based on images measured at different depths within a structure with a very small depth of field; this focal series may also be reconstructed to form a 3D representation of the object.

2.2.3 Other considerations for imaging

Image points in a particular microstructural feature will only be discernible if they have an intensity I_f greater than the intensity I_b of neighbouring points in the background, regardless of the resolution. The contrast C is given by $C = (I_f - I_b)/I_b$. Signals always contain some degree of noise N which will be given by Poisson statistics as $N = \sqrt{n}$, where n is the number of signal counts at a particular point. One criterion for visibility is the Rose criterion, which states that $(I_f - I_b)$ must be greater than 5 times the noise level.

Lenses for light or electrons are not perfect but possess physical and geometrical aberrations that distort the image and lead to a loss in resolution and quality above and beyond the diffraction-limited resolution. These may be summarized as follows:

- *Chromatic aberration* arises because lenses have different focal lengths for different wavelengths of radiation. If the radiation is not monochromatic (of a single λ or energy) then the lens will cause the radiation to be focused over a range of positions; this leads to a point in the object becoming a disc of confusion in the image.

- *Spherical aberration* arises due to the difference in path lengths of rays travelling through different portions of the lens. An axial ray travelling along the optic axis through the centre of a lens will be focused further away from the lens than a non-axial ray travelling through the extremities of the lens. This also leads to a point in the object becoming a disc of confusion in the image.

- *Astigmatism* arises when the focal length of the lens differs in two axial planes perpendicular to each other. For electromagnetic lenses this may be corrected by the use of externally imposed fields known as stigmators.

In light microscopy it is possible to reduce both chromatic and spherical aberrations by using combinations of concave and convex lens elements. However, this is not usually possible for electromagnetic lenses, where the only solution is to reduce the aperture angle, α, which makes the diffraction limited resolution worse. Also it is possible in both light microscopy, SEM and TEM to make the radiation more monochromatic, thereby minimizing chromatic aberrations. Recent developments in instrumentation allow the computer-controlled correction of spherical aberration in electromagnetic lenses in SEM, TEM and STEM; aberration-corrected STEM can produce sub-angstrom probes for imaging, as evinced by the SuperSTEM facility at Daresbury Laboratories, UK.

2.2.4 Light microscopy

As stated the resolution limits for visible or UV light microscopies are insufficient for the direct visualization of nanostructures. However, such techniques may be extremely useful for the analysis of larger-scale devices and architectures constructed from nanoscale elements. One important point concerning light microscopy is that the *detection limit* for dark lines on a white background (white ground/bright field) is considerably less than the resolution (diffraction) limit, while for dark ground/dark field epi-illumination conditions (white lines on a dark background) there is essentially no detection limit other than that of the photon detector. Indeed, epi-illumination fluorescence light microscopy is based on the detection of probes (e.g., DNA markers) at the Angström level (although not their resolution). However, we do not intend to go into great detail on these topics here and the interested reader is referred to the bibliography for a comprehensive description. Key aspects to note are that, owing to the nature of the radiation, the sample can be viewed under ambient conditions and can also be subjected to dynamic experiments whilst being simultaneously observed.

Light microscopy techniques divide into reflection and transmission methods and both may require some sample preparation such as polishing or thin sectioning. Both transmission and reflection light microscopy can exploit the optical anisotropy (birefringence) of specimens by use of incident plane or circularly polarized light and can give colour contrast that depends on specimen anisotropy, orientation absorption and (in the case of transmitted light) thickness.

Two powerful techniques that have arisen in recent years are confocal methods and near-field optical techniques. As stated the diffraction-limited resolution of light microscopy is generally too large for application to the majority of nanostructures, however, scanning optical microscopy (SOM) has had applications. Here a beam of light is scanned across the surface of a specimen. Scanning is performed either by using a spinning disc with a pattern of holes or by using a focused probe such as a laser. The signal is either reflected, fluorescent or transmitted light. To achieve a good resolution, small-diameter spots are required with a large convergence angle which thus limits the depth of field. Confocal scanning light microscopy involves focusing the imaging signal onto a detector aperture in a plane conjugate with the objective focal plane. As shown in Figure 2.4, this

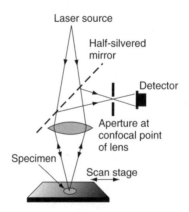

Figure 2.4   Schematic diagram of a confocal scanning light microscope

effectively selects signals from just the focal plane of the objective (i.e., with a very small depth of field). This allows optical sectioning of the image of the specimen, which is useful for quantitative measurements of topography; the sections can be stored and recombined to give an extended focus image.

2.2.4.1 Scanning near-field optical microscopy

Scanning near-field optical microscopy (SNOM) is a scanning optical microscope surface (reflection) technique that provides a limit of resolution beyond the diffraction limit. The sample is illuminated in the near field with light of a chosen wavelength passing through an aperture a few tens of nanometres in diameter formed at the end of a single-mode drawn optical fibre coated with aluminium to prevent light loss, thus ensuring a focused beam from the tip. The probe is usually held fixed while the sample is scanned employing the same piezoelectric drivers used in most commercial scanning probe microscopes (Section 2.5). The tip-to-sample distance may be accurately controlled using a feedback circuit and used to generate both optical images (at a resolution below 100 nm) and topographic images (tip displacement images). The most commonly chosen imaging mode is fluorescence, but other variants include UV-visible, IR and Raman techniques, which can provide chemical information using near-field spectroscopy. Despite the promise that it offers, SNOM remains a less widely used technique than AFM, and is generally perceived as being experimentally challenging.

2.3 ELECTRON MICROSCOPY

2.3.1 General aspects of electron optics

Electron microscopy is an extremely important technique for the examination and analysis of both the surface and subsurface of nanostructured systems (SEM) as well as the bulk structure of thin samples, usually averaged through thickness (TEM and STEM).

In many respects, electron optics may be regarded as analogous to light optics; the main differences may be summarized as follows. Firstly, the wavelength of accelerated electrons is very much smaller than the wavelength of visible or UV photons, which implies a much greater resolution for electron imaging. Secondly, electrons interact much more strongly with matter than photons, thus all optical paths in an electron microscope must usually be under a vacuum of at least 10^{-4} Pa. However a major benefit for nanostructures is that even though the signal is derived from a small amount of material, it will be relatively strong compared to that derived using other incident radiations. Thirdly, electrons are charged particles and may thus be focused by magnetic or electric fields. Furthermore, this focused electron beam may be scanned relatively easily using electrostatic fields. Finally, there is therefore no change in n_r during passage through such a lens, so n_r may be taken as unity, which simplifies the optical formulae. However, in order to reduce lens aberrations when using electromagnetic lenses, the value of α is kept very small and only rays close to the optic axis are employed.

2.3.2 Electron beam generation

The electron beam may be produced either by thermionic emission, where thermal energy is used to overcome the surface potential barrier (work function) of a solid source and so allow extraction of electrons from the conduction band of an emitter, or by field emission, in which an extremely high electric field is employed to reduce the surface potential barrier of an emitter. The finite width of the barrier allows quantum tunnelling of electrons out of the source at room temperature (cold field emission), although higher temperatures have been used more recently (Schottky or thermally assisted field emission), which requires a lower field strength.

A typical thermionic electron gun employs a tungsten cathode filament which is resistively heated to 2800 K in a vacuum of 10^{-4} Pa to give the electrons sufficient energy to overcome the work function of the metal. The emitted electrons are collimated and focused using a Wehnelt cylinder to a beam of typically 50 µm in diameter. The anode potential is typically 1–20 kV in the SEM and 100–200 kV in the TEM.

An important electron gun parameter is the brightness of the source β, the current density per unit solid angle. Increased brightness allows either the use of higher currents, which improve sensitivity and image contrast, or smaller probe areas, which yield higher spatial resolution. The brightness may be increased by using a lanthanum hexaboride (LaB_6) filament that possesses a lower work function than tungsten and a smaller source size (~1 µm), giving an order of magnitude increase in β. Alternatively, a field emission gun (FEG), which employs a high field strength surrounding a pointed tungsten cathode of radius 100 Å, gives a source size as low as 5 nm and a brightness increase of 10^4 over a thermionic tungsten filament. Both sources possess a longer lifetime than tungsten filaments but require a higher vacuum in the gun region. These sources also give a much reduced energy spread of the emitted electrons, leading to improved resolution as a result of reduced chromatic aberration in spectroscopic analysis.

2.3.3 Electron–specimen interactions

Figure 2.5(a) summarises the various signals produced as a result of electron–specimen interactions in SEM and shows the calculated volumes and penetration depths which give rise to the different signals from thick specimens. The *interaction volume* is defined as the volume within which 95% of the electrons are brought to rest by scattering, and has a characteristic teardrop shape (as shown in the figure). The depth and lateral width of electron penetration in the specimen are roughly proportional to V^2 and $V^{3/2}$, respectively, where V is the accelerating voltage. Materials of high (average) atomic number will exhibit reduced electron penetration depths and increased lateral spread relative to those of lower atomic number. For comparison, Figure 2.5(b) shows the signals produced in thin specimens appropriate for TEM imaging. Besides the primary electrons, there are various types of emitted electrons which leave the surface of a bulk sample and may be used for imaging or analysis, as summarized in the following sections.

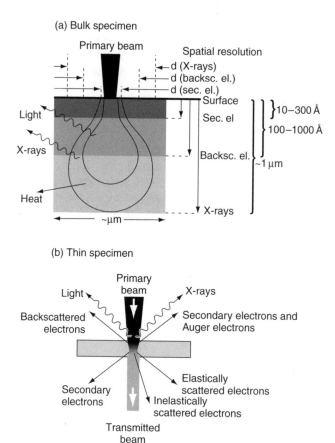

Figure 2.5 Schematic diagram of the beam–specimen interaction in (a) a thick specimen and (b) a thin specimen

2.3.3.1 Secondary electrons

The term 'secondary electrons' is used in general to describe any electrons that escape from the specimen with kinetic energies below about 50 eV. They are most likely to arise from ionized electrons previously associated with atoms close to the surface of the solid which have gained a small amount of kinetic energy and escaped. Alternatively they may be primary electrons which have lost nearly all their energy through scattering and have reached the surface. Secondary electrons are extremely abundant and the secondary electron yield, δ (the number emitted per primary electron), is dependent on the accelerating voltage and can even exceed 1, hence they are extensively used for imaging in SEM.

2.3.3.2 Backscattered electrons

Some primary electrons undergo large deflections and leave the surface with little change in their kinetic energy; these are termed backscattered electrons (BSE). The backscattered electron yield, η, is almost independent of accelerating voltage and is small compared with δ, but backscattered electrons are also used for imaging in SEM as η is strongly dependent on Z (unlike δ) owing to the fact that they arise from Rutherford backscattering from the nucleus. Owing to this nuclear interaction, backscattered electron imaging can be used to differentiate phases of differing average atomic numbers.

2.3.3.3 Auger electrons and emitted X-rays

Atoms which have undergone inner shell ionization and have been promoted into an excited state by the primary electron beam relax when electrons from higher energy levels drop into the vacant inner shells. This process results in the release of the excess energy between the electron energy levels involved in the transitions (ΔE), producing low-energy (100–1000 eV) *Auger* electrons, or X-rays, or visible photons of wavelength $\lambda = hc/\Delta E$. The Auger yield is small, except in the case of light elements, and is extremely useful for surface analysis (Section 2.8). The energies and wavelengths of the emitted X-rays, which are characteristic of the atom involved, are used in elemental analysis (Section 2.8.1).

2.3.4 Scanning electron microscopy

The scanning electron microscope (SEM) is extremely useful for imaging surface and subsurface microstructure. The basic layout of SEM instrumentation is shown in Figure 2.6.

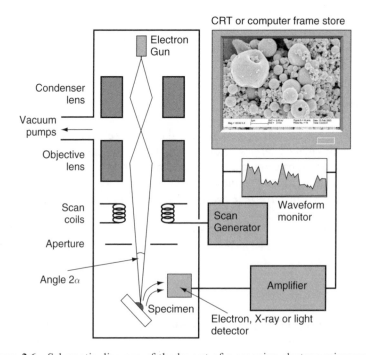

Figure 2.6 Schematic diagram of the layout of a scanning electron microscope

The electron gun usually consists of a tungsten or LaB_6 filament, however FEGs are becoming increasingly necessary for imaging nanostructures at high resolution. The accelerating voltage is usually between 1 and 30 kV. The lower voltages can be used with high-brightness LaB_6 and FEG sources and they are often employed to increase surface detail and obviate sample charging in non-conducting or poorly conducting samples.

In the SEM, two or more condenser lenses are used to demagnify the crossover produced by the gun, while the objective lens focuses the electron probe onto the specimen so that the final probe diameter lies between 2 and 10 nm. The objective aperture limits the angular spread α of the electrons. The focused beam is scanned across the specimen surface in a two-dimensional raster, with the beam passing through the optic axis at the objective lens; meanwhile, an appropriate detector monitors the secondary electrons or other signal, such as backscattered electrons or X-rays, as they are emitted from each point on the surface. Simultaneously, using the same scan generator, a beam is scanned across the recording monitor. The intensity of each pixel on the monitor is controlled by the amplified output of the selected detector and is therefore directly related to the emission intensity of the selected interaction at the corresponding point on the specimen surface. The magnification is simply the ratio of the monitor raster size to the specimen raster size and, as in field ion microscopy, it involves no lenses.

2.3.4.1 Secondary electron imaging

During electron irradiation, a wide spectrum of low-energy secondary electrons are emitted from the surface of the specimen with a high emission cross section, giving images with a high signal-to-noise ratio. They are detected by a scintillator–photo-multiplier system known as an Everhart–Thornley (ET) detector positioned to the side of the specimen chamber. Surrounding the scintillator is a grid (a Faraday cage) which has a small positive bias to attract the low-energy secondary electrons that are travelling in all directions away from the specimen surface. High-energy backscattered electrons travelling directly towards the detector will also contribute to the image, however due to the small detector collection angle, this signal is relatively low. The ET detector will collect nearly all secondary electrons and therefore images appear to look down holes and over hills. However, the signal from sample areas in direct line-of-sight of the detector is slightly higher than the signal from 'unseen' parts of the specimen and consequently they will appear brighter, an effect known as shadowing contrast. The detector and a typical secondary electron image are shown in Figure 2.7.

The shape of the interaction volume relative to the specimen surface, and hence the secondary electron yield, depends on the angle of inclination θ_{inc} between the beam and the specimen surface, varying as $1/\cos \theta_{inc}$. This *inclination effect* means that the edges of a spherical particle will appear brighter than the centre and, for studies of specimen topography, specimens are normally tilted 20–40° towards the detector to maximize the signal. Due to increased field strength at sharp edges and spikes on the specimen surface, the emission of secondary and backscattered electrons increases, and such features will appear bright, known as edge contrast. Consequently, owing to the surface specificity and the contrast effects mentioned above, the major use of secondary electron imaging is for topographic contrast. In general, topographic images obtained with secondary electrons have the same appearance as an object illuminated with diffuse

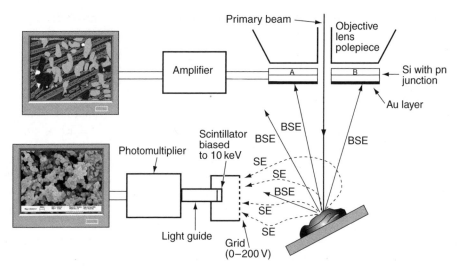

Figure 2.7 Schematic diagram showing a secondary electron detector and a solid-state back-scattered electron detector together with typical images

light, in that there are no harsh shadows. Quantitative information on surface topography may be obtained using stereomicroscopy techniques in which two images are taken under identical magnification conditions at slightly different tilt angles (e.g., $-5°$ and $+5°$). The images may be viewed using a special stereo viewer (which allows the brain to interpret them as a 3D picture). For quantitative work, the relative displacement or parallax between two features on the pair of tilted images allows their height separation to be determined.

Since the detected secondary electrons only originate from the top few nanometres of the specimen surface, the emission diameter will only be slightly bigger than the probe diameter (Figure 2.5). The lateral resolution in secondary electron images is therefore of the order of 1–5 nm for typical specimens but will obviously vary with probe size and signal-to-noise considerations. The depth resolution is a function of the accelerating voltage. Secondary electron imaging is therefore the most useful general signal for imaging nanostructures in the SEM due to the high signal level combined with high lateral and depth resolution.

2.3.4.2 Backscattered electron imaging

BSEs have very high energies and low yields compared to secondary electrons. The ET detector may be used to collect only BSEs by making the grid bias slightly negative so as to exclude secondary electrons. However, only those electrons travelling in line-of-sight towards the detector will contribute to the relatively low signal and therefore images will show very strong shadowing contrast and a low signal-to-noise ratio. More normally a large-area solid-state BSE detector is employed; this consists of a thin, annular-shaped semiconductor device (a reverse-biased pn junction) attached to the bottom of the objective lens, as shown in Figure 2.7. High-energy BSEs incident on this detector cause the formation of electron–hole pairs which, once separated, form a current and may be amplified. The response time is relatively slow, which necessitates the use of slow scan rates.

The resolution in BSE images is typically in the range 25–100 nm, which is significantly worse than in secondary electron images due to the larger penetration depth and thus sampling volume (Figure 2.5). Although the BSE yield η depends sensitively on atomic number Z, it is essential to minimize topographic effects via the use of flat, well-polished specimens if elemental contrast imaging is required.

In a crystalline material, η also depends weakly on the orientation of the crystal with respect to the incident beam. If the beam is parallel to a set of atomic planes, electrons are *channelled* into the crystal structure, giving abnormally large penetration depths, which means that BSEs are less likely to escape to the surface and therefore η is considerably reduced. In a well-polished, undeformed polycrystalline material, grains at different orientations will therefore show different channelling contrasts. Electron backscattered diffraction (EBSD) patterns from a given probe position on the specimen are recorded using a tilted sample and a specially inserted phosphor screen or scintillator to record the angular variation of the BSE yield. When the sample is translated relative to the beam, EBSD may be used to map the relative crystallographic orientations of grains within a microstructure at a resolution ≥ 100 nm.

2.3.4.3 Voltage contrast imaging

Voltage contrast arises since the secondary electron yield is sensitive to the variation in electric fields which may occur at the surface of semiconducting specimens; positive potentials discourage secondary electron emission whereas negative potentials enhance it. Biased semiconductor devices can therefore be imaged in SEM with negatively biased regions appearing bright.

2.3.4.4 Electron beam induced current imaging

Electron beam induced current (EBIC) contrast arises from ionization of valence electrons in the specimen by the incident beam. Normally the electron–hole pairs recombine to produce photons, called cathodoluminescence (Section 2.7.3.2), but in a biased semiconductor specimen the two charge carriers may be separated, giving rise to a current. Measurement of this current allows images to be formed and can give information on conductivity, carrier lifetimes and mobilities. This is useful for the dynamic imaging of integrated devices during operation.

2.3.4.5 Magnetic contrast imaging

Magnetic contrast arises due to the interaction of the external leakage magnetic fields at a specimen surface which can deflect the emitted secondary electrons and BSEs, or via the interaction of the primary beam with the internal magnetic field of the specimen, which will deflect the beam and change emission yields. In principle both mechanisms allow imaging of magnetic domain structures and can also be employed for Lorentz microscopy in the TEM.

2.3.4.6 Environmental scanning electron microscopy

Environmental scanning electron microscopy (ESEM) or variable pressure SEM relies on imaging in a much degraded vacuum in the SEM chamber, but not the electron gun region. Imaging mechanisms in ESEM (particularly secondary electron imaging) rely on the amplification of the signal by the ionization of gas molecules in the chamber. The introduction of gas into the chamber has a number of benefits as well as drawbacks: (a) It is possible to image uncoated non-conducting samples since specimen charging is compensated for by attraction of ionized gas molecules or electrons to the specimen surface. (b) It is possible to image specimens which are in the hydrated state or even reacting in-situ within the SEM. In particular, dehydration of an SEM sample within the vacuum can lead to large changes in specimen morphology. (c) Image resolution is degraded due to interaction of the incident and scattered electrons beams with the gaseous atmosphere.

2.3.5 Transmission electron microscopy

The conventional transmission electron microscope is a key tool for imaging the internal microstructure of ultrathin specimens. The basic TEM instrumentation is summarized in Figure 2.8. The electron gun is usually thermionic tungsten or LaB_6, however FEGs are becoming increasingly common. The accelerating voltage is considerably higher than in an SEM and is typically 100–400 kV, although a number of specialist high-voltage instruments are designed to operate at 1 MV and above. The benefits of high voltage include increased imaging resolution, due to the decreased electron wavelength, and also increased penetration and thus the ability to study thicker samples. Two or more condenser lenses demagnify the probe to typically 1 µm in diameter, although this can be reduced via use of a condenser–objective nanoprobe system. The condenser excitation controls the beam diameter and convergence. The first condenser (C1) controls the demagnification of the crossover (the spot size), whilst the second condenser (C2) controls the size and convergence of the probe at the specimen, and hence the area of the sample that is illuminated. The specimen must be no more than a few hundred nanometres in thickness, and is usually in the form of a 3 mm diameter disc. The specimen is located between the pole pieces of the objective lens. The combination of the objective lens and the projector lens system provides an overall magnification of around 10^6. The *selected area diffraction* (SAD) *aperture* allows selection of a minimum sample area of $\sim$0.1 µm diameter for electron diffraction. Smaller areas (down to a few nanometres in diameter) can be selected using a focused probe rather than an aperture; this is convergent beam electron diffraction (CBED).

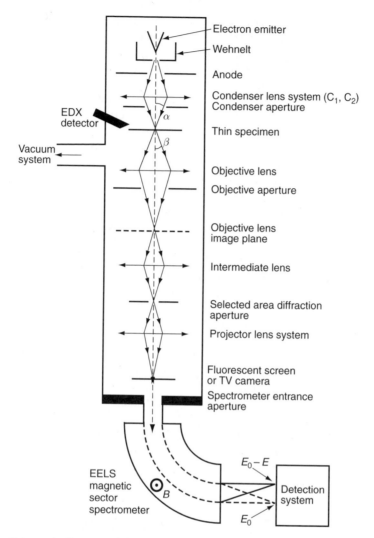

Figure 2.8 Schematic diagram of the layout of an analytical transmission electron microscope

TEM specimens are normally in the form of an ultrathin disc, prepared by cutting, mechanical polishing or chemical dissolution, followed by electropolishing (for conducting materials), chemical polishing (for semiconductors and ceramics) or ion beam milling, including focused ion beam (FIB) machining. Powders may be ground and dispersed onto an amorphous support film on a circular metal grid.

2.3.5.1 Electron diffraction

As described in Section 2.1.2.5, amorphous materials exhibit diffuse diffraction rings related to the average interatomic separations and hence the radial distribution function.

For crystalline materials, the periodic atomic arrangement of atoms scatters the electrons through well-defined angles given by Bragg's law (2.1). Since λ is small (e.g., $\lambda = 0.0037$ nm for 100 keV electrons), the Bragg angles are generally small (of the order of 10^{-3} radians), $\sin\theta \approx \theta$ and therefore $\lambda = 2d\theta$. Hence diffraction occurs only when the planes of atoms are closely parallel to the incident beam, as shown in Figure 2.9.

Since the TEM specimens are so thin, the diffraction conditions along an axis normal to the specimen are significantly relaxed, such that reflections will be excited even when Bragg's law is not exactly satisfied. Thus when the beam is parallel to a major zone axis there will be a large number of (hkl) planes in the zone close to their Bragg angles, giving rise to a large number of diffracted beams. The intensity of a particular diffracted beam, $\mathbf{g} = (hkl)$, is a function of the particular diffracting plane in question, the deviation away from the exact Bragg condition, and the sample thickness.

If the specimen area is polycrystalline, then electron diffraction produces a pattern of concentric rings of radii r which exhibit virtually all possible d spacings, owing to the fact the crystallites are randomly oriented (Figure 2.9). The diffraction pattern may be analysed using the relationship $r/L = 2\theta = \lambda/d$, where L is known as the camera length. Thus r is directly proportional to $1/d$; the constant of proportionality $L\lambda$ is known as the camera constant. For a single-crystal specimen, the diffraction pattern will consist of points, spaced at a distance proportional to $1/d$, aligned in a direction perpendicular to the orientation of the (hkl) planes as shown in Figure 2.9. The electron diffraction pattern essentially represents a scaled section through the reciprocal (Fourier space) lattice normal to the electron beam direction; the beam direction is simply the vector cross-product of two reciprocal lattice vectors in the plane. Analysis of the diffraction patterns obtained along different incident beam directions, which represent differing projections of the reciprocal lattice, allows the extraction of the real space lattice and hence the unit cell of the material.

Any restriction in the dimensions or morphology of the sample, such as plate-like precipitates oriented parallel to incident beam direction, will lead to changes in the form

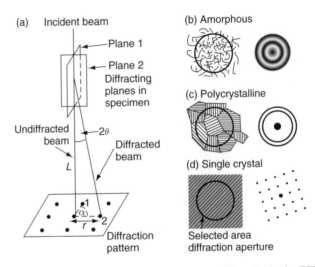

Figure 2.9 Schematic diagram of (a) the geometry of electron diffraction in the TEM and the form of the diffraction pattern for (b) amorphous, (c) polycrystalline and (d) single-crystalline sample regions

of the diffraction pattern such as streaking of the diffraction spots. Furthermore, materials with ultrafine grain sizes will exhibit a degree of broadening in their poly-crystalline ring patterns, as described in Section 2.6.1. Both these effects are highly relevant for the analysis of nanostructured materials.

If the incident electron beam is essentially parallel then, for a crystalline material, sharp spots are obtained in the electron diffraction pattern. As shown in Figure 2.9, the area of the specimen contributing to the diffraction pattern can be defined by the position of the SAD aperture in the TEM image. However, if the SAD aperture is removed and the incident beam is focused onto the specimen (i.e., it is convergent), then there are a range of possible incident beam directions and thus Bragg angles. This constitutes the CBED technique, and leads to a broadening of the normal diffraction spots into discs, the diameters of which are proportional to the beam convergence semi-angle α. For small diffraction angles a plane in reciprocal space is sampled, whereas at larger diffraction angles other reciprocal lattice planes are intersected and appear in the pattern. CBED patterns can give information on the 3D crystal structure, the symmetry of the unit cell, the sample thickness and the exact lattice parameters to an accuracy of 1 part in 10^4. This ability to identify the structure and orientation from individual microstructural regions, either using a SAD aperture or a focused probe, allows the identification of crystallographic orientation relationships between two adjacent nano-structural components (e.g., a layer and a substrate when viewed in cross section or, alternatively, a matrix and a precipitate).

2.3.5.2 TEM imaging

A TEM image is a two-dimensional projection of the internal structure of a material. There are three basic contrast mechanisms which contribute to all TEM images.

Mass thickness contrast

During transmission through a thin TEM sample, electrons undergo a range of scatter-ing processes which change their energy and angular distribution. The intrinsic bore of the microscope column prevents electrons that have been scattered through an angle greater than $\sim 10^{-2}$ mrad from contributing to the image. In the presence of a sample, regions that are thicker or of higher density will scatter the electrons more strongly and hence more electrons will be scattered through an angle greater than 10^{-2} mrad making these areas appear darker in the image. This simple mass–thickness contrast is exhibited by all specimens, whether amorphous or crystalline. Biological samples are often deliberately stained with a heavy metal such as osmium or uranium in order to increase the image contrast from the structural features into which the stain becomes incorp-orated; see Figure 8.10(b).

Contrast may be increased by placing an objective aperture in the back focal plane of the objective lens, which will further limit the angular range of electrons allowed to contribute to the image. If this objective aperture is centred on the optic axis, then in the absence of a specimen the image is bright, hence this is termed *bright-field* imaging (Figure 2.10).

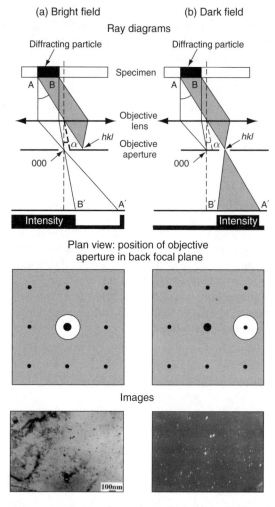

Figure 2.10 TEM ray diagrams for the formation of (a) bright-field and (b) dark-field images together with some typical images of carbide precipitates in a vanadium steel. Image courtesy of Dr Keijan He, Institute for Materials Research, University of Leeds

Diffraction contrast

Diffraction contrast is the major contrast mechanism in crystalline specimens, especially at medium magnifications. It arises from the differing amplitudes of the undiffracted beam and diffracted beams, resulting in intensity variations in the image formed by the different beams. If the objective aperture is centred on the undiffracted beam, then we have a bright-field image as discussed: however, in addition to mass–thickness contrast, any region of the specimen which is in the right orientation (relative to the electron beam direction) for strong diffraction to occur will also appear dark. Correspondingly, if the objective aperture is centred on a diffracted beam, either by displacing the aperture off the optic axis or by tilting the beam so that the diffracted beam travels along the optic axis (giving fewer aberrations in the image) and is used to form the image, then this is known as *dark-field imaging*, also shown in Figure 2.10. Selection of a particular diffraction spot associated with a specific crystal structure can be used, for example, to identify diffracting crystallites. A large range of common microstructural features show strong diffraction contrast.

Bend (extinction) contours

The closer a crystal plane is to a Bragg orientation, the more strongly it diffracts and the darker it appears in the bright-field image (and correspondingly the lighter it appears in the dark-field image). Most thin specimens are buckled, in which case some areas are closer to a major zone axis than others, leading to bright and dark regions known as bend contours. Similar effects allow direct observation of strain fields in TEM images.

Contrast from variations in specimen thickness

As the direct and undiffracted beams propagate through a crystal, they undergo complementary fluctuations in intensity, the periodicity of which (the extinction distance) depends on the specimen, the operative Bragg planes and their deviation from the exact Bragg angle. Hence wedge-shaped crystals or, for example, inclined grain boundaries show corresponding changes in intensity called thickness fringes, from which the specimen thickness can be calculated.

Contrast from crystal defects

When atoms are displaced by a vector $\mathbf{R}$ from their lattice positions due to the strain field of a crystal defect, the diffraction conditions are changed and the defect may appear as a different contrast in bright-field and dark-field images depending on which diffracted beams are used for imaging. Generally the appearance of a defect in a dark-field image using a particular Bragg reflection $\mathbf{g}$ will depend on the magnitude of the product $\mathbf{g.R}$. If this scalar product is zero, then this implies that this particular set of $\mathbf{g}$ planes remain unchanged by the presence of the defect, therefore the diffracted intensity is unaffected and the image is the same as that from a perfect crystal (i.e., the defect is invisible). Other values of this scalar product will lead to some degree of contrast, except for certain special cases. Determination of $\mathbf{R}$ for a particular defect may be achieved by determining two or more $\mathbf{g}$ vectors for which the defect is invisible.

Examples of planar defects include stacking faults and grain boundaries. The defect plane in a stacking fault separates the crystal into two parts, shifted through a vector $\mathbf{R}$ (in addition there might also be a misorientation). Interference between the beams diffracted from the upper and lower parts of the crystal which have undergone different phase changes gives rise to a set of fringes running parallel to the intersection of the fault with the specimen surface. Furthermore, on crossing a grain or phase boundary, the crystal orientation or structure may change and thus the diffraction conditions are unlikely to be similar on both sides of the boundary. If the boundary is inclined, one grain may be strongly diffracting and show thickness fringes whereas the other may show little contrast. If both grains are strongly diffracting, then we may obtain interference between the two sets of fringes producing Moire fringes of a different, usually larger, spacing to the individual interplanar spacings.

An example of a line defect is a dislocation. If a crystal is not in a Bragg condition, then at the core of an edge or screw dislocation the lattice planes are severely bent and must at some point reach the Bragg angle and so diffract. Thus in a bright-field image the dislocation core will normally appear as a dark line.

Examples of three-dimensional defects include precipitates and voids or pores. Incoherent precipitates show diffraction and mass–thickness contrast due to differences in orientation, structure factor or density with respect to the matrix. Coherent precipitates may produce Moire fringes due to interference between two Bragg reflections. Coherent or semicoherent precipitates may also strain the matrix, giving rise to contrast in a similar way to that observed at dislocations. Voids will show some mass–thickness contrast in that they will appear bright. If the void is faceted then effectively the void will produce thickness variations and thus thickness fringes; the void will appear dark on bright thickness fringes, and vice versa.

Phase contrast: lattice imaging

Phase contrast arises when electrons of different phases interfere to form an image. It is present in virtually all TEM images but is generally only visible as a speckled background at high magnifications. Phase contrast is used to form an atomic resolution lattice image by orienting the specimen perpendicular to a major crystallographic zone axis and letting at least two beams pass through the objective aperture; i.e., the undiffracted (000) beam and also the $+\mathbf{g}$ beam and maybe also the $-\mathbf{g}$ beam. The beams interfere and reproduce the periodicity of the object; i.e., a one-to-one correspondence between the observed fringes and the $\mathbf{g}$ lattice planes. If many beams are allowed to pass through the objective aperture then a number of intersecting lattice fringes will be observed, giving rise to spots at the positions of the atomic columns and therefore a structure image (a two-dimensional projection of the atomic structure), as shown in Figure 2.11(d).

2.3.6 Scanning transmission electron microscopy

The general principle of STEM is similar to that of SEM, except that the electron detector is now located below a thin TEM specimen as shown in Figure 2.12. The small probe (of the order of angstroms), produced by a FEG and a condenser–objective lens system, is scanned across the specimen and the signal detected and imaged on a monitor or frame store. The resolution is determined mainly by the probe diameter, which is usually of the order of a nanometre, although recent developments in the correction of spherical aberration in the probe-forming lens have demonstrated sub-angstrom resolution. Many conventional TEMs have a separate scanning attachment; however, dedicated STEMs operate entirely in scanning mode under UHV conditions. Bright-field STEM imaging uses an axial detector to detect electrons scattered through relatively small angles, and images contain diffraction contrast. Dark-field STEM imaging is essentially incoherent and employs an annular detector to detect electrons scattered through higher angles. High-angle annular dark field (HAADF) image intensity is roughly proportional to the square of atomic number (so-called Z contrast) and is therefore extremely useful for mapping phases at ultra-high resolution.

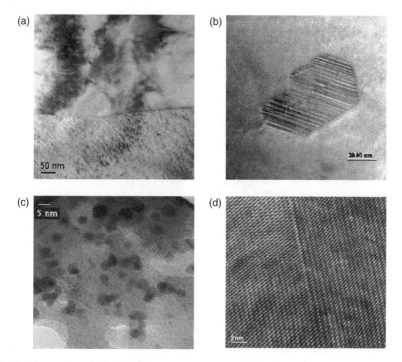

Figure 2.11 Examples of TEM images that exhibit diffraction contrast and phase contrast: (a) ferrite–cementite interface in a steel, (b) twinned titanium boride precipitate in a titanium alloy matrix, (c) nickel catalyst particles on a silica support and (d) grain boundary in zirconia polycrystal. Images courtesy of Drs Andy Brown and Andrew Scott, Institute for Materials Research, University of Leeds

2.4 FIELD ION MICROSCOPY

In the field ion microscope (FIM), the specimen consists of a conducting or semiconducting fine needle-tip of radius r (usually between 5 and 100 nm) prepared by etching or electropolishing techniques, such that individual atoms stand out in a terrace of atomic-scale ledges. The tip is held at a positive potential (between 5 and 20 keV), with respect to a negatively charged conductive fluorescent screen of radius R, in a high vacuum and is maintained at cryogenic temperatures. A low-pressure inert gas, such as hydrogen or helium, is introduced into the chamber, the atoms of which become polarized in the electrostatic field between the screen and the tip and are therefore attracted to the tip by the increased field intensity. The polarized atoms are then ionized and lose electrons (which channel into the tip) either at one point of contact or by 'hopping' over the surface. The resulting positive ions are then accelerated towards the fluorescent screen, giving an instantaneous fluorescent signal. In principle, each fluorescent point on the screen corresponds to an atom in the tip and the high-contrast image consists of a pattern of bright dots on a dark background, which reproduces the atomic arrangement of the tip.

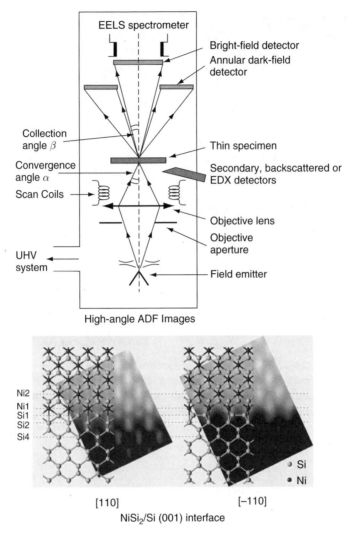

Figure 2.12 Schematic diagram of the layout of a STEM together with HAADF images of a silicon–nickel disilicide interface in two projections with the directly interpretable atomic structure model superimposed. Images courtesy of Dr Andrew Bleloch, Daresbury Laboratories

The image magnification is equal to the ratio of the radius of the screen to that of the tip (R/r) and is of the order of 10^6. More importantly, the limit of resolution is that of an individual atom and the FIM was the first instrument to provide images of atoms and their arrangement at the tip, as well as the location of vacancies and line imperfections. Since no lenses are involved, there are no depth-of-field or diffraction limitations to be considered.

So much for the (potential) advantages of FIM. Apart from the difficulties associated in the preparation of such a fine tip (and the geometrical constraints on a specimen which this imposes) as well as the exclusion of contamination or adsorbed surface species, there is the problem that the field strength at the tip must be sufficiently high to ionize the gas atoms introduced into the chamber, but not so high as to result in field evaporation or emission of atoms from the tip. In practice only those high melting point elements with high cohesive energies (e.g., Fe, Mo, W, Rh and Pt) and their compounds, which can resist field evaporation at the field strengths needed for inert gas ionization, fulfil this requirement. No breakthrough has been found in the case of gases which ionize more easily. On the other hand, this problem is turned to advantage in the atom probe FIM (APFIM), which involves the controlled progressive removal via field evaporation of (ionized) atoms from the tip. Moreover, the desorbed atoms can be individually chemically identified using a time-of-flight mass spectrometer giving atomic-scale resolution maps of elemental distributions. A recent variant, the position-sensitive atom probe (POSAP), allows a three-dimensional reconstruction of the atomic structure in the tip across an area of about 20 nm^2.

In a sense, APFIM has come full circle from the field emission microscope (FEM), invented prior to the FIM, in which the polarities of the tip and the screen are reversed and which, of course, forms the basis of the field emission gun in electron microscopy. In principle the emitted electrons recorded on the conductive fluorescent screen arise from individual atoms projecting from the tip surface as in FIM, but in practice the images suffer from noise and perturbations in the electron trajectories.

2.5 SCANNING PROBE TECHNIQUES

2.5.1 Scanning tunnelling microscopy

Scanning tunnelling microscopy (STM) is the ancestor of all scanning probe microscopies, which allow the real space imaging of surfaces with atomic resolution and employ no illumination and no lenses! The experimental arrangement is shown in Figure 2.13. A sharp conducting tip (often tungsten), acting as the anode, is brought close to the surface of the specimen (the cathode). A bias voltage ranging from 1 mV to 1 V is applied between the tip and the sample. Above the specimen surface there is an electron cloud due to surface atoms, and when the tip is brought within about 1 nm of the sample surface, electrons can quantum mechanically tunnel across the gap, causing a current to flow. The direction of electron tunnelling across the gap depends on the sign of the bias voltage. It is this current that is used to generate an STM image. The tunnelling current falls off exponentially with distance between tip and surface, and if the tip or sample is scanned laterally using piezoelectric drivers, the STM image reflects the variation in the sample surface topography. If the system is carefully damped from mechanical vibrations, the STM image will possess a sub-angstrom resolution vertically and atomic resolution laterally. The tunnelling current also depends on the atomic species present on the surface and their local chemical environment. In principle, no vacuum is required except when studying adsorption of species on surfaces. However, many STMs are operated under UHV as any oxide or contaminant can interfere with the tunnelling current. Furthermore, it is important to note that an STM cannot image insulating materials, except under conditions where they have appreciable conductivity, such as at high temperature.

(a) General scanning probe microscope

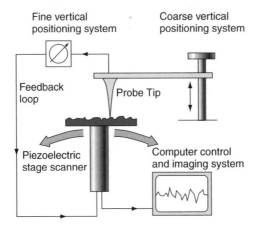

Fine vertical positioning system

Coarse vertical positioning system

Feedback loop

Probe Tip

Piezoelectric stage scanner

Computer control and imaging system

(b) Schematic of tip and sample interaction for STM

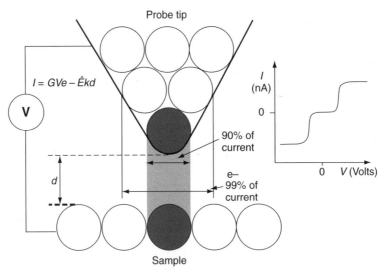

Probe tip

$I = GVe - \hat{E}kd$

V

I (nA)

0

90% of current

e− 99% of current

d

0   V (Volts)

Sample

(c) 3D rendered image of a semiconductor surface

Figure 2.13 Schematic diagram of (a) a general scanning probe microscope, (b) the tip–sample interaction and I–V curve for STM, and (c) a typical STM image. Image (c) courtesy of Thermomicroscopes

An STM can be designed to scan a sample in one of two different modes: constant height or constant current. In constant-height mode, the tip travels in a horizontal plane above the sample and the tunnelling current varies as a function of the surface topography and the local surface electronic states of the sample. The tunnelling current, measured at each location on the sample surface, constitutes the data set, the topographic image. In constant-current mode, the STM uses a feedback system to keep the tunnelling current constant by adjusting the height of the scanner at each measurement point. For example, when the system detects an increase in tunnelling current, it adjusts the voltage applied to the piezoelectric scanner to increase the distance between the tip and the sample. In constant-current mode, the motion of the scanner therefore constitutes the data set. If the system keeps the tunnelling current constant to within a few percent, the tip-to-sample distance will be constant to within <0.01 nm. Comparing the two modes, constant-height mode is faster because the system doesn't have to move the scanner up and down, but it provides useful information only for relatively smooth surfaces. Constant-current mode can measure irregular surfaces with high precision, although the measurement takes considerably more time.

To a first approximation, an image of the tunnelling current provides a map of the topography of the sample. However, since the tunnelling current corresponds to the electronic density of states at the surface, it actually measures a surface of constant tunnelling probability. In fact, STMs can directly measure the number of filled or unfilled electron states near the Fermi level at each particular (x, y) point on the surface, within an energy range determined by the bias voltage. This can cause problems if the sample surface is oxidized (which will dramatically lower the tunnelling current) or contaminated, and for atomic resolution studies the sample is often cooled and examined under UHV conditions.

The ability of the STM to probe the local electronic structure of a surface, in principle with atomic resolution, leads to the various techniques of scanning tunnelling spectroscopy (STS). STS techniques may involve recording 'topographic' (i.e., constant-current) images of the surface using different bias voltages and then directly comparing them, or taking current (i.e., constant-height) images at different heights z. Finally, true STS involves ramping the bias voltage with the tip positioned over a particular surface feature of interest, while recording the variation in tunnelling current; this produces a current versus voltage (I–V) curve characteristic of the surface electronic structure at a specific (x, y) location on the sample surface. STMs can be set up to collect I–V curves at every point in a data set, providing a three-dimensional map of electronic structure. With a lock-in amplifier, dI/dV (conductivity) or dI/dz (work function) versus V curves can be collected directly. STS methods are extremely powerful as they allow the surface electronic properties of a material to be investigated with near atomic resolution. In comparison, many other surface analytical spectroscopies average data from relatively large areas ranging from micrometres to millimetres.

2.5.2 Atomic force microscopy

Atomic force microscopy (AFM) was developed about five years after STM. While STM is limited to the surfaces of conducting specimens, AFM can be used to study nonconducting materials, such as insulators and semiconductors, as well as electrical conductors. As shown in Figure 2.14, AFM again uses a sharp tip, about 2 μm long and down to a minimum of 20 nm in diameter, which is scanned closely over the specimen surface,

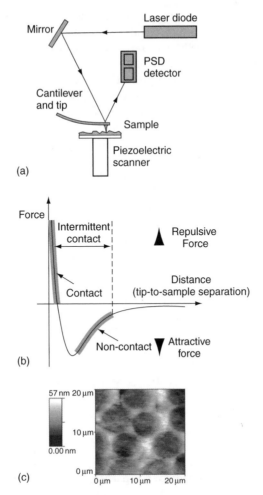

Figure 2.14   Schematic diagram of (a) an atomic force microscope, (b) the tip–sample interaction and (c) a typical AFM image of a carbon fibre–carbon composite. Image (c) courtesy of Professor Brian Rand, Institute for Materials Research, University of Leeds

but in AFM the magnitudes of atomic forces rather than tunnelling currents are monitored as a function of the probe position on the sample surface. AFMs can operate under UHV, ambient air conditions or even with the solid sample and tip submerged in a liquid cell, which is highly useful for biological systems (see Section 9.1.1).

The term 'atomic force microscopy' is used here in its loosest sense. Strictly, most images taken using an AFM are scanning force microscopy (SFM) images. Use of the phrase 'scanning force microscopy' is an acknowledgement that imaging with a scan length of up to 0.1 mm and pixel widths of nearly ∼400 nm cannot be considered to be a probe of individual atomic forces. Indeed, the typical cantilever load of a few nanonewtons, if applied in contact mode through a single atom, would result in a pressure which exceeds the tensile strength of any known material by several orders of magnitude. In fact, contact mode tips have a large radii of curvature (about 50 nm) relative to atomic dimensions. (The only circumstance in which single atom contact occurs is in the use of *non-contact* mode (see below), in which the tip oscillates above the sample, usually in an ultra-high vacuum environment, and the cantilever force is much lower.) In this text, it is perhaps unhelpful to use the term SFM rigorously, because the reader who dips into a particular chapter may be acquainted only with the term AFM, which enjoys much more widespread usage. After some soul-searching, we have therefore used 'AFM' in contexts where 'SFM' would be considered technically more appropriate.

The AFM tip is located at the free end of a cantilever that is 100 to 200 μm long and forces between the tip and the sample surface cause the cantilever to bend, or deflect. A detector is used to measure the cantilever deflection (usually just the displacement in the z direction although lateral force microscopy is also possible) as the tip is scanned over the sample, or alternatively, as the sample is scanned under the tip. The measured cantilever deflections allow a computer to generate a map of surface topography (see Figure 1.36).

The position of the cantilever is most commonly detected optically using a focused laser beam reflected from the back of the cantilever onto a position-sensitive (photo)-detector (PSD). Any bending of the cantilever results in a shift in the position of the focused laser spot on the detector, and the PSD can measure such displacements to an accuracy of < 1 nm. However, the ratio of the path length between the cantilever and the detector to the length of the cantilever itself, produces an additional mechanical amplification and, as a result, the overall system can detect sub-angstrom vertical movements of the cantilever tip. Other methods of detecting cantilever deflection rely on optical interference, or even an STM tip to read the cantilever deflection above a gold contact. If the cantilever is fabricated from a piezoresistive material, this obviates the need for a laser and PSD system as any deflection will result in strain, thereby changing the cantilever's resistivity, which can be monitored directly using an integrated electrical circuit.

Analogous to the case of STM, topographic AFM data sets are generated by operating in either constant-height or constant-force mode. In constant-height mode, the scanner height is fixed during the scan and the spatial variation of the cantilever deflection is recorded. Constant-height mode is employed for recording real-time images of changing surfaces, where high scan speeds are essential. It is also often used for taking atomic-scale images of atomically flat surfaces, where the cantilever deflections and thus variations in applied force are small. Conversely, in constant-force mode the deflection of the cantilever can be used as input to a feedback circuit that moves the scanner up and down in the z direction, responding to the topography by keeping the cantilever deflection (and thus also the total force applied to the sample) constant. Constant-force mode data sets are therefore generated from the scanner's motion in the z direction. In constant-force mode, the speed of scanning is limited by the response time of feedback circuit, but the total force exerted on the sample by the tip is well controlled. Constant-force mode is generally preferred for most applications.

Several forces typically contribute to the deflection of an AFM cantilever, the most common being relatively weak, attractive van der Waals forces as well as electrostatic repulsive forces. The general dependence of these forces upon the distance between the tip and the sample is shown schematically in Figure 2.14. When the tip is less than a few angstroms from the sample surface, the interatomic force between the cantilever and the sample is predominantly repulsive, owing to the overlap of electron clouds associated with atoms in the tip with those at the sample surface; this is known as the contact regime. In the non-contact regime, the tip is somewhere between ten to a hundred Angströms from the sample surface, and here the interatomic force between the tip and sample is attractive, largely as a result of long-range attractive van der Waals interactions.

In contact-AFM mode, the AFM tip makes soft, physical contact with the sample. The tip is attached to the end of a cantilever, which has a lower spring constant than the effective spring constant holding the atoms of the sample together. Hence as the scanner gently traces the tip across the sample, the repulsive contact force (which is a very steep function of tip–sample separation) causes the cantilever to bend to accommodate changes in topography rather than forcing the tip atoms closer to the sample atoms. If the cantilever is very stiff (i.e., it has a larger spring constant) then the sample surface is likely to deform during contact and this can be used for nanopatterning of surfaces (Chapter 1).

In addition to the electrostatic repulsive forces described above, two other forces are generally present during contact-mode AFM operation. Firstly, an attractive capillary force (the magnitude of which is about 10^{-8} N depending on the tip–sample separation) exerted by the thin water layer often present between the tip and the sample in an ambient environment and, secondly, the force exerted by the cantilever itself. The force exerted by the cantilever can be either repulsive or attractive and depends on the deflection of the cantilever and upon its spring constant. Overall the total force that the tip exerts on the sample is the sum of the capillary plus cantilever forces (typically 10^{-8} to 10^{-6} N) and, in the case of contact-mode AFM, will be balanced by the repulsive electrostatic force.

In non-contact AFM mode, a stiff cantilever is vibrated near the surface of a sample close to its resonant frequency. The spacing between the tip and the sample is of the order of tens to hundreds of Angströms and this is similar to the vibration amplitude of the tip. As the tip approaches the surface, this is detected as a change in the resonant frequency or vibration amplitude of the tip oscillation, and this can provide sub-Angström vertical resolution in the topographic image. Generally the detection system monitors the resonant frequency or vibrational amplitude of the cantilever and keeps it constant with the aid of a feedback system that moves the scanner up and down, ensuring that the average tip-to-sample distance also remains constant.

Non-contact AFM mode provides a means of measuring sample topography with minimum contact between the tip and the sample, thus removing any possibility of contamination or degradation of the surface by the tip, as can occur during contact AFM scanning. Furthermore, the total force between the tip and the sample in the non-contact regime is very low, approximately 10^{-12} N, which is beneficial for studying soft or elastic samples. Generally both contact and non-contact topographic images of rigid samples will appear similar. However, the presence of a few monolayers of liquid (e.g., water) on the surface of a sample will result in non-contact AFM mode imaging the surface of the liquid layer, whereas operating in contact mode will penetrate the liquid layer to image the underlying surface.

A further AFM variant is intermittent-contact or 'tapping' AFM mode. This is similar to non-contact mode except the vibrating cantilever tip is brought closer to the sample so that at the bottom of its travel it just touches the sample. Tapping mode is less likely to damage the sample than contact AFM because it eliminates lateral forces (friction or drag) between the tip and the sample. Furthermore, in comparison with non-contact mode, tapping mode can image larger scan sizes that can include greater variations in sample topography.

AFM may also be used to measure the local elastic properties of a point on the surface or even a whole surface. Here the force on an AFM cantilever tip in the z direction is monitored as a function of the z position of the piezoelectric scanner tube.

This force versus distance curve will vary with changes in local elastic properties of the surface, or the presence of contaminants and lubricants.

Generally AFM can only be used to provide information about the surface of a sample; any subsurface information about the morphology and structure must be obtained by scattering techniques (such as SEM or reflectometry) or sputter depth profiling (dynamic SIMS). It is often possible to reveal structural information of mixtures by selectively dissolving one component with an appropriate solvent. AFM is performed on the surface of, say, a film of a two-component mixture before exposing that film to a solvent which dissolves only one component of the mixture. The change in morphology will tell us where that component resided in the film. However, on certain samples it is possible to reveal three-dimensional structural information with nanometre resolution. This technique, termed nanotomography by its inventor, involves imaging the surface of a film before etching only a small amount in a plasma. The same part of the film is imaged again using AFM before another etching step is performed. The images can be stacked together to produce the final three-dimensional volume image. Image reconstruction is complicated by the difficulty in obtaining the same scan location *exactly* each time, and also by the fact that different components etch at different rates, hence the need for a careful alignment of images (registration) similar to that used in magnetic resonance imaging. Figure 2.15 shows an example of a nanotomography scan of a block copolymer.

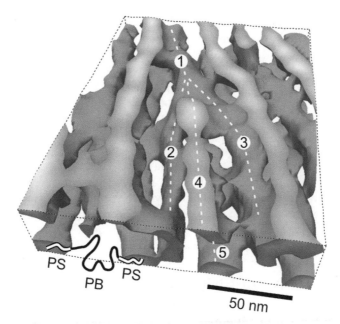

Figure 2.15 Nanotomography image of a triblock copolymer of polystyrene and polybutadiene (PS–PB–PS). The polystyrene component is shown in this image (clear regions indicate the location of polybutadiene). This method of imaging a material reveals information that would probably be impossible using other techniques; for example, notice how one of the polystyrene 'cylinders' (marked 1 in the figure) splits into four arms (marked 2, 3, 4 and 5). Reproduced with permission from R. Magerle, Nanotomography, *Phys. Rev. Lett.* **85**, 2749–2752 (2000). Copyright 2000 by the American Physical Society

2.5.3 Other scanning probe techniques

There are an increasing number of variations on the standard STM and AFM techniques that can provide additional information to surface topography. However, these may well be at a lower spatial resolution depending on the particular interaction involved. Force modulation microscopy works in contact mode and involves applying a periodic oscillation to the tip or the sample and monitoring the amplitude and sometimes phase lag of the cantilever oscillation; it provides information on the elastic properties of the sample. Lateral force microscopy (LFM) detects lateral deflections such as twisting of the cantilever that arise from forces on the cantilever parallel to the plane of the sample surface. LFM studies are useful for imaging variations in surface friction that can arise from surface inhomogeneities.

Magnetic force microscopy (MFM) is used to image the spatial variation of magnetic forces on a sample surface (such as domain structures in magnetic materials) and employs a tip coated with a ferromagnetic thin film. The system operates in non-contact mode, detecting changes in the resonant frequency of the cantilever induced by the dependence of the long-range magnetic field on tip-to-sample separation. The image also contains information on sample topography, particularly when the tip is close to the sample surface, however this can be separated from magnetic effects by recording images at different tip heights. Further details are given in Chapter 4.

In a similar fashion, local charge domains on a sample surface (such as a microprocessor) may be imaged using electrostatic force microscopy (EFM) where a voltage is applied between the tip and the sample whilst the cantilever vibrates in non-contact mode. Scanning capacitance microscopy (SCM) images spatial variations in capacitance, again with a voltage applied to the tip and a special circuit to monitor the capacitance between the tip and sample. SCM can be used to map the dielectric properties of a surface, which may be influenced by layer thickness or the presence of subsurface charge carriers.

Scanning thermal microscopy (SThM) measures the thermal conductivity of the sample surface using a resistively heated Wollaston wire as the probe. Changes in the current required to keep the scanned probe at constant temperature produce a thermal conductivity map of the surface (or thermal diffusivity as the temperature is oscillated). Conversely, changes in the probe resistance as a constant current is applied can generate local surface temperature maps. In addition, with the probe at a particular point on the surface, a localized microthermal analysis can be performed monitoring calorimetric or mechanical changes as a function of temperature (Section 2.9.4).

2.6 DIFFRACTION TECHNIQUES

2.6.1 Bulk diffraction techniques

As we have seen, diffraction techniques provide an insight into the crystallography of a material and, in some cases, the structure of a sample system or device. We have already mentioned electron diffraction in relation to TEM of thin specimens, and here we discuss bulk diffraction techniques which refer principally to X-ray diffraction (XRD) and neutron diffraction (ND) in which relatively large volumes of material, greater than about 0.1 mm^3, need to be sampled. This need arises because, unlike electrons, X-rays and neutrons cannot easily be focused. They can usually only be collimated and the practical limit of collimation (in order to provide diffracted beams of sufficient intensity) is of the order of 0.5 mm diameter for X-rays and 10 mm diameter for neutrons. In practice, much wider beams than this are employed.

X-ray diffraction is of course of paramount importance in determining the structures of crystals. From the pioneering work of the Braggs in 1913–15, who solved the first crystal structures, to the determination of the structure of DNA by Watson, Crick, Franklin and Wilkins in 1953, XRD has unlocked the door to the transmission of the gene and the evolution of life itself. The application of XRD to nanocrystalline solids, powders, single-crystal thin films or multilayers may be less spectacular, but apart from the standard procedures for the determination of their crystalline structures (from an analysis of the directions and intensities of diffracted beams), XRD can provide information on nanocrystal size (strictly, coherence length) and microstresses and microstrains (from an analysis of line broadening), macrostrains (from an analysis of line shifts), repeat distances (or superlattice wavelengths) and total thicknesses in multilayer films (from an analysis of low-angle Bragg peaks) and orientation distributions or 'texture' (from an analysis of pole figures or orientation distribution functions).

In all these applications monochromatic X-rays are employed, either by the use of crystal monochromators or filters which pass only the strong characteristic $K\alpha_1$ components of the whole spectrum from an X-ray tube, or by crystal monochromatized radiation from a synchrotron source (and which has the advantage that the wavelength can be tuned to minimize unwanted fluorescent excitation in the sample). The original Laue technique, which used the whole white or bremsstrahlung radiation (plus a whole range of characteristic wavelengths in the spectrum) is inapplicable to polycrystalline solids because the two variables λ and θ (arising from many orientations) would, by simple inspection of Bragg's law (2.1), give rise in effect to very many superimposed diffraction patterns which would be quite impossible to sort out.

The X-ray technique of greatest importance is therefore the powder diffractometer in which the monochromatic beam is incident at an angle θ on a specimen of about 10 mm^2 in area and (for polycrystalline or powder specimens) a minimum thickness of about 20 μm, mounted on a support film that does not give rise to interfering reflections. The detector is set to receive reflections at an angle θ (the Bragg–Brentano symmetrical arrangement shown in Figure 2.16(a)) and this is varied over the angular range of interest (typically 1–6° for low-angle reflections and 6–80° for high-angle reflections), either by keeping the incident beam direction fixed and rotating the specimen and detector (the detector at twice the angular velocity) or by keeping the specimen fixed and rotating the incident beam and detector in opposite senses. In both cases this instrumental set-up preserves the symmetrical arrangement.

Reflected beams, it should be noted, are only recorded from those planes which lie parallel, or nearly parallel, to the specimen surface. In order to record reflections from planes which lie at an angle to the specimen surface, as required for texture or macrostrain determination, the specimen must be tilted or rotated to the required angle(s); i.e., away from the symmetrical Bragg–Brentano arrangement. The texture goniometer is essentially a specimen holder arrangement in which such tilts or rotations are carried out automatically: the incident beam and detector directions are fixed to receive a specific reflection (d spacing) of interest and the variations in intensity are recorded.

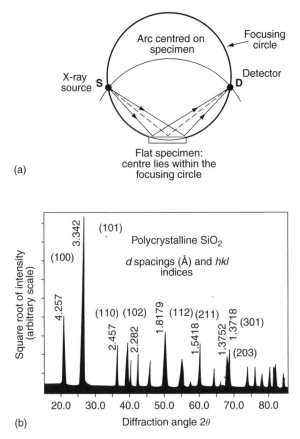

(a)

(b)

Figure 2.16   Schematic diagram of (a) the geometry in an XRD powder diffractometer and (b) an example of an XRD powder pattern from crystalline silicon oxide

In the diffractometer, the angles (d spacings) and intensities of the high-angle reflected beams serve as a 'fingerprint' for the crystal structure. This may be identified by comparing the XRD pattern (e.g., Figure 2.16(b)), with a database such as the Powder Diffraction File (PDF), which lists some 140 000 inorganic structures. The identification is not always easy, especially when the specimen contains two or more phases, or when the reflections are broadened or shifted as a result of small grain size or strain. The broadening β due to crystallite grain size t arises from the limited number of diffracting planes within a diffracting object and is given by the Scherrer equation

$$\beta = \frac{K\lambda}{t\cos\theta},$$ (2.6)

where K is a shape factor (≈ 1). This is analogous to the broadening of points in the reciprocal lattice that is encountered in electron diffraction of thin TEM samples. The peak broadening due to the presence of internal elastic microstrains (varying between different crystals in the samples) is given by

$$\beta = \frac{-2\varepsilon}{\cot \theta}, \tag{2.7}$$

where ε is the elastic strain. The deconvolution of these contributions, as well as the instrumental broadening arising from such factors as detector slit width, is also not a simple task and a common procedure is to compare the broadening from, for example, large grain size and nanograin size specimens.

The low-angle reflections give information on the thickness of thin films and multi-layers, as well as the mean size and size distributions of scattering objects such as micelles, colloidal particles and macromolecules. Since the scattered intensity is low, this is most commonly performed using high-intensity synchrotron radiation. The reflections, or rather diffuse diffraction maxima, from amorphous or semicrystalline materials such as glasses can provide information on short-range structural coherency and average spacings between atoms or ions.

The diffractometer is not appropriate or useful when only small quantities of material are available. In such situations, for high-angle work, the Debye–Scherrer powder camera and Seeman–Bohlin camera still retain their usefulness, particularly with the advent of improved area X-ray detectors. For very low angle work, the diffraction pattern (or rather radial diffracted intensity distribution) may be recorded from the beams transmitted through a small thickness or volume of material, the incident beam being very finely collimated using carefully aligned knife-edges; this is the basis for the small-angle X-ray scattering (SAXS) technique.

As mentioned in Section 2.1.2.5, neutrons are largely scattered by the nuclei, rather than the atomic electrons, and the scattering amplitudes vary in an irregular way with atomic number. The large scattering amplitudes of, for example, hydrogen and oxygen atoms, in comparison with heavy metal atoms, allow hydrogen and oxygen atoms to be located within the structure. Moreover, since the nuclear cross sections are very small, the interference effects, which in the case of X-rays lead to a decrease in scattering amplitude with angle, are also small and neutron scattering amplitudes do not decrease rapidly with angle. Finally, since neutrons possess spin, they interact additionally with the unpaired electrons in magnetic materials and hence neutron diffraction finds applications in the study of ferro-, antiferro- and ferrimagnetic materials.

Neutrons emerge from a high-flux nuclear reactor with a range of velocities corresponding (via de Broglie's equation) to wavelengths which peak typically in the range 0.1–0.2 nm; i.e., close to the characteristic Kα X-ray wavelengths from metal anodes of X-ray tubes. Single-wavelength beams are achieved, as with X-rays, through the use of crystal monochromators, hence the diffraction or reflection angles are similar to those for X-rays. Neutron diffraction is geometrically similar to X-ray diffraction except that the physical size of the apparatus needs to be much larger because of the lower intensities of neutron beams. In summary, therefore, neutron diffraction is very much a bulk technique, and apart from structural or magnetic changes which may occur in large assemblies of nanoparticles or nanocrystalline solids, it currently offers limited additional benefits for the analysis of nanostructures. Reflectometry experiments using neutrons are discussed in Section 2.8.4.

2.6.2 Surface diffraction techniques

In addition to grazing incidence (angle) techniques using X-rays and to some extent neutrons, which limits the penetration depth of the incident radiation, there are two important techniques for the structural analysis of surfaces, both of which employ electrons. These are reflection high-energy electron diffraction (RHEED) and low-energy electron diffraction (LEED), shown in Figure 2.17. Both methods can be used to determine the periodic, two-dimensional arrangement of atoms at surfaces. Since electrons are scattered much more strongly by matter than either X-rays or neutrons

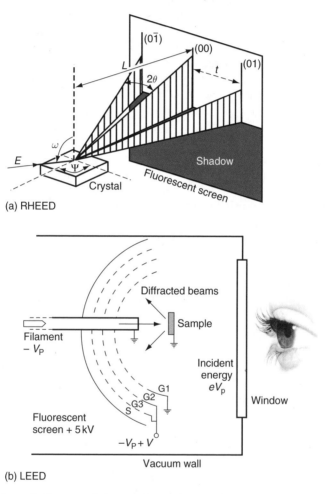

Figure 2.17   Schematic diagram of the experimental set-up for (a) RHEED and (b) LEED. In (a) the electron beam is of energy E and the direction relative to the sample is defined by the two angles ω and Ψ. The camera length L defines the observed spacing t of the diffraction streaks separated by twice the Bragg angle θ. In (b) the filament is held at a potential $-V_P$ and the diffracted beams are accelerated between the grid G_1 and the screen S. Grids G_2 and G_3 are used to reject inelastically scattered electrons

(Section 2.1.2.5) and penetrate less deeply, they constitute a very important probe of surface structure as, in general, the surface-specific signal can often be extremely weak relative to the bulk. To prevent surface contamination both techniques require UHV conditions and also methods for cleaning surfaces in situ, either by heating, low-energy ion sputtering or cleaving the sample in situ.

In RHEED a high-energy (typically 100 keV) beam of electrons is incident on a plane surface at a grazing angle, thus preventing excessive electron penetration. This experimental arrangement results in a set of diffraction streaks normal to the shadowed edge of the sample the form of which is dependent on Bragg's law and the electron beam direction. RHEED can determine the symmetry and dimension of the surface unit mesh (the repeat unit) by varying the incident beam direction. The grazing incidence geometry is very convenient for observing the change in surface structure whilst a material is being deposited on a surface. During monolayer deposition, such as occurs during MBE (Section 1.4.2.3), the intensities of the RHEED beams oscillate, which allows this technique to be used for monitoring atomic layer epitaxial growth.

In LEED a low-energy beam of electrons (between 10 and 1000 eV) is incident normal to a surface and diffracted. The elastically backscattered electron beams are accelerated towards a fluorescent screen or camera and the inelastically scattered electrons are rejected by grids held at a slightly positive potential. A single crystal surface will produce a spot pattern which can directly give information about the surface symmetry and the surface unit mesh. In principle LEED (and also RHEED) intensities can be modelled to obtain information on where atom types are located within the surface unit mesh.

2.7 SPECTROSCOPY TECHNIQUES

Generally in all types of spectroscopic measurements (e.g., optical, UV, IR, X-ray and electron spectroscopies) we are concerned with three types of processes: absorption (transmission), reflection or emission (luminescence). The different experimental methodologies are shown in Figure 2.18. Such experiments probe the energy differences between electronic, vibrational and, in some cases, rotational quantum energy states of atoms as well as the lifetime of excited states and their energy relaxation channels. The latter studies are often aided by using a pulsed radiation source.

Besides the spatial resolution of the radiation incident upon and propagating within the sample (which determines the volume of analysis), a further important experimental parameter is the wavelength or energy resolution of both the incident radiation and the detected radiation, which determines the overall spectral resolution. The intensity of the reflected, transmitted (absorbed) or emitted radiation may be recorded as a function of the wavelength of the incident radiation; alternatively, the wavelength of the incident radiation is fixed and the distribution in intensity of the reflected, transmitted or emitted signals are spectroscopically analysed as a function of wavelength.

A further experimental variable in such spectroscopic experiments is the polarization of the incident and/or reflected, transmitted or emitted radiation. The absorption of electromagnetic radiation involves the resonant response of electrons in an atom or molecule to the oscillating electric field of the radiation, which is a vector quantity and is orthogonal to the propagation direction and the oscillating magnetic field vector of the

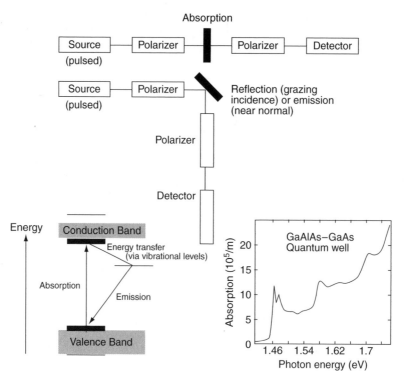

Figure 2.18 Schematic diagram of absorption, reflection and emission spectroscopic experiments together with an energy level diagram and a typical absorption spectrum from a quantum well

radiation. If the radiation is linearly polarized, we restrict the electric field vector of the radiation to one plane (plane polarized radiation) and this can be used to probe the directional properties of, in this case, the electronic transitions of atoms within the sample and hence the structure of the sample with respect to the plane of polarization. There are other forms of polarization, known as circular and elliptical polarization, which possesses a polarization vector that describes either a left-or right-handed helical path as a function of time or position and can be of use for studying optically active materials or magnetic transitions. Briefly, circular polarization involves the electric and magnetic fields of the polarization each being described by two equal orthogonal components which are out of phase (by 90°), while elliptical polarization involves the electric and magnetic fields of the polarization each being described by two unequal orthogonal components which are out of phase. In addition to electromagnetic radiation, it is also sometimes possible to perform similar experiments with electrons.

2.7.1 Photon spectroscopy

Generally the lateral spatial resolution of photon-based spectroscopies is limited by diffraction, and can attain probe sizes of at best a few micrometres, unless near-field techniques are used (Section 2.2.4.1), whereas X-ray synchrotron sources have demonstrated spot sizes of 50–100 nm by focusing using Fresnel plates. Lateral resolution is therefore limited and essentially the techniques provide an ensemble-averaged signal from a nanostructured sample. Furthermore, most photon-based spectroscopies probe bulk rather than surface properties, however there do exist various techniques for localizing the signals from surfaces (e.g., grazing incidence studies) which are of use for analysing thin film nanostructures. Photon-based spectroscopies have been revolutionized in the X-ray regime by the use of synchrotron radiation and, in the optical/UV regime, by the use of lasers that provide coherent, intense and often tunable radiation (over a large range of wavelengths). Furthermore, lasers can often be pulsed over sub-picosecond timescales.

Despite the limited spatial resolution, extremely important information can be derived from photon-based spectroscopies. As discussed in Chapter 1, all electronic energies in nanostructures are affected by quantum confinement, in addition to the effects of the increased contribution of surfaces. These changes are most apparent for transitions between the highest occupied molecular orbitals (HOMO) in the valence band and the lowest unoccupied molecular orbitals (LUMO) in the conduction band of a solid, both of which are directly influenced by changes in the bonding. Such transitions often lie in the optical region of the electromagnetic spectrum, where λ is typically 400–700 nm (sometimes expressed as a wavenumber $(1/\lambda)$ in cm^{-1}). Photon-induced transitions follow a set of dipole selection rules for the quantum numbers of the initial and final electronic states of the transition, which causes only selected transitions to be observed.

Colloidal metallic nanoparticles have long been known to show colours that depend on their size and shape. Semiconductor nanoparticles also show a distinct quantum size effect, whereby the absorption edge shifts from the infrared to the visible region as the particle size decreases and the band gap correspondingly increases. Generally semiconductor nanoparticles are luminescent, however the surface properties of the particles can change the emission characteristics significantly via the introduction of defect states. Quantum well structures often exhibit distinct steps in absorption spectra, rather than a continuous increase as observed in the bulk, corresponding to the quantized energy levels associated with the localized well. This is shown schematically in Figure 2.18. There are further effects due to the fact that the electron–hole pair (an exciton) created during the energy absorption process may be constrained within the well, leading to an increase in the exciton binding energy and the appearance of sharp exciton peaks at the edges of the steps in the absorption spectrum.

Further effects due to nanodimensionality can be seen in the longer-wavelength, infra-red and far-infra-red (1–20 µm) portion of the electromagnetic spectrum which are associated with changes in vibrational (and hence thermal) properties arising from changes in structure and/or composition.

2.7.1.1 Optical measurements

The two simplest optical measurements are reflectivity and transmission/absorption. Both record either the reflectance or the transmittance/absorbance as a function of wavelength. For transmittance measurements in the wavelength range 185–3000 nm (UV, optical, IR region) there are a large number of commercially available dual-beam spectrophotometers.

Simplistically, absorption techniques involve a measurement of the attenuation of the intensity of a monochromatic light beam as it traverses the sample. At a given wavelength or frequency, the relative absorption intensity I/I_0 generally follows the Beer–Lambert law:

$$I/I_0 = \exp(-Kx), \qquad (2.8)$$

where K is the absorption or extinction coefficient and x is the path length through the sample. There are a large number of variations on this simple experiment, for instance using polarized light, external applied (electric or magnetic) fields or an external applied stress. The light source can be a tungsten lamp, a gas discharge lamp, a highly mono-chromatic continuous laser or even a pulsed laser. Some experiments probe the change in absorption following the absorption of a laser pulse. The absorption can be measured by monitoring the intensity of the transmitted beam directly (i.e., using the photoelectric effect to measure a current) or by measuring the absorbed energy in the form of heat deposited in the sample. Samples for optical spectroscopic measurements are often cooled to low temperatures (e.g., liquid helium at 4.2 K), essentially to depopulate vibrational energy levels and ease the interpretation of the spectra in terms solely of excited electronic states.

Modulation spectroscopy involves the measurement of small changes in optical properties (typically absorption or reflectivity) of a material that are caused by a modification of the measurement conditions. The modification is generally an external perturbation such as the application of an electric field, a light pulse, a magnetic field, a heat pulse or uniaxial stress. In some cases this can give rise to transitions to states that are not evident in normal absorption spectra. Two techniques of use for studying nanoparticles embedded in a matrix of another material are electro-absorption and photo-absorption. Here the wavelength is tuned to a region of the spectrum where the host matrix does not absorb radiation. In electro-absorption the absorptivity is altered by application of an external AC field, the so-called Stark effect. This can provide details on the nature of the excited state in nanocrystals depending on how the absorp-tion is modified by the polarization of the incident radiation. Photo-absorption involves the measurement of the change in transmission resulting from a periodic light pulse that can cause filling and also splitting of electronic levels which will modify the absorption spectra. This can reveal the presence of non-linear optical effects, where the properties of the absorbing or reflecting species or particle are altered by the incident radiation.

Generally the optical properties of materials are expressed using a complex quantity known as the dielectric function, ε, which may be separated into real and imaginary parts, $\varepsilon = \varepsilon_1 + i\varepsilon_2$. The dielectric function represents the response of electrons in the solid to the electrical and magnetic field of the incident radiation. Reflectance and transmittance measurements allow determination of the index of refraction, n, and the extinction/absorption coefficient, K, of the sample. Under certain conditions these quantities can be used to derive the behaviour of the real and imaginary parts of the dielectric function as a function of frequency or wavelength. This allows identification of the contribution of resonant plasmon-type oscillations as well as interband transi-tions in the absorption or reflection spectrum. Measurement of the optical transmit-tance, and hence the absorption coefficient, allows a direct determination of the imaginary part of the dielectric function. When plane polarized light is reflected from a solid surface the reflected wave is, in general, found to be elliptically polarized.

Ellipsometry is a useful technique when one is interested in the real part of the dielectric function and involves the measurement of the ellipticity (the amplitudes of the two components of the oscillating electric field and their phase difference) as a function of incidence angle, the plane of polarization of the incident light and its wavelength.

As highlighted in Chapter 1, metallic nanoparticle systems show an absorption behaviour that is dependent on the size and shape of the particles (Figure 1.17). The original explanation was provided by Mie, who solved Maxwell's equations for the case of spherical particles. In the visible region this usually corresponds to excitation of surface plasmons by the incident electromagnetic radiation. Surface plasmons are collective oscillations of valence electrons associated with surface atoms, and while they are insignificant in the spectra obtained from bulk materials, they are an important phenomenon in nanometre-sized particles which have a large surface area to volume ratio. The technique forms the basis for the technique of surface plasmon resonance spectroscopy. If the particles are not spherical but rod-like, for example, then the surface plasmon absorption will split into tranverse and longitudinal modes corresponding to the electron oscillations perpendicular and parallel to the major axis of the rod.

2.7.1.2 Photoluminescence

Emission or luminescence studies are complementary to optical and IR absorption techniques in that they also provide details of the spectral distribution of electron energies and their polarization behaviour, in particular low energy states which may be obscured due to vibrational transitions. A comparison between the luminescence peak and the lowest absorption peak can provide information on the nature of the emitting state.

Photoluminescence (PL) measurements are made by exciting a sample with a laser tuned to a particular wavelength, collecting the scattered light from the front surface and dispersing it as a function of wavelength then measuring its intensity. The polarization dependence of the emitted intensity can be studied by inserting a linear polarizer in the detection pathway. A variation on this technique is photoluminescence excitation (PLE) spectroscopy, which monitors the intensity of a single luminescence band while the excitation wavelength is varied. Since the size of a nanoparticle affects the absorption wavelength, then PLE can be used to study size distributions. Imaging luminescence from individual nanoparticles has also been achieved via dilution of nanoparticle solutions to levels of 1nanocrystal/μm^2 and coating as thin films. The fluorescence is then filtered and imaged on a two-dimensional CCD camera. Spectroscopy from single nanocrystals can also be achieved in a similar fashion.

Extremely important information comes from the time dependence and efficiency of luminescence. Once an excited state has been created, it can either emit or decay non-radiatively into thermal energy; alternatively the excitation may migrate to a different site within the lattice, which may itself either emit or decay non-radiatively. The migration process is known as energy transfer and can reveal how the two sites are coupled. An impurity, defect or surface site may act as a trap for the mobile excitation. The use of pulsed laser sources and gated detection in time-resolved PL can provide details of the energies and lifetimes of these trap states. For example, nanoparticles with different surface modifications (e.g., surfactants and stabilizers) will show different time-resolved PL spectra.

2.7.1.3 Infrared and Raman vibrational spectroscopy

The energy of most molecular vibrations corresponds to the infrared region of the electromagnetic spectrum and these vibrations may be detected and measured in an infrared (IR) or Raman spectrum, which typically covers a wavelength range of $2-16\,\mu m$ (more commonly expressed as $400-5000\,cm^{-1}$).

For IR the spectrometer system consists of an IR source emitting throughout the whole frequency range; the beam is split, with one beam passing either through the sample (transmission) or alternatively reflected (for studies of surfaces) while the other beam is employed as a reference beam. Samples may be a vapour or solution contained in a special NaCl cell, a liquid between NaCl plates, or a solid ground together with excess KBr and pressed into a disc. Fourier transform IR (FTIR) spectroscopy works by allowing the transmitted (or reflected) beam to recombine with the reference beam after a path difference has been introduced. The interference pattern so produced is Fourier transformed to produce the spectrum of material being analysed as a function of wavelength or wavenumber of the incident radiation.

If a molecular vibration is excited in the sample, the molecule absorbs energy of the particular frequency and this is detected as absorption relative to the reference beam. As shown in Figure 2.19, different functional groups have characteristic vibration frequencies arising due to stretching, bending, rocking and twisting of bonds which allows the particular functional group to be identified. These may change slightly when the functional group is incorporated into a solid. However, not all possible vibrations are excited in IR spectroscopy, as a dipole moment must be created during the vibration. In molecules possessing a centre of symmetry, vibrations symmetrical about the centre of symmetry are IR-inactive.

The attenuated total reflection spectrometer (ATR-FTIR) is an extension of the FTIR technique and consists of an infrared-transparent block; silver chloride, zinc selenide and germanium are suitable materials. An infrared beam passes through this block at a glancing angle and is totally reflected several times at the top and bottom before exiting at the other end. The sample can be adhered to the block, changing the IR spectrum. In total internal reflection there is no power transmission outside of the block but this does not imply that the IR wave is completely confined within the block; in fact, the beam penetrates a very small distance into the contacting sample layer (this is the evanescent wave) and this can be utilized for infrared absorption. In the IR case, the evanescent wave penetrates about a micron into any sample clamped onto the sample cell. If this sample has strong IR absorption lines, these will be revealed in the final spectrum. In fact, ATR-FTIR is not restricted to solid materials attached to the ATR block; the block can also be immersed in solution to study adsorption to the block. This is a particularly useful way of studying nanoscale conformational events such as protein folding and unfolding, which might take place at surfaces.

A complementary vibrational spectroscopic technique is Raman spectroscopy; this usually employs a laser source and here the scattered light is analysed in terms of its wavelength, intensity and polarization. The parent line is simply formed by absorption and re-emission of the light, whereas other weaker lines involve not only absorption/emission but also vibrational excitation or de-excitation (Raman lines). The vibrational frequency is just the difference between the parent line and the Raman line. Generally IR-inactive vibrations are Raman active, which makes the techniques complementary for analysis of functional groups or bonds (chemistry). Additionally, low-frequency

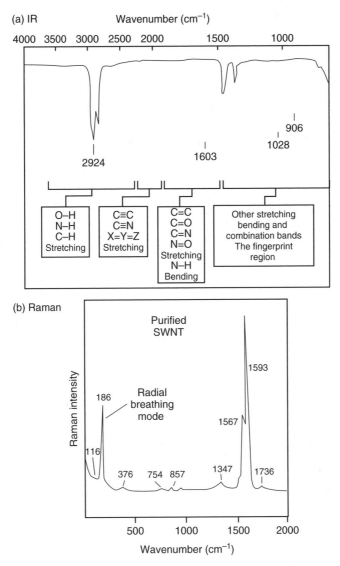

Figure 2.19 (a) Diagram of the IR spectral regions for different chemical bonds in organic materials. (b) A schematic example of a Raman spectrum from single-walled carbon nanotubes; the radial breathing mode at 186 cm^{-1} is characteristic of the entire nanostructure

Raman scattering can probe elastic vibrations of an entire nanoparticle, which most often shows a distinct size dependence; it is used extensively to analyse single- and multi-walled carbon nanotubes.

Both IR and Raman microscopy are possible using a specially designed confocal scanning laser microscope (CSLM) that provides a laser spot size, hence lateral spatial resolution, of around 1 μm. The confocal nature of the microscope allows different depths of up to 2 μm below the surface to be sampled. Typically the laser produces polarized red light, the scattered light is filtered to remove the parent line and the remaining Raman lines are analysed with a monochromator.

2.7.1.4 X-ray spectroscopy

X-rays from a high-intensity source with a controllable wavelength, such as a synchrotron, can also be used to perform absorption or reflectivity studies. X-ray absorption spectroscopy (XAS) involves the ionization of atoms by high-energy photons causing electron transitions to occur from deep inner shell atomic levels to the empty conduction band. X-ray ionization edges occur at energies characteristic of both the element and the inner shell involved and show two types of fine structure: X-ray absorption near-edge structure (XANES), which involves strong intensity oscillations within about 40 eV of the absorption edge onset, and weaker oscillations at higher energies known as extended X-ray absorption near-edge structure (EXAFS). XANES is directly related to the unoccupied density of electronic states and has characteristic features and energy positions for different bonding environments of the atoms in the solid, whereas EXAFS oscillations can be used to determine bond lengths and coordination numbers in terms of a radial distribution function around the ionized atom. This is shown schematically in Figure 2.20. Both these techniques have electron equivalents in the technique of EELS in the TEM (Section 2.7.3.3).

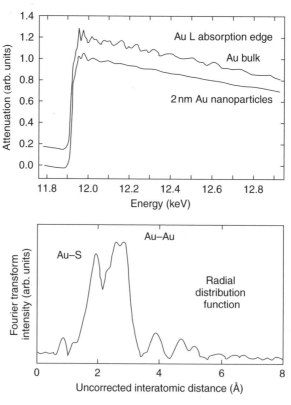

Figure 2.20 Schematic comparison between the L X-ray absorption edge in bulk gold and nanoparticulate gold together with the radial distribution function obtained via Fourier transformation of the EXAFS oscillations

Surface studies (often called SEXAFS or NEXAFS) can be performed using glancing angle incident radiation, whereas polarized X-ray sources can provide information on the orientation of species (using linearly polarized radiation) as well as magnetic dichroism (using circularly polarized radiation).

2.7.2 Radio frequency spectroscopy

Atomic nuclei contain protons and neutrons, both of which possess an intrinsic spin just like the electron (i.e., the spin quantum number s). In a nucleus the proton and neutron spins combine to give an overall nuclear spin (I). Certain nuclei, such as ^{1}H, ^{2}H, ^{7}Li, ^{11}B, ^{13}C, ^{14}C, ^{14}N, ^{15}N, ^{17}O, ^{19}F, ^{23}Na, ^{27}Al, ^{29}Si, ^{31}P and ^{35}Cl, have a nuclear spin that is non-zero and therefore they possess a magnetic moment and behave like a bar magnet. If a nucleus of spin 1/2 (e.g., ^{13}C) is placed in a magnetic field, it may orient itself with the field (in a low energy state) or it may align itself against the field (in a higher energy state). This is known as Zeeman splitting and the energy difference between these two states is proportional to the applied magnetic field multiplied by the magnetic moment of the nucleus. At equilibrium, the lower energy state will be slightly more populated, however in a field of 10–100 kGauss, transitions between the two energy states can be induced by radio frequency (RF) radiation ($\Delta E = h\nu$, where $\nu = 50-500$ MHz). Such transitions will absorb radiation and then re-emit upon relaxation; overall, radiation will be absorbed since there are initially very slightly more nuclei in the lower energy state. This forms the basis for the technique of nuclear magnetic resonance (NMR) spectroscopy, essentially the absorption of RF radiation by nuclei in a strong magnetic field, which can provide information on the local environment of the nuclei and hence the local bonding of atoms in a sample.

The experimental set-up for NMR involves an RF transmitter, a strong homogeneous magnetic field and an RF receiver. In a large molecule or solid, the presence of different chemical bonds may lead to the different magnetic nuclei of a particular atom species experiencing slightly different electronic environments. Hence the actual resonance frequency may vary between different nuclei and, to accommodate this, either the frequency or the magnetic field is swept through a range of values typically spanning the parts per million (ppm) range. As the detected signal is usually quite low, the radio frequency is often pulsed, which causes it to behave like a range of frequencies. As the signal decays following the pulse, it is Fourier transformed to obtain the separate resonance frequencies that are related to the differing electronic and hence chemical environments. After many such RF pulses, the spectral noise will average out, so improving statistics.

The actual value of the resonance frequency depends on a number of factors. Firstly, there is the chemical shift interaction caused by the fact that different nuclei have different chemical environments. Here the individual nuclei are shielded from the applied field to different extents by the electron cloud surrounding the nucleus, providing a means for identifying the local atomic environment; the actual value of the chemical shift is often referenced to a particular molecule, such as H or Si in tetramethylsilane (TMS) for ^{1}H or ^{29}Si NMR. In a molecule containing OH, for example, the electronegative oxygen atom essentially pulls the electron away from the hydrogen atom, which is largely unshielded, giving rise to a smaller chemical shift than would be expected for other bonds (e.g., C–H) where the electron will behave so as to oppose the applied magnetic field. Secondly, the coupling of spins between neighbouring nuclei within a molecule, or even between different molecules, can change the observed NMR frequencies, leading to splitting and broadening of peaks. The magnitudes of these effects are dependent on the orientation of the magnetic nuclei with respect to the magnetic field.

The measured NMR frequencies and splittings can be used to characterize the chemical environments of the constituent atoms in a molecule, while the individual signal intensities are proportional to the number of nuclei in each particular environment. As a result, this technique is frequently used on solution species. However, in a disordered solid or a polycrystalline material, the orientation of the magnetic nuclei relative to the field is largely random. This anisotropy affects the chemical shifts and the dipolar couplings; it leads to a range of resonance peaks due to the large range of possible molecular or crystallite orientations. In low-viscosity liquids these problems disappear because the rapid molecular motion averages all these anisotropies to zero. However, in solids, translational movement is not possible and the resonances all overlap, leading to very broad peaks. This anisotropy problem can be overcome by spinning the sample around an axis inclined at an angle of 54.7° (the magic angle which trisects the three Cartesian axes) to the axis of the magnetic field. The powdered solid sample is typically contained in a plastic or ceramic rotor, and the rotor is spun by an air turbine around an axis oriented at exactly the magic angle using a spinning rate of a few kilohertz, which is comparable to the frequency spread of the anisotropy. The technique is known as magic angle spinning NMR (MASNMR) and produces a single, much narrower isotropically averaged NMR peak from solids, as in spectra obtained from solutions. A schematic MASNMR spectrum is shown in Figure 2.21. A variety of pulse sequences and magnetization changes (referred to as cross-polarization techniques) are often used to enhance signals from certain nuclei within a system. The MASNMR signal is obviously an average over the whole sample volume.

Dynamical information can also be obtained by NMR spectroscopy by measuring relaxation times. The interaction of a nucleus with its environment or with other (identical) nuclei can be obtained by measuring the longitudinal (spin–lattice) or transverse (spin–spin) relaxations, respectively. Nuclei precess about the applied magnetic field, B_0, at their *Larmor* frequency. By applying a circularly polarized magnetic field orthogonal to B_0 at the Larmor frequency, the nuclear spins will essentially follow a gyroscopic motion precessing about two different axes simultaneously, which allows one to isolate the nuclei of interest. This is because, if one removes the circularly polarized field after only 90°, the nucleus precesses around an axis away from the other nuclei. This orthogonality means that the behaviour of these nuclei can be monitored without interference from all the other nuclei with different Larmor frequencies precessing about B_0. Of course, these nuclei will relax back to precession about B_0, but then that is generally the sort of information that one is interested in. This $\pi/2$ spin flip is of central importance to dynamical measurements, but it must be used in parallel with one π spin flip (spin–lattice experiments) or multiple π spin flips (spin–spin experiments). By isolating individual nuclei it is possible to map their location throughout the sample, enabling a two-dimensional image to be constructed. Indeed it is possible to scan such two-dimensional sections through a solid and reconstruct 3D images. This is essentially a tomographic technique which forms the basis for magnetic resonance imaging (MRI).

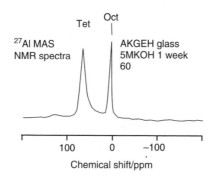

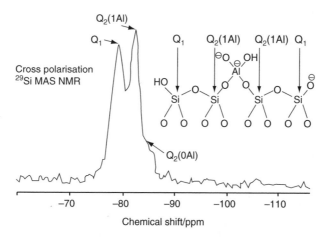

Figure 2.21 Schematic diagram of ^{27}Al and cross-polarized ^{29}Si MASNMR spectra from a Portland cement sample revealing the different chemical environments; inset is the derived chemical structure of the calcium silicate hydrate chains within the cement nanostructure. Data courtesy of Dr Ian Richardson and Dr Adrian Brough, School of Civil Engineering, University of Leeds

In biology and medicine, NMR images are constructed from ^{1}H NMR spectra of proton-bearing species such as water and fatty acids. However, the spatial resolution of these techniques is limited to at best microns or tens of microns, depending on the magnetic homogeneity of the sample. It is also possible to use the spin–spin relaxation to measure molecular diffusion coefficients using NMR. This is often called spin-echo NMR.

Another related spectroscopy worthy of mention is Mössbauer spectroscopy, which employs a γ-ray source containing an isotope with the same nuclear energy as the species of interest, for example, a ^{57}Co source is used to analyse ^{57}Fe, so as to excite transitions between nuclear states in the sample. The source is moved relative to the fixed sample so as to vary the incident γ-ray wavelength via the Doppler effect. Transitions between the ground and excited states are modified by interactions between the nucleus and the surrounding electronic and magnetic fields, termed hyperfine interactions, and these can be analysed to determine the valence states of atoms in the sample. This technique is restricted to elements possessing isotopes with suitable nuclear transitions such as ^{57}Fe, ^{119}Sn and ^{197}Au.

Finally, electron paramagnetic resonance (EPR) is a microwave spectroscopy which employs a variable magnetic field to analyse crystals containing molecules or ions with unpaired electronic spins. The magnetic field causes Zeeman splitting of electronic energy levels, and transitions between these levels are induced by the absorption of microwave radiation.

2.7.3 Electron spectroscopy

Initially we concentrate on electron-induced spectroscopies in the electron microscope, either the SEM or TEM, in particular the analysis of the X-rays emitted by the sample as a result of electron beam ionization of atoms in the sample. True electron spectroscopy is discussed in Section 2.7.3.3.

2.7.3.1 X-ray emission in the SEM and TEM

As discussed in Section 2.3.3.3, following ionization of atoms in a sample by an electron beam, one possible de-excitation process is X-ray emission. The energy of the X-ray photon emitted when a single outer electron drops into the inner shell hole is given by the difference between the energies of the two excited states involved. A set of dipole selection rules determine which transitions are observed. Due to the well-defined nature of the various atomic energy levels, it is clear that the energies and associated wavelengths of the set of emitted X-rays will have characteristic values for each of the atomic species present in the specimen. By measuring either the energies or wavelengths of the X-rays emitted from the top surface of the sample, it is possible to determine which elements are present at the particular position of the electron probe; this is the basis for energy-dispersive and wavelength-dispersive X-ray analysis (EDX and WDX). WDX spectrometers use crystal monochromators to disperse the emitted X-ray spectrum in terms of diffraction angle and hence wavelength. Detectors then move along the arc of a circle centred on the specimen and collect the spectrum serially. EDX detectors collect X-rays in a near-parallel fashion and rely on the creation of electron–hole pairs in a biased silicon crystal; the number of electron–hole pairs and hence current is directly proportional to the energy of the incident X-ray. Fast electronics allows separate pulses of X-rays to be discriminated and measured. EDX detectors often have some form of window which, depending on the material, may reduce sensitivity to light elements ($Z < 11$). Compared to EDX, WDX is slow but has increased resolution and hence sensitivity to all elements, particularly light elements. WDX is confined to dedicated analytical SEMs known as electron probe microanalysers (EPMA), whereas EDX detectors may be fitted as an add-on attachment to most SEMs and TEMs.

Figure 2.22 shows a typical electron-generated X-ray emission spectrum from molybdenum oxide. The Mo Kα, Kβ and Lα X-ray lines as well as the O Kα line are superimposed on the bremsstrahlung background. The bremsstrahlung X-rays are not characteristic of any particular atom but depend principally on specimen thickness. To a first approximation, peak intensities are proportional to the atomic concentration of the

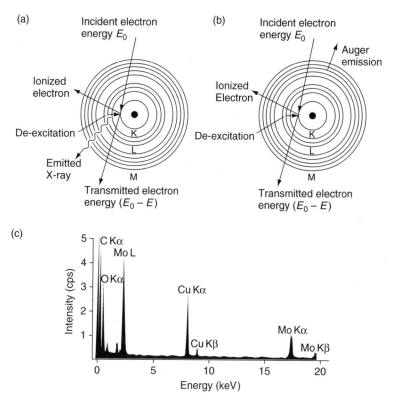

Figure 2.22 Schematic diagram of (a) de-excitation by X-ray emission, (b) de-excitation by Auger electron following ionization by an electron of incident energy E_0. (c) Schematic energy-dispersive X-ray emission spectrum from a thin specimen of molybdenum oxide on a carbon support film; the peaks are labelled with standard X-ray notation as discussed in the text; the peaks due to copper are from the specimen holder

element and, with careful measurement, EDX and WDX can detect levels of elements down to 0.1 at%.

When EDX is used in the TEM as opposed to the SEM, the reduced sample thickness and probe size lead to much higher spatial resolution. At 100 kV, a 100 nm thick sample typically gives a beam broadening of the order of 10 nm. EDX analysis will collect all X-rays produced isotropically within the beam-broadened volume, and this is of obvious import-ance for the analysis of nanostructures. EDX maps of multilayers and nanoscale precipitates have demonstrated resolutions of at best a few nanometres in ultrathin sample areas.

2.7.3.2 Cathodoluminescence in the SEM and STEM

Cathodoluminescence (CL) is generated by the recombination of electron–hole pairs produced by electron beam induced ionization of valence electrons in a semiconducting or insulating sample. The exact CL energies (and therefore wavelengths) depend on the width of the band gap of the material. Any changes in temperature, crystal structure (e.g., different polymorphs), impurity levels or defect concentrations will modify the band gap and thus the CL wavelength. Spectroscopic analysis of the CL emission can therefore provide information on the microstructure of the sample. The emitted photon intensity is low and therefore low scan rates and high probe currents (large probe diameters) are required, limiting the resolution to 1–10 µm in the SEM. In the STEM the resolution can be improved considerably, making the technique viable for nano-structural analysis.

2.7.3.3 Electron energy loss spectroscopy

Electron energy loss spectroscopy (EELS) in a TEM or STEM involves analysis of the inelastic scattering suffered by the electron beam via measurement of the energy distribution of the transmitted electrons. The technique allows high-resolution elemental mapping and the measurement of local electronic structures for the determination of the local chemical bonding, such as that present at an interface or defect.

The transmitted electrons are dispersed according to their energy losses using a magnetic sector spectrometer (see Figure 2.8). EELS spectra can be recorded in parallel using a scintillator and diode array detection system, with the TEM in either diffraction or image mode; spatially resolved spectra are obtained using a small, highly focused probe in a STEM. The transmitted EELS signal is highly forward-peaked and the collected signal can be well defined using a suitable aperture that can significantly improve the ultimate spatial resolution of the analysis compared to EDX, where beam broadening will limit the ultimate (lateral) spatial resolution for microanalysis. EELS spectra can be obtained from individual atomic columns.

The various energy losses observed in a typical EELs spectrum are shown in Figure 2.23a, which gives the scattered electron intensity as a function of the decrease in kinetic energy (the energy loss E) of the transmitted fast electrons and essentially represents the response of the electrons in the solid to the disturbance introduced by the incident electrons. In a specimen of thickness less than the mean free path for inelastic scattering (roughly 100 nm at 100 keV), by far the most intense feature is the zero-loss peak at 0 eV energy loss, which contains all the elastically and quasi-elastically (i.e., phonon) scattered electron components. Neglecting the effect of the spectrometer, the full width at half maximum (FWHM) of the zero-loss peak is usually limited by the energy spread inherent in the electron source. In a TEM the energy spread will generally lie between ~0.1 and 3 eV, depending on the type of emitter, and this parameter often determines the overall spectral resolution.

The *low-loss* region of the EELs spectrum, extending from 0 to about 50 eV, corresponds to the excitation of electrons in the outermost atomic orbitals which are often delocalized due to interatomic bonding and extend over several atomic sites. This region therefore reflects the solid-state character of the sample. The smallest energy losses (10–100 meV) arise from phonon emission, but these are usually subsumed in the zero-loss peak. The dominant feature in the low-loss spectrum arises from collective, resonant oscillations of the valence electrons known as plasmons. The energy of the plasmon peak is governed by the density of the valence electrons, and the width by the rate of decay of the resonant mode. In a thicker specimen ($\geq$100 nm) there are additional peaks at multiples of the plasmon energy, corresponding to the excitation of more than

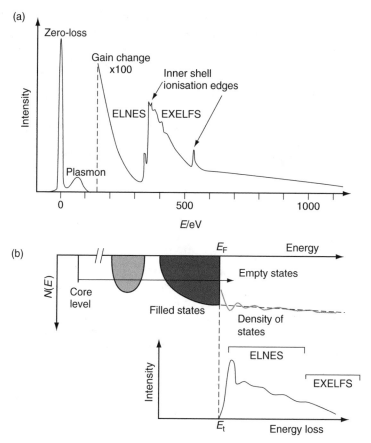

Figure 2.23 Schematic diagram of (a) an EELS spectrum and (b) the ELNES intensity which reflects the unoccupied DOS above the Fermi level

one plasmon; the intensities of these multiple plasmon peaks follow a Poisson distribution. A further feature in the low-loss spectra of insulators are peaks, known as interband transitions, which correspond to the excitation of valence electrons to low-energy unoccupied electronic states above the Fermi level. These single-electron excitations may lead to a shift in the energy of the plasmon resonance. The low-loss region is used mainly to determine the specimen thickness and to correct for the effects of multiple inelastic scattering when performing quantitative microanalysis on thicker specimens. In a more detailed analysis the overall shape of the low-loss region may be related to the dielectric response function of the material, which allows a correlation with optical measurements, including reflectivity and band gap determination in insulators and semiconductors.

The *high-loss* region of the EELs spectrum extends from about 50 eV to several thousand electron volts and corresponds to the excitation of electrons from localized orbitals on a single atomic site to extended, unoccupied electron energy levels just above the Fermi level of the material. This region therefore reflects the atomic character of the specimen. As the energy loss progressively increases, this region exhibits steps or edges superimposed on the monotonically decreasing background intensity; the intensity at 2000 eV is typically eight orders of magnitude less than at the zero-loss peak and so, for clarity, in Figure 2.23 a gain change has been inserted in the linear intensity scale at 150 eV. These edges correspond to excitation of inner shell electrons and are known as ionization edges. The various EELS ionization edges are classified using the standard spectroscopic notation; e.g., K excitation for ionization of 1s electrons. Since the energy of the edge threshold is determined by the binding energy of the particular electron subshell within an atom, the atomic type may be easily identified. The signal under the ionization edge extends beyond the threshold, since the amount of kinetic energy given to the excited electron is not fixed. The intensity or area under the edge is proportional to the number of atoms present and this allows the technique to be used for quantitative analysis. EELS is particularly sensitive to the detection and quantification of light elements ($Z < 11$) as well as transition metals and rare earths.

If electrons are scattered via inelastic collisions with K-shell electrons of free atoms (e.g., gases) the core-loss edges are sharp, sawtooth-like steps displaying no features. Other core-loss excitations in free atoms display a variety of basic edge shapes that are essentially determined by the degree of overlap between the initial and final state wavefunctions. In solids, however, the unoccupied electronic states near the Fermi level may be appreciably modified by chemical bonding, leading to a complex density of states (DOS), and this is reflected in the electron energy loss near-edge structure (ELNES), which modifies the basic atomic shape within the first 30–40 eV above the edge threshold. The ELNES effectively represents the unoccupied DOS (above the Fermi level) in the environment of the atom(s) being ionized, as shown in Figure 2.23b, hence it gives information on the local structure and bonding. Beyond the near-edge region, superimposed on the gradually decreasing tail of the core-loss edge, a region of weaker, extended oscillations is observed, known as the extended energy loss fine structure (EXELFS). As in EXAFS (Section 2.7.1.4), the period of the oscillations may be used to determine bond distances, while the amplitude reflects the coordination number of the particular atom.

Energy-filtered TEM (EFTEM) involves the selection of a specific energy loss value, or narrow range of energy losses, from the transmitted electron beam via use of an energy-selecting slit after the spectrometer. Using only zero-loss (elastically) scattered electrons to form images and diffraction patterns increases contrast and resolution, allowing easier interpretation than with unfiltered data. Chemical mapping may be achieved by acquiring and processing images formed by electrons which have undergone inner shell ionization events. An alternative approach, used when mapping is performed with an EELs spectrometer and detector attached to an STEM, is to raster the electron beam across the specimen and record an EELS spectrum at every point (x,y) – this technique being known as 'spectrum imaging'. The complete data set may then be processed to form a 2D quantitative map of the sample using either standard elemental quantification procedures or the position and/or intensity of characteristic low-loss or ELNES features. The inherently high spatial resolution of the EELS technique has allowed maps of elemental distributions to be formed at subnanometre resolution. In addition, the possibility of plasmon or ELNES chemical bonding maps has also been demonstrated.

2.8 SURFACE ANALYSIS AND DEPTH PROFILING

One basic form of surface analysis, highly relevant for nanostructured materials, is the determination of the specific surface area (in $m^2 g^{-1}$) of a sample. This is usually performed by measurement of the volume of gas adsorbed onto a specific mass of sample, as a function of gas pressure or more usually relative gas pressure (pressure/saturation vapour pressure). This measurement is known as an adsorption isotherm. Samples are carefully outgassed under vacuum and cooled to low temperatures, prior to controlled physical adsorption of a suitable gas, such as nitrogen, which is small enough ($\sim$0.4 nm) to access even the smallest nanopores. The form of the resulting adsorption isotherm depends on the nature of the adsorbent and the adsorbate and may be analysed using a number of theoretical models, all of which make assumptions about the nature of the gas–solid interaction and the types of pores present. In essence the models attempt to identify when one monolayer of gas molecules has been adsorbed at the surface so as to determine the surface area. In addition to the total surface area, by modelling the data it is also possible to calculate the average pore size and the pore size distribution within a porous material, although it is important to realise that such experiments will only access open (not closed) porosity within a sample.

In addition to the basic adsorption methods, surface chemical analysis techniques directly probe the elemental composition and chemical state of the outermost atomic layers (i.e., 0.1–10 nm below the surface) of all types of solid materials. The principal techniques used for probing surface chemistry are based on electron, mass and vibrational spectroscopy. In terms of nanostructured materials and devices, the surface is obviously an important consideration in terms of properties and material performance, as well as the nature of the interaction between the sample and the environment. Properties of adhesion, colour, catalytic activity, biocompatibility and lubrication, for example, can all be strictly controlled by the surface of a material. However, if samples are of reduced dimension laterally this can present particular problems as high lateral spatial resolution on the surface is then also required.

Due to the extreme surface sensitivity of most surface analysis methods, the amount of information available decays rapidly with degrading vacuum since, as highlighted above, ambient gases readily adsorb onto the surface of the specimen. Most surface analysis methods therefore require analysis to be performed under UHV conditions (10^{-10} mbar) where the pressure is an order of magnitude lower than for, say, electron microscopy. For this reason instrument design has evolved around a substantial evacuation system involving a two-chamber apparatus incorporating metal-on-metal vacuum seals for maintaining a good vacuum during analysis. The first preparation chamber is for sample introduction and high-vacuum experimentation (e.g., gas adsorption experiments, sample heating or ion beam etching), while the second chamber is the analysis chamber containing the various radiation sources and detection systems. Figure 2.24 is a schematic diagram of a surface analysis system.

Depth profiling provides surface analysis coupled with subsurface information. Techniques such as neutron reflectometry can provide information about the structure of samples many microns below the surface, without the often prohibitive UHV requirements. Other techniques such as ion beam analysis can also provide similar information. The key to depth profiling is the ability to have a probe which can interrogate deep into the sample. Neutrons, X-rays, and high-energy ions are generally

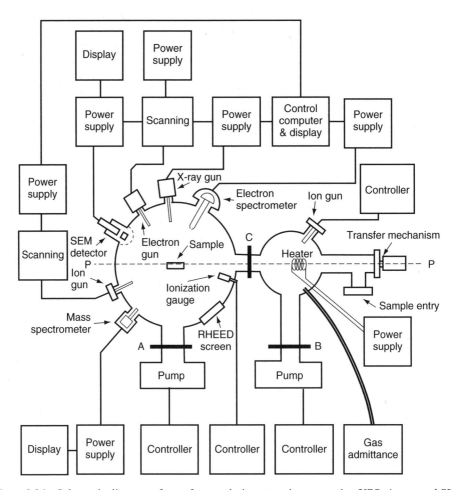

Figure 2.24   Schematic diagram of a surface analysis system incorporating XPS, Auger and SIMS

very effective at this. We shall consider depth profiling with ions and neutrons below, but first we turn to the surface-specific electron spectroscopy and mass spectrometry techniques.

2.8.1   Electron spectroscopy of surfaces

X-ray photoelectron spectroscopy (XPS) and Auger electron spectroscopy (AES) have shown the greatest applicability to the widest range of materials surfaces. XPS and AES are conveniently linked as techniques since they are both dependent on the analysis of low-energy electrons emitted from surfaces, in the range 10–3000 eV. As a result, both techniques can use the same electron spectroscopy instrumentation, although they tend to use different excitation sources.

In XPS the sample is irradiated with a beam of usually monochromatic, low-energy X-rays. Photoelectron emission results from the atoms in the specimen surface, and the kinetic energy distribution of the ejected photoelectrons is measured directly using an electron spectrometer. Each surface atom possesses core-level electrons that are not directly involved with chemical bonding but are influenced slightly by the chemical environment of the atom. The binding energy of each core-level electron (approximately its ionization energy) is characteristic of the atom and specific orbital to which it belongs. Since the energy of the incident X-rays is known, the measured kinetic energy of a core-level photoelectron peak can be related directly to its characteristic binding energy. A typical XPS spectrum is displayed in Figure 2.25, which shows photoelectron peaks superimposed on a background due to inelastically scattered photoelectrons that have lost energy before emerging from the surface. The binding energies of the various photoelectron peaks (1s, 2s, 2p, etc.) are well tabulated and XPS therefore provides a means of elemental identification which can also be quantified via measurement of integrated photoelectron peak intensities and the use of a standard set of sensitivity factors to give a surface atomic composition. The low binding energy region of the XPS spectrum is usually excited with a separate ultraviolet photon source, such as a helium lamp, (ultraviolet photoelectron spectroscopy, UPS) and provides data on the valence band electronic structure of the surface.

Atomic orbitals from atoms of the same element in different chemical environments possess slightly different but measurable binding energies within the range 0.1–10 eV. These chemical shifts in the binding energies of the photoelectron peaks arise because of the variations in electrostatic screening experienced by core electrons as the valence and conduction electrons are drawn towards or away from the specific atom in question. Differences in oxidation state, molecular environment and coordination number will all provide different chemical shifts that can be measured in a high-resolution XPS spectrum. Using deconvolution and peak synthesis routines it is possible to identify and also quantify the relative occupations in these differing environments as indicated in Figure 2.25.

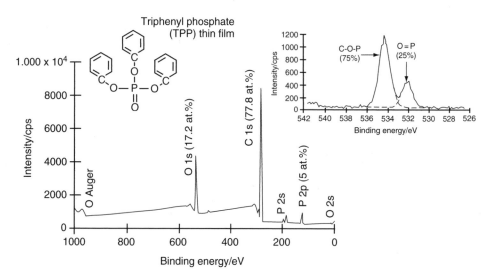

Figure 2.25 Schematic XPS spectrum from a triphenyl phosphate polymer film: (a) survey scan showing the elemental composition (in at %) and (b) high-resolution scan of the oxygen 1s peak showing two chemical states of oxygen and their relative proportions

After photoemission of a core electron, the ion is left in an excited state and must decay back to the ground state. Energy is released when an electron drops back into the core hole and this energy can escape as an X-ray photon (X-ray fluorescence) or, alternatively, it can eject a third weakly bound outer shell electron (Auger electron) with a certain kinetic energy (Figure 2.22). The characteristic kinetic energy of the Auger electron is dependent only on the binding energies of the core levels within the target atom. Auger electron bands are designated using classical X-ray notation (e.g., K, L, M) referring to the electron energy levels involved in the relaxation process.

The sole requirement for the ejection of an Auger electron is the formation of an initial core electron level hole and this can be generated by either an electron beam, X-rays, ions or even by thermal energy. Auger peaks are therefore also observed in photoelectron spectra, as seen in Figure 2.25. However, traditionally Auger electron spectroscopy has been studied by excitation with primary, low-energy electron sources of a few keV. Here the electron-induced Auger peaks are superimposed on an intense background due to the large numbers of emitted secondary electrons. Auger peaks, particularly those of light elements, may be identified and quantified to give surface compositions, in a similar manner to XPS. A major benefit of using electron beams, rather than photons, for the production of Auger spectra is that it is relatively easy to focus low-energy electron beams from field emission sources to a dimension of 10–50 nm. This gives much improved lateral spatial resolution at the surface of a sample, which is of great applicability for the analysis of nanodimensional samples. As a result, Auger electron spectroscopy (AES) has been combined with scanning electron microscopy (SEM) to produce the imaging technique of scanning Auger microscopy (SAM). In contrast, X-ray photon beams (needed to generate X-ray photo-electrons and Auger electrons for XPS) are much more difficult to focus and are hence much broader, being as large as 1 mm to 1 cm in laboratory instruments. Small-area XPS instruments can achieve analysis from areas as small as 10 μm by the use of area-selecting apertures before the photoelectron spectrometer. It is now also possible to perform XPS mapping of a surface by using a photoelectron microscope arrangement to image the position of surface species using the emitted photoelectrons.

In principle AES is also capable of providing information on the chemical state of a particular element. In practice the observed chemical shifts are small compared to the broader widths of Auger peaks and this reduces the quality of the chemical information. Nevertheless, in some cases, distinct changes in Auger peak shape are observed for different chemical states. In an XPS spectrum the difference between (or sum of) the XPS and Auger peaks, known as an Auger parameter, is also used widely for its chemical sensitivity. This parameter also enjoys independence from any sample-charging problems often experienced in XPS and AES, as peak separations and not absolute energies are measured.

Although X-rays can penetrate up to several microns within a solid, XPS and Auger electrons only escape, without energy loss, from the outermost atomic layers of the sample. This is because these low-energy electrons have very short inelastic mean free paths (IMFPs) in solid materials. The electron IMFP is dependent on the electron energy as well as the density and nature of the material through which the electrons pass. In the energy range of XPS and AES (100–2000 eV), the electron IMFP in virtually all solids is between 1 and 10 monolayers, roughly equivalent to 0.3–3 nm. This sampling depth is dependent also on the electron take-off angle relative to the detector and is maximized for the detector normal to the sample surface. By varying the geometry in angle-resolved XPS measurements, it is possible to determine the nature of the surface structure such as the presence of thin overlayers on substrates. This technique can also be used to determine the orientation of certain molecules at surfaces. Alternatively, in XPS or AES, depth information can also be achieved using sputter ion depth-profiling techniques where the surface is eroded using a low-energy, inert gas ion source. However, for many materials this approach is known to significantly damage the chemical integrity of the surface under investigation and does not have sufficient depth resolution to probe the outermost surface layers. In this case, angle-resolved XPS provides a convenient method for acquiring depth information in a non-destructive manner and thus enables molecular structure (as indicated by the chemical shift) as well as the elemental distribution to be determined.

2.8.2 Mass spectrometry of surfaces

In the same way that analysis of compounds by mass spectrometry provides information about chemical and structural properties in analytical chemistry, surface mass spectrometry is of key importance in the surface analysis of materials. The principal method of signal generation is by sputtering ionized particles from the specimen surface. Mass discrimination can involve the use of alternative mass spectrometers that are specially configured for particular aspects of surface analysis.

In the sputtering process, the specimen surface is bombarded by a primary beam of particles having a fixed energy. These particles (commonly ions or atoms) induce a series of hard sphere collision cascades along pathways of up to 10–25 nm into the surface region, whereupon the majority of their energy is lost. When the collision cascade returns to the surface and has sufficient energy to break chemical bonds, then atomic and clustered fragments of surface atoms, representative of the surface chemistry, are ejected into the vacuum.

The vast majority of species which are ejected in the sputtering process are electrically neutral. However, a finite portion is ionized during the process of bond fracture and this ionized portion of the emitted species can be extracted directly for analysis in a mass spectrometer. This is the basis of surface analysis by secondary ion mass spectrometry (SIMS); a typical spectrum is shown in Figure 2.26. Extraction of the ionized species is often achieved by placing a ring or cone of high potential in close proximity to the sample surface. By altering its polarity both positive and negative ions can be drawn into the mass spectrometer for analysis. SIMS is not usually a directly quantitative technique. It can, however, be used accurately in a semiquantitative manner in conjunction with a series of suitable calibration samples. In sputtered neutral mass spectrometry (SNMS) an attempt is made to ionize the portion of neutral species that are emitted in the sputtering process. Since these species constitute the vast majority of the sputtered 'plume' of particles ejected from the surface then, in principle, SNMS should be able to provide surface mass spectral information that is more directly quantitative than information derived from SIMS. In addition to ion sources, lasers can be used to desorb or volatilize surface species and generate mass spectra from the surfaces of materials.

Generally the amount of material removed during sputtering is governed by the energy and atomic mass of the primary beam and its angle of incidence to the sample surface. Removal of a large volume of surface atoms from the analysis area is undesirable for

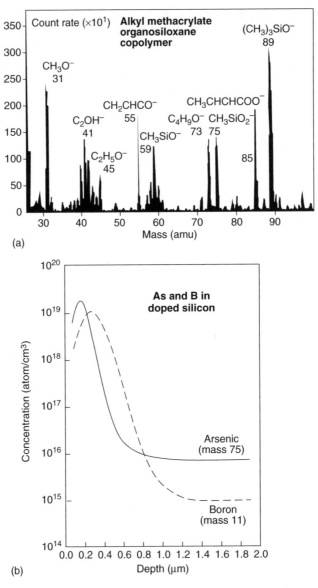

Figure 2.26 Schematic diagram of (a) a static SIMS spectrum from a copolymer film and (b) an example of a depth profile measured using dynamic SIMS

surface analysis and so minimal primary particle fluxes are preferred for the most surface-sensitive analysis. This mode of SIMS is called static SIMS, whereby the surface chemical state can be considered to remain unaltered, or static, as a consequence of analysis. Static SIMS generally employs high mass resolution time-of-flight (ToF) mass spectrometers to provide detailed and often complex spectra with many mass fragment peaks evident, as seen in Figure 2.26(a). ToF-SIMS is capable of resolving very high mass fragments compared to the more conventional and cheaper quadrupole mass spectrometers.

If the removal rate is precisely controlled, at higher primary beam fluxes it is possible to monitor the SIMS signal from successive layers of a material as a function of removal rate. This mode of analysis is termed dynamic SIMS, whereby the surface moves or is eroded in real time as a consequence of analysis. This is the mode of SIMS used for depth profiling materials with high elemental sensitivity of the order of parts per million, as seen in Figure 2.26(b). In static and dynamic SIMS it is common practice to raster scan the focused primary ion beam across the surface region of interest. For static (scanning) SIMS, this will increase sensitivity and decrease beam-induced damage effects compared to point analysis. When scanning SIMS is performed with the mass spectrometer preset to detect the intensity of masses of interest at each scan point, chemical mapping of the distribution of secondary ions over the area of interest can be performed. The resolution of the SIMS image is determined by the size of the focused ion beam, which can reach 50–100 nm or below. The use of scanning and dynamic SIMS for 3D sample imaging has also been demonstrated.

2.8.3 Ion beam analysis

SIMS is not the only use of ions to analyse the surface and subsurface properties of materials. Ion beam analysis refers to a variety of techniques that use ions to interact with the sample under consideration. These ions are generally produced in an accelerator, often van der Graaff accelerators adapted from nuclear experiments in the post-war years. The ions typically have energies between 1 and 2 MeV and can penetrate deep (several microns) into the sample. Some ions will be scattered by the components of the material or even undergo nuclear reactions with these components. By a careful consideration of the energetics of the scattered products, it is possible to build a profile for the composition of a particular component in a material as a function of depth. As in dynamic SIMS, this form of material analysis is known as depth profiling, but it does have certain advantages. Firstly, sample preparation is easier; there is no need for a surface coating on the sample (this is required in some dynamic SIMS experiments to control the sputtering rate before the sample is reached). Secondly, the variation of scattered products as a function of depth is easier to ascertain, which makes for easier data analysis (changing composition can have a strong effect on sputtering rates in dynamic SIMS experiments). Finally, some ion beam experiments will cause minimal sample destruction compared with dynamic SIMS.

Surface and subsurface composition can be determined using high-energy ions of low mass (e.g., H^+ or He^{2+}) produced in an accelerator. Interaction of these ions with a sample results in considerable elastic backscattering of the incident ion flux. Analysis of the energy loss spectrum of the backscattered ions forms the basis for Rutherford backscattering spectrometry (RBS). An RBS spectrum exhibits steps at energies which are characteristic of each backscattering element present and whose widths depend on the elemental depth distribution in both the surface and subsurface regions. Although the lateral spatial resolution of the technique is poor, RBS can provide a non-destructive means of depth-resolved subsurface elemental analysis, but any light (e.g., organic) components are likely to be destroyed by the high-energy ion beam. A related technique is proton-induced X-ray emission (PIXE), which allows the determination of low elemental concentrations.

The scattering of high-energy low-mass ions in RBS works because these ions will be backscattered from heavier elements in the material. Lighter elements (mainly hydrogen) will be ejected from the sample. However, if one considers labelling some of the materials with deuterium (to provide contrast with hydrogen), then one can use this to analyse scattering in the forward geometry. A high-energy α-particle beam will eject both H and D from a material and the energies of these ejected ions will depend on their original location in the material. The expelled hydrogen and deuterium are detected in a *forward-scattered* geometry in contrast with the backscattered geometry of RBS. This technique goes by two names: forward-recoil spectrometry (FReS) and elastic-recoil detection analysis (ERD or ERDA). The choice of name generally depends on the community doing the research.

Another method of obtaining a depth profile using ion beams is the technique of nuclear reaction analysis (NRA). In this case an ion such as $^{15}N^+$ reacts with a target nucleus (e.g., deuterons) in the sample, a nuclear reaction ensues and the products can be detected. The energy of the incident ion can be correlated to the depth in the sample where the reaction occurred. This technique is less commonly used because of the low product yield and also because the ion energy needs to be varied carefully. Another technique, ^{3}He NRA (an example is shown in Figure 2.27(a)) has a broader reaction cross section than the other methods, so it can obtain a depth profile with one incident ion energy; this overcomes the difficulties caused by a narrow reaction cross section. However, the low yield is still a problem in many circumstances.

2.8.4 Reflectometry

Neutrons and X-rays are powerful probes of material structure, but they also provide a means of obtaining a depth profile from a film or multilayer structure. Neutron and X-ray reflectometry can only provide limited information on a one-component film, such as thickness, density or roughness. However, for mixtures or multilayer structures, reflectometry can be a very powerful technique provided contrast exists between the two components. Neutrons are more powerful than X-rays, because deuteration or other isotopic labelling can be used to provide material contrast. With X-ray reflectometry experiments, a heavy metal is generally needed to provide the large electron density required for sufficient contrast with any lighter elements present. In the brief discussion that follows, we consider neutron reflectometry, but X-rays would not be treated all that differently.

A reflectometry experiment generally consists of a neutron beam incident at a glancing angle (typically $\sim 1°$) to the film. The reflectivity at the specular peak (i.e., the same reflected angle) is detected as a function of angle or wavevector. At low wavevectors, neutrons are usually totally reflected (neutron scattering length densities work in the same way as refractive indices for light, so this means that the neutron scattering length density is greater at the film surface than in air or the surrounding medium). At higher wavevectors the reflectivity decays dramatically in a fashion that relates to the material structure. Bragg and Kiessig fringes in the data reflect dominant length scales from internal structure and the total film thickness, respectively (Section 2.6.1). In the example data shown in Figure 2.27(b), Bragg fringes from the deuterated layer at the front and the back of the sample are clearly visible in the spectrum. Kiessig fringes are difficult to observe in neutron experiments for films thinner than ~ 250 nm.

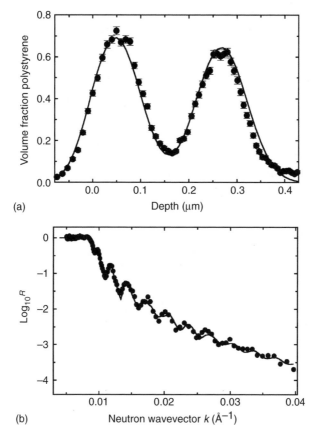

(a)

(b)

Figure 2.27 (a) Helium-3 nuclear reaction analysis (NRA) data for a film of polystyrene mixed with polybutadiene. The polystyrene is deuterium labelled to provide contrast with the polybutadiene. In this example a polystyrene-rich phase self-assembles at the front and back of the film, with polybutadiene in the centre. The solid line is not a fit to the NRA data but the best fit to neutron reflectometry data convolved with the resolution function of the ion beam experiment. The neutron reflectivity (R) data are shown in (b) as a function of the perpendicular component of the neutron wavevector; note the Bragg fringes from the front and back deuterated layers in the film. Here the solid line is a best fit to the data. This work is reported in M. Geoghegan *et al.*, Lamellar structure in a thin polymer blend film, *Polymer* **35**, 2019–2027 (1994)

Analysing reflectivity data is not trivial; typically a scattering length density–depth profile needs to be modelled and fitted to the data, adjusting any fit parameters as necessary. The difficulty arises because neutron reflection, like diffraction, occurs in **k**-space, and phase information is lost. As a result, there is no one-to-one correlation between reflectivity and the depth profile, and a priori information is needed before fitting can reveal any meaningful results; this information may be obtained from what we know about the sample in its preparation, or from a complementary experiment such as ion beam analysis. An example of this complementarity is shown in Figure 2.27. A few tricks can be used to simplify the fitting process but, in truth, great advances in reflectometer design over the past 15 years have not been mirrored in data analysis, except that increased computational power has helped considerably.

An advantage of neutron reflectometry is that it is suitable for working with liquids; indeed neutrons can penetrate many different media, including silicon substrates, should one need to access the sample from underneath (this is not possible with X-rays). Both neutrons and X-rays provide subnanometre depth resolution, if there are sharp interfaces present in the material. However, considering neutrons in terms of resolution is unwise because, although they might be able to resolve a 0.2 nm interface, they usually would not be able to resolve an interface with a width of 1 μm. To understand this, remember that the neutron wavevector is related to its wavelength λ by $(2\pi/\lambda) \sin \theta$. Even if it were possible to obtain a neutron wavelength comparable to micron length scales, the relevant value of the wavevector would be so low that, in all probability, total external reflection would render data analysis impossible. Neutron reflectometry is generally non-destructive but the timescale for obtaining data usually precludes kinetic experiments, although a new generation of neutron sources in America and Japan should alleviate these problems considerably.

2.9 SUMMARY OF TECHNIQUES FOR PROPERTY MEASUREMENT

Many sample properties, such as strength or electrical conductivity, are a strong function of the material's microstructure, including features such as the nature and distribution of phases, the crystallite grain size and defect concentration. Nanostructured systems often exhibit very different material properties owing to the step change in the nature and distribution of such features. However, accurate property measurement, particularly in a spatially resolved manner, can present considerable challenges for nanoscale science.

2.9.1 Mechanical properties

The important mechanical properties are strength, ductility, toughness, resilience and hardness. Generally all measurements are determined by applying a load to a specimen (either in tension, compression or shear) and the dependence between the applied stress (the force per unit area) and the measured strain (the fractional change in dimension) is monitored. In many cases the mechanical behaviour is dependent on the rate at which the specimen is stressed and may also be a function of whether the material undergoes a series of cyclic loadings and unloadings.

A typical tensile stress–strain curve is shown in Figure 2.28. In tension and at low levels of stress and strain, the stress–strain relationship is linear and the material deforms elastically. The proportionality constant in this initial portion of the stress–strain curve is known as the Young's modulus or elastic modulus and may be thought of as stiffness. Resilience is the capacity of the material to absorb energy when deformed elastically and is given by the area under the stress–strain curve up to the yield point, as shown in Figure 2.28.

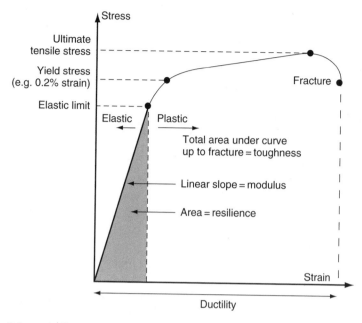

Figure 2.28   Schematic diagram of a typical stress–strain curve indicating the important quantities

At higher stresses, the stress–strain curve departs from linearity and plastic deform-ation occurs. The stress value at this departure point gives a measure of the elastic limit and, slightly above this, the yield strength for a given strain; the overall maximum in the curve is known as the tensile strength. In reality the cross-sectional area may change during deformation, leading to the concept of true stress. Ductility is the degree of plastic deformation that has occurred at fracture and is given by the percentage elongation of the material. Toughness is the ability of the material to absorb energy up to fracture; it is very dependent on specimen geometry and the rate at which the strain is applied. A strong, ductile material will generally also be tough. Hardness is a measure of the resistance of a material to localized plastic deformation and is generally determined by indenting the surface of a sample with a controlled load using a known indenter geometry at a known rate of indentation.

Many mechanical tests are difficult to apply to systems of reduced dimensionality owing to the small diameters and length scales involved, making conventional tensile testing difficult where, say, fibres or filaments need to be gripped without sliding. In recent years, nanoindentation techniques have been developed and they have allowed some spatially resolved studies of mechanical behaviour. Figure 2.29 shows how a stiff indenter with a well-defined geometry can be positioned with submicron accuracy within the microstructure of a sample using a sample stage controlled by piezoelectric drivers. Very small loads can be applied to the indenter while monitoring its displace-ment, hence the depth of penetration below the sample surface. The loading–unloading versus displacement curve (inset) can give information on both Young's modulus and hardness at that position on the sample. Accurate manipulation of nanostructures is also extremely difficult and specialized testing techniques have had to be developed using, for example, scanning probe tips. By deflecting one end of a nanofibre with an

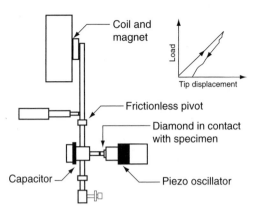

Figure 2.29 Schematic diagram of a nanoindenter together with a typical loading–unloading curve

AFM tip and holding the other end fixed, the mechanical strength has been calculated by correlating the lateral displacement of the fibre as a function of applied force. Other techniques have included the development of nanomanipulators for use inside the SEM or TEM which allow the mechanical loading and bending of nanostructures, such as the use of oscillating voltages in the TEM to mechanically resonate a nanotube and so determine its elastic modulus.

The adsorption of proteins, DNA and other polymers at surfaces can be studied by using an AFM in one dimension. The molecular force probe (MFP) involves the measurement of AFM cantilever deflection as a function of distance from the sample. The cantilever deflection can be correlated to atomic forces of the order of piconewtons. On approaching a polymer adsorbed at the surface, the tip experiences a van der Waals attraction to the polymer. On contact with the polymer, the tip is removed. Force (again of the order of piconewtons) is required to remove the tip (and polymer) from the surface. Since not all of the polymer is adsorbed to the surface, energy can be released when this part of the polymer is pulled from the surface. Such pull-off events reveal information about the nature of the chain conformation on the surface (Figure. 2.30), including parameters such as the polymer contour length.

2.9.2 Electron transport properties

The electronic conductivity σ of a material is given by

$$\sigma = nq\mu, \tag{2.9}$$

where n, q and μ are the concentration, charge and mobility of the charge carriers, respectively. Charge carriers can be electrons, holes or sometimes ions. Measurement of the conductivity or resistivity ($\rho = 1/\sigma$) as a function of temperature can often provide an insight into conduction mechanisms. Conductivity measurements rely on measuring the current passed through the test sample on applying a known voltage. However, this is not straightforward as a contact applied to a sample will develop a contact resistance,

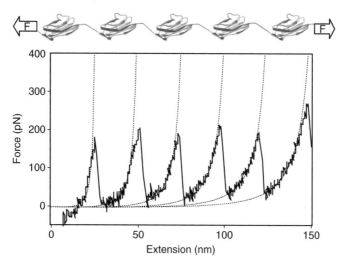

Figure 2.30 Typical mechanical protein unfolding data obtained using the atomic force micro-scope (AFM). The applied force is plotted as a function of the extension of a multimeric protein system engineered to comprise five identical repeated domains of the 27th immunoglobulin domain of the giant muscle protein titin (shown schematically at the top of the figure). At the start of the experiment all five domains are folded. At a critical applied force ($\sim$180pN) a domain unfolds releasing a length of unfolded polypeptide chain into the system which allows the AFM cantilever to return almost to its rest position and the applied force falls almost to zero. Further extension of the system results in a second and third domain unfolding and so on until all five domains in the multimer have unfolded. Each domain introduces exactly the same length of unfolded polypeptide into the system as expected. The final feature in the data is the stretching of the entire polypeptide chain comprising five unfolded domains until it becomes detached from the AFM tip. The dashed lines are of the force-extension data using a model of polymer elasticity known as the 'worm-like chain' model. Figure reproduced courtesy of D. Alastair Smith, David Brockwell, and Sheena Radford

as in general it is non-ohmic. The choice of contact electrode, based on the work function, is therefore critical. The arrangement of the electrodes is also important; for high resistance materials ($>$100 Ohms) a two-probe technique is adequate, whereas for small values of resistance the residual contact resistance and the impedance of the leads become significant. In this latter case a four-probe technique is required, which uses two probes to pass a current through the material while the potential difference between the other two probes is measured using a high-impedance voltmeter.

To deconvolve the conductivity into the two components, n and μ, a Seebeck experiment is often used. Here a known thermal gradient is set up across a uniform sample area, a voltage develops across the sample due to the flow of carriers induced by the differing electrochemical potentials. If the Seebeck voltage is measured, this provides a measure of the carrier concentration. A Hall experiment can also provide a direct measure of n and μ. Here a known current, either alternating current (AC) or direct current (DC), is passed through a sample at right angles to a magnetic field; the moving carriers experience a resultant force, which leads to a Hall voltage being set up across the sample.

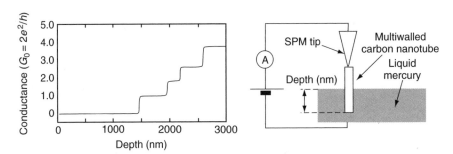

Figure 2.31 Schematic data showing ballistic quantum conductance in a multiwalled carbon nanotube; a possible experimental set-up is also shown

Besides DC methods, AC conductivity can provide data on the impedance of the system. The impedance is the ratio of the applied modulated voltage to the resultant current modulation and has both a magnitude and a phase. The dependence of the impedance on the frequency of the applied AC is known as impedance spectroscopy and this technique can separate out electrical responses from separate nanostructural features; e.g., grain boundaries, bulk grains and interfaces with electrodes, all with different response or relaxation times.

For electrical measurements on systems of reduced dimensionality, such as the system in Figure 2.31, unless the total sample is large and one is seeking an average electrical response, a major problem is in contacting small probes to samples. Some measurements can be performed in situ within an SEM or TEM, for example, and techniques are also now available for the preparation of nanostructured electrodes. However, in recent years STM (Section 2.5.1) has allowed current–voltage responses to be obtained in addition to the normal topographic imaging data. Here the tip is held in a constant position and the tunnelling current measured as a function of the tip bias voltage, producing an *I–V* curve at the local tip position. Furthermore, there are a number of related SPM techniques that can directly probe electrical properties such as scanning capacitance microscopy (SCM) mentioned in section 2.5.3.

2.9.3 Magnetic properties

Magnetic properties are discussed fully in Chapter 4. However, as an introduction, materials are broadly classified by their response to an inhomogeneous magnetic field. For a uniform field, a diamagnetic substance will tend to move towards the weakest region of the field, whereas a paramagnetic substance will tend to move towards the strongest region. Usually diamagnetic substances have all the electrons spin-paired while if unpaired electrons are present then the substance may be paramagnetic. Cooperative interionic interactions may lead to long-range magnetic ordering into domains, resulting in ferromagnetism, antiferromagnetism or ferrimagnetism.

The ratio of the sample magnetization to the field strength is known as the volume magnetic susceptibility. Experimental techniques such as the Gouy and Faraday methods generally rely on measuring the force exerted on a sample in a magnetic field gradient in order to derive the sample susceptibility. Magnetic force microscopy (MFM) using a ferromagnetic scanning probe tip is a spatially resolved extension of such methods. Other experimental methods are based on changing the magnetic flux of a solenoid and involve measuring this change upon the introduction of the sample. Changes in flux may be measured by direct current, alternating current or induction methods (such as the Meissner effect in superconductors) in the presence of a static magnetic field, an oscillating magnetic field or both. Finally a vibrating sample magnetometer is a device where the sample is vibrated in a uniform magnetizing field and the magnetization of the sample is detected as an electrical signal in a pick-up coil.

2.9.4 Thermal properties

Thermal analysis can be defined as the measurement of changes in the physical properties of a material as a function of temperature whilst the material is subjected to a controlled temperature programme. Typically, the temperature programmes used are either heating or cooling at a constant rate, isothermal holds or a combination of the two. Sample property measurements include temperature, enthalpy, mass, dimensions or mechanical properties and the techniques are termed differential thermal analysis (DTA), differential scanning calorimetry (DSC), thermogravimetry, thermodilatometry, and thermomechanical analysis, respectively. However, most of these techniques can only be performed on bulk samples. Other less common thermo-optical, thermomagnetic, thermoelectrical and thermo-acoustic measurements are also occasionally employed. These measurements may be made either in absolute terms, or as differential or derivative measurements. Output is in the form of a thermal analysis curve and features of this curve (peaks, discontinuities or changes in slope) are related to thermal events in the sample such as phase transitions, chemical reactions or microstructural changes. Sometimes two or more of these techniques may be employed simultaneously on the same sample and any of these techniques may be combined with evolved gas analysis by spectroscopy, chromatography or mass spectroscopy. Use of a scanning thermal microscope allows spatially resolved micro-DTA measurements to be undertaken (Section 2.5.3).

In DTA the temperature difference between the sample and a reference sample (e.g., Al_2O_3 or SiC) is recorded using a thermocouple, as both are subjected to the same temperature programme within say a furnace, possibly under a controlled atmosphere, as shown in Figure 2.32. The temperature difference, rather than the sample temperature, is measured since this gives a greater sensitivity. During a temperature programme (heating or cooling), the temperature of the sample will differ from that of the reference due to the difference in heat capacity between the sample and the reference. If an endothermic thermal event occurs in the sample, the sample temperature will lag behind the temperature of the reference, and vice versa for an exothermic event. Plotting the temperature difference versus the reference temperature, as shown in Figure 2.32, will produce a trough for an endotherm and a peak for an exotherm; each thermal event is best characterised by the onset temperature where the thermal curve first deviates from the baseline. The area under the peak is related to the enthalpy change during the transformation, although the classical DTA method is not well suited to the determination of enthalpies owing to difficulties with calibration.

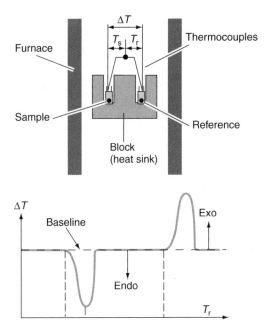

Figure 2.32   Schematic diagram of a differential thermal analyser and a typical DTA trace

In the alternative technique of DSC, the sample and reference are maintained at the same temperature during the temperature programme by varying the power supplied to the two sample heaters. Hence the thermal analysis curve from a DSC is a plot of energy difference between the supplies to the sample and reference furnaces versus programme temperature. Different conventions apply to DTA and DSC plots in that an endotherm is a peak in DSC. DSC is capable of determining accurate enthalpies and heat capacities.

BIBLIOGRAPHY

Whilst we have attempted to cover most techniques applicable to the characterization of nanostructures, the following should prove useful for further study.

C. N. Banwell, *Fundamentals of Molecular Spectroscopy*, 4th edition, McGraw-Hill, London (1994).

S. Bradbury and B. Bracegirdle, *Introduction to Light Microscopy*, Bios, Oxford (1998).

D. Briggs and M. P. Seah, *Practical Surface Analysis*, Vols 1 and 2, John Wiley & Sons, Ltd, Chichester (1990).

R. Brydson, *Electron Energy Loss Spectroscopy*, Bios, Oxford (2001).

W. D. Callister, *Materials Science and Engineering*, John Wiley & Sons, Inc., New York (1994).

A. K. Cheetham and P. Day (eds), *Solid State Chemistry Techniques*, Oxford University Press, Oxford (1987).

W. K. Chu, J. W. Mayer and M. A. Nicolet, *Backscattering Spectrometry*, Academic Press, New York (1978).

P. J. Goodhew, F. J. Humphreys and R. Beanland, *Electron Microscopy and Analysis*, 2nd edn, Taylor and Francis, London (2000).

S. J. Gregg and K. S. W. Sing, *Adsorption, Surface Area and Porosity*, Academic Press, London (1967).

C. Hammond, *The Basics of Crystallography and Diffraction*, Oxford University Press, Oxford (1997).

R. J. Keyse, A. J. Garrett-Reed, P. J. Goodhew and G. Lorimer, *Introduction to Scanning Tranmission Electron Microscopy*, Bios, Oxford (1998).

M. K. Miller, *Atom Probe Tomography*, Kluwer, Dordrecht (2000).

M. Prutton, Introduction to Surface Physics, Oxford University Press, Oxford (1994).

L. Reimer, *Scanning Electron Microscopy*, Springer, Berlin (1985).

C. J. R. Sheppard and D. M. Shotton, *Confocal Laser Scanning Microscopy*, Bios, Oxford (1997).

L. Solymar and D. Walsh, *Lectures on the Electrical Properties of Materials*, Oxford University Press, Oxford (1995).

J. Stohr, *NEXAFS Spectroscopy*, Springer-Verlag, Berlin (1992).

J. F. Watts, *Introduction to Surface Analysis by XPS and AES*, John Wiley & Sons, Ltd, Chichester (2003).

D. B. Williams and C. B. Carter, *Transmission Electron Microscopy*, Plenum Press, New York (1997).

3

Inorganic semiconductor nanostructures

3.1 INTRODUCTION

The information technology revolution of the previous decades has been based firmly on the development and application of inorganic semiconductors. Silicon forms the basis of the vast majority of electronic devices, whilst compound semiconductors such as gallium arsenide (GaAs) are used for many optoelectronic applications. Conventional devices utilise bulk semiconductors in which charge carriers are free to move in all three spatial directions. However, the formation of a nanostructure, in which the dimensions along one or more directions are reduced below ~10 nm, dramatically modifies the carrier properties, which become governed by the laws of quantum mechanics. Transistors in current-generation microprocessors have feature sizes as small as 50 nm, with a reduction to less than 10 nm predicted by 2016. Such devices will soon enter deep into the nanoscale regime, where their behaviour will no longer follow a simple extrapolation from larger devices. In this case the advent of quantum mechanical behaviour may be seen as a problem to be overcome but, as will be seen in this chapter, inorganic semiconductor nanostructures exhibit a wide range of new and unusual properties, which can be employed to fabricate improved and novel electronic and electro-optical devices.

This chapter is organised in the following manner. After a brief review of basic semiconductor physics, the modified electronic properties of semiconductor nano-structures are considered. The different techniques developed to fabricate inorganic semiconductor nanostructures are then discussed, including a comparison of their relative advantages and disadvantages. Novel physical processes which occur in semiconductor nanostructures are considered and the experimental techniques used

Nanoscale Science and Technology   Edited by R. W. Kelsall, I. W. Hamley and M. Geoghegan
© 2005 John Wiley & Sons, Ltd

to probe their structural, electronic and optical properties are described. A number of representative applications are discussed and possible future developments are considered. The treatment in this chapter is relatively non-mathematical and the emphasis is on breadth rather than depth. Suggestions for further reading are given, these will hopefully allow the reader to explore a particular topic in greater detail.

3.2 OVERVIEW OF RELEVANT SEMICONDUCTOR PHYSICS

3.2.1 What is a semiconductor?

Semiconductors behave as insulators at absolute zero temperature ($T = 0$) but at non-zero temperatures ($T > 0$) exhibit a relatively small electrical conductivity, the size of which increases rapidly with increasing temperature. Furthermore, their electrical conductivity can be increased by adding small amounts of certain impurities (dopants) or by illumination with particular wavelengths of light. These properties contrast strongly with those of good conductors (metals), whose electrical conductivity is many orders of magnitude larger, decreases relatively weakly with increasing temperature and, to a good approximation, is not affected by small levels of impurities or illumination.

The electronic band theory of solids has been described in Chapter 1, and Figure 3.1 summarises the main features of the band structure of a semiconductor. At absolute zero temperature all states in the valence band are occupied by electrons and all states in the conduction band are empty. Under these conditions, electrical conduction cannot occur. As the temperature is increased, electrons are excited from the valence band across the band gap E_g into the conduction band. Electrical conduction is now possible via the small number of electrons in the conduction band and the large number of electrons which remain in the valence band, but whose motion is limited because there are only a small number of vacancies. Although electrical conduction in the valence band is due to the movement of the large number of electrons, it is more convenient instead to consider this contribution to the electrical conductivity in terms of the much smaller number of vacancies. These vacancies, which are termed holes, move in the opposite direction to the electrons and hence they behave as carriers of opposite charge sign.

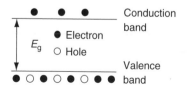

Figure 3.1   The electronic band structure of a semiconductor. Electrical conduction occurs in the conduction band by a small number of electrons and in the valence band by a small number of vacancies or holes

3.2.2 Doping

Electrons and holes created by thermal excitation across the band gap are known as intrinsic carriers. For these carriers the density of electrons in the conduction band n equals the density of holes in the valence band p. Although n and p increase rapidly with increasing temperature (resulting in increased electrical conductivity), their absolute values are relatively small. For example, in Si ($E_g = 1.12\,\text{eV}$) $n = p \sim 10^{15}\,\text{cm}^{-3}$ at 300 K, many orders of magnitude less than the density of electrons in the conduction band of a typical conductor ($\sim 10^{22}\,\text{cm}^{-3}$). Consequently, a semiconductor's intrinsic electrical resistance, which is inversely related to its conductivity, is relatively high; potentially a serious problem for many device applications.

Fortunately, the electron or hole densities in a semiconductor can be increased significantly, and in a controllable manner, by the addition of small amounts of certain impurities, a process known as doping. If atoms with one additional valence electron are added to the host semiconductor, such as phosphorus to silicon, the impurity atoms form a series of new states within the forbidden band gap, located slightly below the bottom of the conduction band. At $T = 0$ these states are occupied by the additional electrons but for $T \neq 0$ these electrons may be thermally excited into the conduction band, increasing the free electron density. Because the impurity states are relatively close to the conduction band, this excitation requires relatively little thermal energy, and at moderately high temperatures all the impurity atoms will lose their electrons to the conduction band. Therefore n is increased by an amount approximately equal to the density of impurity atoms, which are known as donors; this process is called n-type doping. Similarly, the introduction of impurity atoms with one fewer valence electron than the host semiconductor, such as boron to silicon, forms a series of levels slightly above the top of the valence band, which are unoccupied at $T = 0$. For $T \neq 0$ electrons may be excited into these states, leaving free holes in the valence band. This is called p-type doping and increases the free hole density by an amount approximately equal to the density of the impurity atoms, which are known as acceptors. Electrons or holes produced by doping are known as extrinsic carriers and for a doped semiconductor $n \neq p$. The concept of doping is summarised schematically in Figure 3.2.

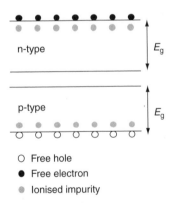

O Free hole
● Free electron
◉ Ionised impurity

Figure 3.2   n- and p-type doping. Impurity states are formed just below the bottom of the conduction band and just above the top of the valence band for n- and p-type doping, respectively. The close proximity of these states to the respective band edges means the thermal excitation of extrinsic carriers is much more probable than intrinsic carrier excitation across the band gap

3.2.3 The concept of effective mass

Electrons and holes in a semiconductor are not free particles, possessing, in addition to kinetic energy, potential energy due to their electrostatic interaction with the charged ions. Particles with potential energy are considerably more difficult to describe mathematically than free particles, but in a solid this problem can be simplified by using the concept of effective mass. In this model the electrons and holes are treated as free particles by assigning them a modified mass, the effective mass, which combines their potential and kinetic energies into a single kinetic-like energy. The effective mass must be used in all equations describing the dynamical properties of carriers in a solid. The symbol for effective mass is m^* with, in general, a subscript e or h to indicate the effective mass of the electron or hole, respectively. Effective masses are expressed as a multiple of the free electron mass, and holes and electrons typically have different effective masses. As an example, the semiconductor GaAs has an electron effective mass $m_e^* = 0.067m_e$ and a hole effective mass $m_h^* = 0.35m_e0$, where m_e0 is the free electron mass.

3.2.4 Carrier transport, mobility and electrical conductivity

An externally applied voltage produces an electric field within a semiconductor and this results in an electrostatic force that acts on the charge carriers. This force produces an acceleration and hence motion of the carriers along the field direction; it is this motion which constitutes an electrical current. Carriers will be accelerated by the electric field until they hit an obstacle, at which point their velocity is randomised. Following this collision, the acceleration recommences. This process is shown schematically in Figure 3.3. Although the time between any two collisions is random, there will be a well-defined average scattering time, τ, between collisions and this allows a mean velocity, the carrier drift velocity v_d, to be defined. The electrical conductivity is directly proportional to v_d, which is a measure of how easily the carriers are able to move through the semiconductor. At low fields v_d is proportional to the size of the electric field, hence a measurement-independent quantity can be obtained by dividing v_d by the field; the resultant quantity is the carrier mobility μ. It can be shown that the mobility is related to the scattering time by $\mu = e\tau/m^*$, where e is the electronic charge.

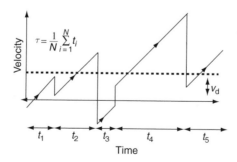

Figure 3.3 The dynamics of a charge carrier subjected to a constant electric field. A mean scattering time, τ, can be defined by averaging over a large number of events, leading to an average drift velocity v_d

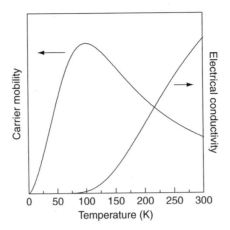

Figure 3.4   The temperature variation of the electrical mobility and electrical conductivity for a semiconductor

The mean time between collisions is dependent on the nature of the collisions. A carrier travelling through a periodic crystal lattice only experiences a collision if there is a local departure from the crystal periodicity. This can result from the presence of an impurity atom, such as a dopant atom, or by thermal vibrations of the lattice, the quanta of which are termed phonons. Scattering by impurity atoms is important at low temperatures but decreases with increasing temperature. In contrast, scattering by phonons increases with temperature, reflecting the increasing amplitude of the lattice vibrations. The combined effect of these two processes is to give a mobility which, at low temperatures, increases with increasing temperature, followed by a decrease at high temperatures. This behaviour is shown schematically in Figure 3.4. Section 3.6.1 shows that in certain semiconductor nanostructures it is possible to turn off the scattering by impurity atoms, resulting in very high carrier mobilities at low temperatures. The temperature variation of the electrical conductivity is also shown in Figure 3.4. For the case where electrical conduction is dominated by one type of carrier, this is given by Equation 2.9; for the more general case it is necessary to include two terms representing the contributions of both electrons and holes. The electrical conductivity increases monotonically with temperature, the high-temperature decrease in μ being overwhelmed by the very rapid increase in n.

3.2.5 Optical properties of semiconductors

The application of semiconductors in electro-optical devices relies on their ability to efficiently emit or detect light. If photons of energy greater than or equal to the band gap are incident on a semiconductor, they may excite an electron from the valence band to the conduction band. In this process the photon is destroyed (absorbed) and an electron and hole are created. In the reverse process an electron in the conduction band may return to the valence band and recombine with a hole; the energy lost by the electron creating a photon. As the energies of the electron and hole will generally be very close to the bottom of the conduction band and the top of the valence band respectively, the emitted photon will have an energy approximately equal to the band gap of the semiconductor.

3.2.6 Excitons

The band gap of a semiconductor represents the energy required to create an electron and hole when there is no final interaction between the two carriers. However, the negatively charged electron and positively charged hole may interact to form a hydrogen-atom-like complex in which the two carriers orbit each other, a system known as an exciton. The electrostatic interaction between the electron and hole reduces their energy compared to the non-interacting case, resulting in a series of energy levels just below the conduction band edge. These excitonic states have discrete energies

$$E_n = E_g - E_b/n^2 \qquad (n = 1, 2, 3, \ldots, \infty), \qquad (3.1)$$

where for $n \to \infty$ they merge into the continuum states of the conduction band. The binding energy of the exciton E_b is the energy difference between the lowest exciton state ($n = 1$) and the conduction band edge ($n = \infty$). In addition to the states formed below the conduction band edge, there is a modification of the states above the band edge, referred to as the Sommerfeld enhancement. Absorption into excitonic states is possible, and the inset of Figure 3.5 shows how the absorption of a bulk semiconductor is modified by the inclusion of excitonic effects. Although there are an infinite number of exciton states, their absorption strength and separation both decrease rapidly with increasing n, and hence experimentally only absorption into the $n = 1$ state is generally observed. Figure 3.5 shows absorption spectra for the semiconductor gallium arsenide (GaAs) at low temperature and room temperature. At low temperature, excitonic effects are clearly visible in the spectrum as an enhanced absorption close to the band gap. However, because the exciton binding energy in GaAs is only 4.2 meV, at room temperature there is sufficient thermal energy ($k_B T = 25$ meV) to ionise the majority of excitons. Hence excitonic effects are absent, or at most extremely weak, in GaAs and

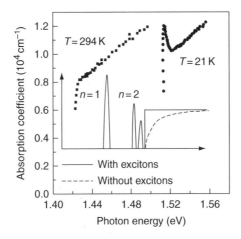

Figure 3.5 Low-temperature and room temperature absorption spectra of bulk GaAs. The energy shift between the two spectra results from the temperature variation of the band gap. The inset shows the density of states with and without the inclusion of excitonic effects. Reproduced from M. D. Sturge, *Phys. Rev.* **127**, 768 (1962). Copyright 1962 by the American Physical Society

most other bulk semiconductors at room temperature. Section 3.6.6 shows that it is possible to significantly increase the exciton binding energy in a nanostructure, allowing the observation of excitonic effects at higher temperatures.

3.2.7 The pn junction

The majority of semiconductor devices are based on the pn junction, which is formed at the interface between two regions, one doped n-type the other p-type. In equilibrium a potential step is formed at the interface which prevents the net movement of electrons from the n-type region into the p-type region, and vice versa for holes. In addition, free carriers are absent from regions either side of the junction, forming a depletion region. A schematic diagram of a pn junction under equilibrium conditions is shown in Figure 3.6(a). If an external voltage of the correct sign is applied (Figure 3.6(b)) the potential step is reduced, allowing electrons and holes to move across the junction, a process known as injection. In a pn junction designed for optical applications, an undoped or intrinsic (i) region may be placed between the n- and p-type regions to form a p-i-n structure. Electrons and holes meet in the intrinsic region, where they recombine to produce photons. A nanostructure may be incorporated within the intrinsic region, providing a convenient and efficient mechanism for injection of electrons and holes into the nanostructure.

The bias condition for current injection is referred to as forward bias. Changing the polarity of the applied voltage produces reverse bias. In this case the potential step is increased and there is negligible current flow. However, electrons and holes created in the intrinsic region by photon absorption may be swept out into the n- and p-type

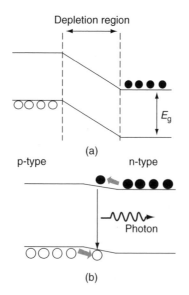

Figure 3.6 Schematic band diagrams of a pn junction (a) under equilibrium conditions and (b) with an external voltage applied to reduce the potential step, resulting in carrier injection across the junction

regions, respectively, resulting in an electrical current that can be measured by an external circuit. This process allows a semiconductor to act as a photon detector.

3.2.8 Phonons

Carriers in a solid may lose or gain energy by emitting or absorbing a phonon. Figure 3.7 shows the phonon dispersion – the frequency-wave vector relationship – calculated for a one-dimensional chain consisting of atoms of two alternating masses. This model provides a good approximation to real semiconductors. Of the two calculated branches, the lower or acoustic branch corresponds to the propagation of sound, and has a frequency which tends to zero for small wave vectors. The upper or optical branch corresponds to phonons which can interact with electromagnetic radiation, and has a frequency which remains non-zero for small wave vectors. For three-dimensional solids each branch consists of three sub-branches, corresponding to the three possible directions of the lattice vibrations with respect to the propagation direction, two transverse and one longitudinal. For GaAs and related semiconductors the strongest carrier–phonon interaction occurs for longitudinal optical (LO) phonons, and it is these phonons that are preferentially emitted as carriers lose energy.

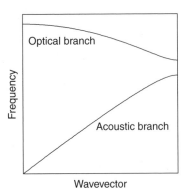

Figure 3.7 Schematic diagram of the phonon dispersion relationship for a one-dimensional linear chain consisting of atoms of alternating mass

3.2.9 Types of semiconductor

The majority of purely electronic devices are based on the elemental semiconductor silicon (Si). However Si has an indirect band gap, with the lowest energy state in the conduction band occurring at a different wavevector to the lowest energy state in the valence band. When an electron and hole in Si recombine, this wavevector difference must be conserved, in addition to energy conservation. A photon is unable to conserve both energy and wavevector, so a second particle, usually a phonon, must be created in addition to the photon. This two-particle, photon plus phonon, recombination process occurs relatively slowly, hence it allows other processes to occur in which a photon is not created.

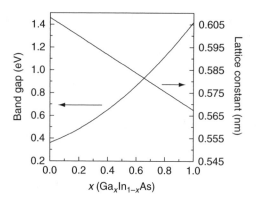

Figure 3.8 The compositional variation of the lattice constant and band gap of the ternary alloy semiconductor $Ga_xIn_{1-x}As$

For example, the electron may return to the valence band by relaxing via phonon emission through a series of impurity states formed within the band gap, or it may transfer its energy to a second electron which is excited to a higher state in the conduction band. As a result of these non-radiative processes the majority of electrons and holes recombine without the emission of a photon; consequently, the light production efficiency of Si is very poor, making it unsuitable for many electro-optical applications.

Light production efficiency is much greater in direct band gap semiconductors where the recombining electron and hole have the same wavevector, and only a photon is required to satisfy energy conservation. For electro-optical applications, binary semiconductors consisting of elements from columns three and five of the periodic table are typically used, the majority of which have direct band gaps. Examples of III–V semiconductors include gallium arsenide (GaAs), indium phosphide (InP) and gallium nitride (GaN). It is also possible to form semiconductors by combining elements from columns two and six, although these II–VI semiconductors, which include cadmium telluride (CdTe) and zinc selenide (ZnSe), are technologically less important. Furthermore, it is possible to combine two semiconductors to form an alloy semiconductor. For example, InAs and GaAs can be combined to form the ternary semiconductor gallium indium arsenide ($Ga_xIn_{1-x}As$), where the variable x ($0 \leq x \leq 1$) indicates the relative proportions of InAs and GaAs. The properties of an alloy semiconductor are approximately equal to the appropriate weighted average of the constituent semiconductors. Figure 3.8 shows the variation of the band gap and lattice constant of $Ga_xIn_{1-x}As$ as a function of x. Both quantities vary smoothly between the values for InAs and GaAs. One important practical application of alloy semiconductors is that a specific band gap can be obtained by a suitable choice of composition.

3.3 QUANTUM CONFINEMENT IN SEMICONDUCTOR NANOSTRUCTURES

In a bulk semiconductor, carrier motion is unrestricted along all three spatial directions. However, a nanostructure has one or more of its dimensions reduced to a nanometre length scale and this produces a quantisation of the carrier energy corresponding to motion along these directions. In this section the nature of this quantisation for nanostructures of different dimensionalities is considered.

3.3.1 Quantum confinement in one dimension: quantum wells

Consider initially an isolated, thin semiconductor sheet of thickness L. Carrier motion is unrestricted along the two orthogonal directions within the plane of the sheet, but is quantised perpendicular to the plane, forming a one-dimensional quantum well. The resultant quantised energy levels are found by solving the one-dimensional form of the time-independent Schrödinger equation (1.3):

$$-\frac{\hbar^2}{2m^*}\frac{d^2\psi_n(x)}{dx^2} + V(x)\psi_n(x) = E_n\psi_n(x), \tag{3.2}$$

where $V(x)$ is the potential and $\psi_n(x)$ and E_n are the wavefunction and energy of the nth confined state. For the present case, $V(x)$ is zero within the semiconductor (which extends from $x = 0$ to $x = L$) and is infinite elsewhere; this is the infinite-depth potential well model. Solving the Schrödinger equation and applying the boundary condition that the wavefunctions must be zero at the edges of the sheet, results in the following energies and wavefunctions:

$$E_n = \frac{h^2 n^2}{8m^* L^2} \qquad \psi_n(x) = \sqrt{\frac{2}{L}}\sin\left(\frac{n\pi x}{L}\right) \qquad (n = 1, 2, 3, \ldots, \infty) \tag{3.3}$$

Figure 3.9 shows the energies and wavefunctions for the first three confined states ($n = 1, 2$ and 3) of an infinite-depth potential well.

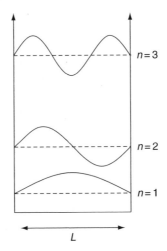

Figure 3.9 The energies and wavefunctions of the first three confined states of an infinite-depth quantum well

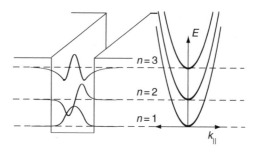

Figure 3.10   Energies and wavefunctions of the confined states in a finite-depth quantum well (left-hand side). The dispersion resulting from the unrestricted motion of the carriers in the plane of the well is shown to the right

A thin, free-standing semiconductor sheet would possess negligible mechanical strength, and practical quantum wells are formed by sandwiching a thin layer of one semiconductor between two layers of a second, larger band gap semiconductor, which form the barriers. This results in a finite-depth potential well as shown in Figure 3.10. The wavefunctions and energies of the confined states are again determined by the solution of the Schrödinger equation with the appropriate potential, which now remains finite outside of the well. The left-hand side of Figure 3.10 shows energies and wavefunctions for a finite-depth well. In contrast to the infinite-depth well, there are now only a finite number of confined states and the wavefunctions penetrate out of the well and into the barriers. For a finite-depth well it is not possible to obtain analytical forms for the confined energies and the Schrödinger equation must be solved numerically. However, for many applications the energies and wavefunctions of an infinite-depth well can be used as reasonable approximations, particularly for states that lie close to the bottom of the well.

For a semiconductor quantum well both the electron and hole motion normal to the plane will be quantised, resulting in a series of confined energy states in the conduction and valence bands (inset of Figure 3.11). One consequence of this quantum confinement is that the effective band gap of the semiconductor E_g^{ef} is increased from its bulk value by the addition of the electron and hole confinement energies corresponding to the states with $n = 1$:

$$E_g^{\text{ef}} = E_g + \frac{h^2}{8m_e^* L^2} + \frac{h^2}{8m_h^* L^2}. \tag{3.4}$$

This effective band gap will determine, for example, the energy of emitted photons, and can be altered by varying the thickness of the well. Figure 3.11 shows an example of this behaviour where the emission spectrum of a structure containing five quantum wells of different widths is shown. Each well emits photons of a different energy; the energy increasing as the width of the well decreases, in agreement with the predictions of Equation (3.4).

Although the carrier energy is quantised for motion normal to the well, within the plane of the well the motion is unrestricted. The total energy of a carrier is given by the sum of the energy due to this unrestricted motion plus the quantisation energy. The in-plane

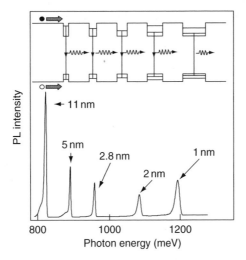

Figure 3.11   Emission spectrum of a quantum well structure containing five wells of different thicknesses. The wells are $Ga_{0.47}In_{0.53}As$ and the barriers are InP. The inset shows the electronic structure and the nature of the optical transitions

motion is characterised by a wavevector, $k_{\parallel}$, which corresponds to the combination of the wavevectors for motion along the two mutually orthogonal in-plane directions. If the z-axis is taken as being perpendicular to the plane of the well, then the two in-plane directions are x and y and

$$k_{\parallel} = \sqrt{k_x^2 + k_y^2}. \tag{3.5}$$

From the relationship between momentum, $p = m^*v$, and wave vector, $p = \hbar k$, where $\hbar = h/2\pi$, and the definition of kinetic energy

$$E = \frac{1}{2}m^*v^2 = \frac{p^2}{2m^*}, \tag{3.6}$$

the energy corresponding to in-plane motion can be written in the form

$$E = \frac{\hbar^2 k_{\parallel}^2}{2m^*}. \tag{3.7}$$

The total energy for a carrier in the nth confined state is therefore given by

$$E_{n,k_{\parallel}} = \frac{h^2 n^2}{8m^* L^2} + \frac{\hbar^2 k_{\parallel}^2}{2m^*}. \tag{3.8}$$

Since $k_{\parallel}$ is unrestricted, equation (3.8) gives a continuum of energies for each value of n, as shown in the right-hand side of Figure 3.10; these energy bands are known as *subbands*.

3.3.2 Quantum confinement in two dimensions: quantum wires

A quantum wire consists of a strip of one semiconductor confined within a second, larger band gap barrier semiconductor. Unrestricted carrier motion is now only possible along the length of the wire and is quantised along the two remaining orthogonal directions. For simple wire shapes (square or rectangular cross sections) it is possible to calculate the quantisation energies for the two directions independently. These two quantisation energies are then added to the energy resulting from the unrestricted motion along the wire. Using the infinite-depth well approximation for the quantised energies, the total energy for a carrier in a quantum wire with z and y dimensions L_z and L_y respectively is

$$E_{n,m,k_x} = \frac{h^2 n^2}{8m^* L_z^2} + \frac{h^2 m^2}{8m^* L_y^2} + \frac{\hbar^2 k_x^2}{2m^*} \qquad (n, m = 1, 2, 3, \ldots). \qquad (3.9)$$

The total energy depends on the two quantum numbers n and m and the wave vector for free motion along the wire k_x. For each confined state, given by a particular combination of n and m, there will be a subband of continuous states resulting from the unrestricted values of k_x. In section 3.5 it will be seen that real quantum wires have complex cross sections. This prevents the confined energies from being calculated by separating them into terms corresponding to the two directions perpendicular to the axis of the wire. Instead the confined energies of a quantum wire must be obtained from a numerical solution of the appropriate Schrödinger equation.

3.3.3 Quantum confinement in three dimensions: quantum dots

A quantum dot consists of a small region of one semiconductor totally surrounded by a second, larger band gap barrier semiconductor. Carrier motion is now quantised along all three spatial directions and there remains no unrestricted carrier motion. For a simple shape such as a cube or cuboid, confinement for the three spatial directions can be considered separately. In the infinite-depth well approximation, the total energy for a carrier in a cuboid-shaped dot of dimensions L_z, L_y and L_x is a function of three quantum numbers n, m and l:

$$E_{n,m,l} = \frac{h^2 n^2}{8m^* L_z^2} + \frac{h^2 m^2}{8m^* L_y^2} + \frac{h^2 l^2}{8m^* L_x^2} \qquad (n, m, l = 1, 2, 3, \ldots). \qquad (3.10)$$

The energy is now fully quantised and the states are discrete, in a manner similar to those of an atom. The shapes of real quantum dots are more complex than simple cuboids and a calculation of the confined energy levels requires a numerical solution of the relevant Schrödinger equation.

3.3.4 Superlattices

It is possible to fabricate a structure consisting of many quantum wells, with each well separated from neighbouring wells by a barrier. If the barriers are sufficiently thick, carriers located in different wells are essentially isolated and the structure behaves identically to a single well, although some properties (e.g., the absorption) will increase linearly with the total number of wells. In this case the structure is known as a multiple quantum well. However, if the thickness of the barriers is reduced, carriers in neighbouring wells may interact via the part of their wavefunctions which penetrates into the barriers. Significant interaction will occur if the wavefunctions (which decay exponentially into the barriers) overlap strongly, requiring relatively thin barriers. For strong interaction, originally identical states in different wells couple together to form a miniband of closely spaced states, a system known as a superlattice. Figure 3.12(a) and (b) show the electronic structures of a multiple quantum well and superlattice, respectively. Figure 3.12(c) demonstrates how the widths of the minibands increase as the barrier width decreases, allowing an increasing interaction between states in different wells. The formation of a superlattice allows carrier motion to occur normal to the plane of the quantum wells, with the resultant structure exhibiting properties intermediate between a bulk semiconductor (3D) and a true quantum well (2D). By extension it is possible to visualise a structure consisting of repeated quantum wires (exhibiting quasi 1D to 2D behaviour) or repeated quantum dots (exhibiting quasi 0D to 1D behaviour).

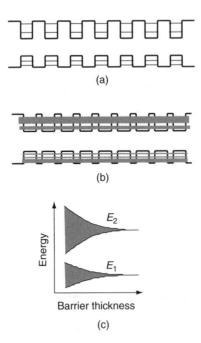

(a)

(b)

(c)

Figure 3.12 The form of the confined energy states in (a) a multiple quantum well and (b) a superlattice; (c) shows how the discrete states in a multiple quantum well evolve into superlattice minibands as the barrier thickness is reduced

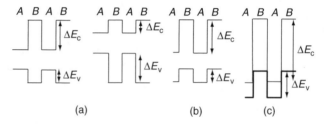

(a)   (b)   (c)

Figure 3.13   Possible band offsets for a semiconductor quantum well: (a) type I systems with conduction and valence band minimum energies both in semiconductor A; (b) type II system with conduction and valence band minimum energies in A and B, respectively; and (c) an extreme type II system where the minimum energy in the conduction band of semiconductor A lies below the minimum valence band energy of semiconductor B

3.3.5 Band offsets

The interface between two different semiconductors is referred to as a heterojunction, and at this point a discontinuity in the energies of both the conduction and valence bands occurs. The numerical sum of these two discontinuities – known as the band offsets: ΔE_c for the conduction band and ΔE_v for the valence band – equals the difference between the band gaps of the two semiconductors. However, there are an infinite number of possible combinations of the offsets and Figure 3.13 shows some examples with reference to a quantum well. In these diagrams the electron energy increases when moving vertically up the page, the hole energy increases moving vertically downwards. Hence in both examples shown in Figure 3.13(a) the minimum energy for both electrons and holes occurs in semiconductor A. This configuration is known as a type I system and is the one most commonly encountered. In Figure 3.13(b) the minimum energy for electrons occurs in semiconductor A but the minimum energy for holes occurs in semiconductor B. This configuration is known as a type II system and results in spatially separated electrons and holes. An extreme example of a type II system is shown in Figure 3.13(c). Here the conduction band of semiconductor A lies below the valence band of semiconductor B, allowing electrons from the latter to transfer to the former. The result is a relatively high density of electrons in the conduction band of semiconductor A and the system exhibits semimetallic like properties with a relatively large electrical conductivity.

The magnitudes and signs of the band offsets are important parameters, relevant to a wide range of properties of a nanostructure. However, their accurate theoretical prediction and experimental determination for a given semiconductor combination is relatively difficult.

3.4 THE ELECTRONIC DENSITY OF STATES

The concept of the density of states was introduced in Chapter 1, where the density of states for a bulk material was shown to be proportional to $E^{1/2}$. The density of states is strongly affected by reductions in the system dimensionality, because of the corresponding reduction of degrees of freedom in wavevector space. Simple theory, for purely 3D, 2D and 1D systems, gives the density of states to be proportional to k^{n-2}, where n

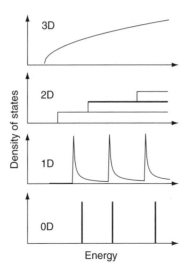

Figure 3.14 The electronic density of states for a bulk semiconductor (3D), a quantum well (2D), a quantum wire (1D) and a quantum dot (0D)

is the number of dimensions present, and for parabolic bands this translates into a dependence on $E^{(n-2)/2}$. For a quasi low-dimensional system, such as a semiconductor quantum wells or quantum wire, in which a series of confined subbands is formed, the density of states takes the $E^{(n-2)/2}$ dependence in each subband, as shown in Figure 3.14. For a quantum dot, with quantum confinement in all three dimensions, there is no continuous distribution of states and the density of states takes the form of a spectrum of discrete energy values, similar to that found for individual atoms, as shown in the bottom graph on Figure 3.14.

Many of the optical and electronic properties of a nanostructure depend critically on the density of states, hence they exhibit a strong dependence on dimensionality. For example, in electrical transport the density of states determines the number of states available for the motion of the charge carriers, and their scattering time is dependent on the number of available states into which they can be scattered. In addition, the strength of an optical transition is proportional to the density of states at the initial point in the valence band and the final point in the conduction band. This combination is known as the joint density of states and has the same functional form as the individual electron and hole density of states. The energy dependence of the absorption follows the joint density of states and therefore exhibits a very different form for nanostructures of different dimensionality.

3.5 FABRICATION TECHNIQUES

In this section the range of techniques developed for the fabrication of semiconductor quantum wells, wires and dots are considered. It begins with established epitaxial techniques capable of producing high-quality quantum wells then moves on to more specialised techniques suitable for fabricating wires and dots. To help compare the relative advantages and disadvantages of the different techniques, a summary of the requirements for ideal semiconductor nanostructures is first provided.

3.5.1 Requirements for an ideal semiconductor nanostructure

The following lists the main requirements for an ideal semiconductor nanostructure. In practice the relative importance of these requirements will be dependent upon on the precise application being considered.

- *Size*: for many applications the majority of electrons and holes should lie in their lowest energy state, implying negligible thermal excitation to higher states. The degree of thermal excitation is determined by the ratio of the energy separation of the confined states and the thermal energy, $k_B T$. At room temperature $k_B T \approx 25\,\mathrm{meV}$ and a rule of thumb is that the level separation should be at least three times this value ($\sim 75\,\mathrm{meV}$). As the spacing between the states is determined by the size of the structure, increasing as the size decreases (Section 3.3), this requirement sets an upper limit on the size of a nanostructure. For electrons in a cubic GaAs quantum dot, a size of less than 15 nm is required. However, below a certain quantum dot size there will be no confined states and this places a lower limit on the dot size.

- *Optical and structural quality*: semiconductors produce light when an electron in the conduction band recombines with a hole in the valence band – a radiative process. However, electron–hole recombination may also occur without the emission of a photon in a non-radiative process. Such processes are enhanced by the presence of certain defects which form states within the band gap. If non-radiative processes become significant then the optical efficiency – the number of photons produced for each injected electron and hole – decreases. For optical applications, nanostructures with low defect numbers are therefore required. Poor structural quality may also degrade the carrier mobility.

- *Uniformity*: devices will typically contain a large number of nanostructures. Ideally each nanostructure should have the same shape, size and composition.

- *Density*: dense arrays of nanostructures are required for many applications.

- *Growth compatibility*: the epitaxial techniques of MBE and MOVPE are used for the mass production of electronic and electro-optical devices. The commercial exploitation of nanostructures will be more likely if they can be fabricated using these techniques.

- *Confinement potential*: the potential wells confining electrons and holes in a nano-structure must be relatively deep. If this is not the case then at high temperatures significant thermal excitation of carriers out of the nanostructure will occur.

- *Electron and/or hole confinement*: for electrical applications it is generally sufficient for either electrons or holes to be trapped or confined within the nanostructure. For electro-optical applications it is necessary for both types of carrier to be confined.

- *p-i-n structures*: the ability to place a nanostructure within the intrinsic region of a p-i-n structure allows the efficient injection or extraction of carriers.

3.5.2 The epitaxial growth of quantum wells

The epitaxial techniques of molecular beam epitaxy (MBE) and metallorganic vapour phase epitaxy (MOVPE) are described in Chapter 1. Using these techniques it is possible to deposit semiconductor films as thin as one atomic layer, starting with a suitable substrate material. In this way, quantum well structures can be fabricated with almost perfectly abrupt interfaces. In addition, the doping can be modulated so that only certain layers are doped. MBE and MOVPE are used extensively for the commercial growth of a number of electronic and electro-optical devices, including semiconductor lasers and high-speed transistors. They also form the basis of a number of techniques developed for the fabrication of quantum wires and dots, the most important of which are described in the following sections.

3.5.3 Lithography and etching

An obvious fabrication technique for quantum dots or wires is to start with a quantum well, which provides confinement along one direction, and selectively remove material to leave ridges or mesas, forming wires or dots, respectively. Material removal is achieved by the use of electron beam lithography followed by etching. The advantage of this technique is that any desired shape can be produced, although because the electron beam has to be scanned sequentially over the surface, writing large-area patterns is a very slow process. A more serious problem arises from surface damage which results from the etching step. An optically dead surface region is formed, within which the dominant carrier recombination is non-radiative and the importance of which increases as the size of the nanostructure decreases. A suitable choice of semiconductors can minimise this problem (the GaInAs–GaAs system is a common one) but it can never be entirely eliminated. Hence although it is possible to fabricate structures with dimensions as small as $\sim 10\,\mathrm{nm}$, structures with acceptable optical properties are considerably larger ($\gtrsim 50\,\mathrm{nm}$).

3.5.4 Cleaved-edge overgrowth

This technique starts with the growth of a quantum well in an MBE reactor. The wafer is then cleaved in situ along a plane normal to the well. The sample is subsequently rotated through 90° and a second quantum well and barrier are deposited on the cleaved surface. This growth sequence is shown in Figure 3.15.

The two quantum wells form a T-shaped structure. At the intersection of the wells, the effective well width is increased slightly, resulting in a reduced potential which traps both electrons and holes. As this potential minimum extends along the intersection of the wells, a quantum wire is formed. The initial growth of multiple wells followed by the overgrowth of the final well allows the formation of a linear array of wires. In addition a second cleave, followed by a further overgrowth step can be used to produce a quantum dot.

The cleaved surface is atomically flat and clean, in contrast to the damaged surfaces formed after etching. As a consequence, cleaved edge overgrowth dots and wires have a high optical quality. With careful optimisation, reasonably deep confinement potentials

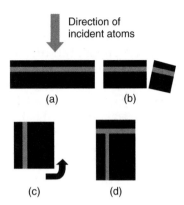

Figure 3.15   The growth sequence used to produced cleaved edge overgrowth quantum wires: (a) growth of initial quantum well, (b) in situ cleaving, (c) rotation of the structure and (d) growth of a second quantum well on the cleaved surface

are possible; values ~50 meV for the sum of the electron and hole confinement energies have been achieved. However, the separation between the confined states is significantly less than $3k_BT$ and only a single layer of wires can be formed, preventing the fabrication of dense two-dimensional arrays. In addition, the cleaving step is a difficult, non-standard process that requires a significant modification of the MBE reactor.

3.5.5 Growth on vicinal substrates

The periodic, crystalline nature of a semiconductor results in flat surfaces only for certain orientations. For the majority of orientations, the surface consists of a periodic series of steps in one or two dimensions. The step periodicity is determined by the orientation of the surface but is typically ~20 nm or less. Although epitaxial growth is generally performed on flat surfaces, growth on stepped or vicinal surfaces provides a technique for the fabrication of quantum wires.

Figure 3.16 shows the main steps in the growth of vicinal quantum wires. Starting with the stepped surface, the semiconductor that will form the wire is deposited by MBE or MOVPE. Under suitable conditions, growth occurs preferentially at the step corners where there is the highest density of unterminated atomic bonds. The growth, consisting of a single atomic layer, proceeds laterally from the corner of the step. When approximately half of the step has been covered, growth is switched to the barrier material which is used to cover the remainder of the step. Growth is then switched back to the initial semiconductor to increase the height of the wire. This growth cycle is repeated until the desired vertical thickness is obtained. Finally the wire is overgrown with a thick layer of the barrier material.

Although very thin wires can be produced using this technique, the growth has to be precisely controlled so that exactly the same fraction of the step is covered during each cycle. In addition, the coverage on different steps may vary and it can be difficult to ensure that the original steps formed on the surface of the substrate are uniform. As a result, quantum wires formed by the vicinal technique tend to exhibit poor uniformity.

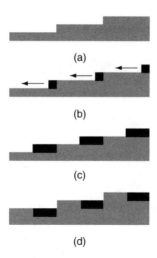

Figure 3.16 The growth of quantum wires on a vicinal surface: (a) initial stepped surface, (b) initial growth of the wire semiconductor in the corners of the steps, (c) growth proceeds outwards along the steps and (d) second half of step completed with growth of the barrier semiconductor. The figure greatly exaggerates the angle of the surface

3.5.6 Strain-induced dots and wires

Applying stress to a semiconductor results in a distortion of the atomic spacing – a strain – which, if of the correct sign, reduces the band gap. If the strain occurs over a small region then a local reduction of the band gap occurs, resulting in the formation of a wire or dot. A strain can be produced by depositing a thin layer of a different material, for example carbon, on the surface of a semiconductor. Because of their different lattice spacings, the materials distort near to their interface in order to fit together. This distortion constitutes a strain, which extends a short distance into the bulk of the semiconductor. If the carbon layer is patterned by lithography and then etched to leave only stripes or mesas, the resulting localised strain fields generate wires or dots in the underlying semiconductor. The remaining isolated regions of carbon are known as stressors. The strain only provides in-plane confinement and so it is necessary to place a quantum well near to the surface to provide confinement along the third direction.

Although this technique involves an etching step, it is only the optically inactive carbon layer that is etched, with the optically active quantum well spatially separated from any surface damaged region. Stressor-induced dots therefore exhibit high optical efficiency. However, the strain field produces only a weak modulation of the band gap, hence the confinement potential is relatively shallow. In addition, fluctuations in the sizes of the stressors results in a distribution of wire and dot sizes. This inhomogeneity is particularly significant in a variant on this technique where the stressors, in the form of small islands, form spontaneously during the deposition of the carbon. Because of the quasi-randomness of this self-assembly processes (Section 3.5.12) there is a relatively large distribution of stressor, and hence underlying quantum dot, size.

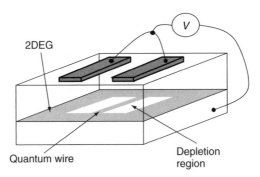

Figure 3.17   A schematic diagram of a split gate quantum wire. The electrons remaining below the gap between the gates form a quantum wire. A 2DEG remains in the regions away from the gates

3.5.7 Electrostatically induced dots and wires

It will be shown in Section 3.6.1 that the technique of modulation doping results in two-dimensional sheets of electrons (or holes) exhibiting very high mobilities at low temperatures. The electron density of a two-dimensional electron sheet or gas (2DEG) can be modified by depositing a metal layer, known as a Schottky gate, on the surface of the semiconductor, and this gate can be patterned using electron beam lithography followed by etching. The application of a negative bias voltage to the gate repels the electrons of the 2DEG from the region immediately below the gate, forming a region depleted of free electrons. If two parallel but spatially separated metal gates (a split gate) are formed on the surface, then biasing these gates depletes the regions below the gates but leaves a long, thin undepleted region directly below the gap between the gates. This is shown schematically in Figure 3.17. This undepleted central region forms a quantum wire, consisting of a one-dimensional strip of free electrons. With increasing gate voltage the depleted regions extend horizontally outwards from the regions directly below the gates, allowing the width of the quantum wire to be varied, a very useful experimental parameter. More complicated structures, in which constrictions are added to the gap between the gates, allow the formation of quantum dots. Further complexity involves the use of gates on both the top and bottom surfaces of the structure, allowing the properties of two parallel and interacting 2DEGs to be varied.

Electrostatically induced nanostructures form clean systems as only the metal is etched, not the semiconductor. Their very high low-temperature carrier mobility makes them excellent structures for studying transport processes in low-dimensional systems. However, the shallow potential minima and small energy-level spacing, a result of their relatively large size, limits their use to cryogenic temperatures. In addition, only electrons or holes are confined in a given structure, hence they are not suitable for optical applications.

3.5.8 Quantum well width fluctuations

The width of a quantum well is in general not constant but exhibits spatial fluctuations which form during growth. Potential minima for electrons and holes are formed at points where the well width lies above its average value, spatially localising the carriers within the plane of the well. With the well providing confinement along the growth direction, the carriers are confined in three dimensions, forming a quantum dot. Although dots formed by well width fluctuations have very good optical properties, their confining potential is very small, as are the spacings between the confined levels. The in-plane size of the dots is difficult to control – the well width fluctuations are essentially random – and the spread of dot sizes is large. Although it is possible to study zero-dimensional physics in these dots, they are not suitable for device applications.

3.5.9 Thermally annealed quantum wells

Starting with a GaAs–AlGaAs quantum well, grown using standard epitaxial techniques, a very finely focused laser beam is used to locally heat the structure. This produces a diffusion of Al from the AlGaAs barrier into the GaAs well, resulting in a local increase of the band gap. By scanning the beam along the edges of a square, a potential barrier is produced surrounding the unannealed centre. Electrons and holes within this square are confined by this potential barrier, with confinement along the growth direction provided by the quantum well potential. The carriers are confined in all three directions, forming a quantum dot. Quantum wires may be formed by scanning the laser beam along the edges of a rectangle. Because the size of the focused laser beam is ~1 μm the minimum dot size is fairly large (~100 nm), resulting in very closely spaced energy levels. In addition, the annealing processes can affect the optical quality of the semiconductor by introducing defects. The technique is relatively slow and hence is not suitable for the production of large arrays of dots or wires. It also requires specialised, non-standard equipment.

3.5.10 Semiconductor nanocrystals

Very small semiconductor particles, which act as quantum dots, can be formed in a glass matrix by heating the glass together with a small concentration of a suitable semiconductor. Dots with radii between 1 and about 40 nm are formed; the radius being a function of the temperature and heating time. Although the dots have excellent optical properties, the insulating glass matrix in which they are formed prevents the electrical injection of carriers.

3.5.11 Colloidal quantum dots

These II–VI semiconductor quantum dots are formed by injecting organometallic reagents into a hot solvent. Nanoscale crystallites grow in the solution, with sizes in the range 1 to about 10 nm. Subsequent chemical and physical processing may be used to select a subset of the crystallites which display good size uniformity. The dots can be coated with a layer of a wider band gap semiconductor which acts to passivate the surface, reducing non-radiative carrier recombination and hence increasing the optical efficiency. Organic capping groups are also used to provide additional surface passivation. Colloidal quantum dots exhibit excellent optical properties and can be deposited on to a suitable surface to form close-packed solid structures. Electrical contact to the dots is more difficult, although they can be deposited on silicon wafers that have been prepatterned with metal electrodes. In addition colloidal dots, either incorporated into a polymer film or as a thin film deposited by spin casting, can be formed into layered devices that allow the injection of both electrons and holes.

3.5.12 Self-assembly techniques

Using self-assembly techniques, quantum dots or wires form spontaneously under certain epitaxial growth conditions. The resulting structures have high optical quality, and self-assembled quantum dots, in particular, are suitable for a wide range of electro-optical applications. Much of the current interest in the physics of semiconductor nanostructures and the development of device applications is based on self-assembled systems. The underlying fabrication techniques are therefore described in detail in this section.

The first form of self-assembly involves growth on prepatterned substrates. The main steps of this technique for the fabrication of quantum wires can be understood with reference to Figure 3.18. Starting with a flat semiconductor substrate, an array of parallel stripes are formed in a layer of etch resist by optical holography or electron beam irradiation. The structure is then etched in an isotropically acting acid, which attacks different crystal surfaces at different rates, resulting in the formation of an array of parallel V-shaped grooves. The patterned substrate is then cleaned and transferred to a growth reactor.

Quantum wires formed by this technique are typically GaAs with AlGaAs barriers. Initially AlGaAs is deposited, which grows uniformly over the whole structure, sharpening the bottom of the grooves which, after the etching step, have a rounded profile. Next a thin layer of GaAs is deposited which grows preferentially at the bottom of the grooves due to diffusion of Ga atoms from the side walls; the diffusion length for Ga atoms is greater than that of Al atoms. A spatially modulated quantum well is formed, with an increased well thickness at the bottom of the groove. Carriers in this thicker region have a lower energy, so a quantum wire is formed with a crescent-shaped cross section

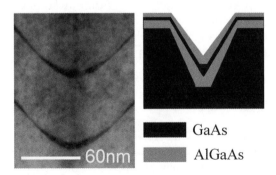

Figure 3.18 A schematic cross section of a V-groove quantum wire and a transmission electron microscope image of two stacked GaAs–AlGaAs quantum wires. Image courtesy of Dr Gerald Williams, QinetiQ, Malvern

and running along the length of the groove. The shape of the wire's cross section reduces the sharpness of the growth surface at the upper surface of the wire. However, following the wire growth, a second AlGaAs barrier is deposited, which re-establishes the sharpness of the groove. This resharpening permits the growth of further, nominally identical wires. Figure 3.18 also shows a cross-sectional TEM image of a V-groove quantum wire structure containing two wires.

V-groove quantum wires are spatially separated from the original etched surface, hence they exhibit excellent optical quality. By careful optimisation of the growth conditions it is possible to achieve a separation between the wire transitions of 80 meV. However, this energy represents the sum of the electron and hole confinement energies. Because of the larger hole effective mass, the main contribution will be due to the electrons, with the spacing between the confined hole subbands being relatively small ($\sim$10 meV). In addition, the optical spectra of V-groove quantum wires are subject to inhomogeneous broadening resulting from wire width fluctuations; these are probably related to the original groove quality. Furthermore, numerous emission lines are observed, arising from the different spatial regions of the structure. These regions include the quantum wires and quantum wells formed on the sides of the groove (the side-wall wells) and quantum wells formed in the region between the grooves (the top wells). A vertical well is also formed in the AlGaAs at the centre of the groove where the diffusion of Ga to the groove centre produces a Ga rich region. All these regions may capture carriers, reducing the fraction that recombine in the wire (Section 3.7.1). Although the top wells and a fraction of the side wells can be removed by a suitable short wet etch, this requires a further fabrication step followed by a return to the growth reactor to complete a p-i-n structure if required. Alternatively, high-energy ions incident at a shallow angle may be used to degrade the optical efficiency of the top wells and upper section of the side wells.

It is also possible to form quantum dots by a refinement of this technique in which the substrate is initially patterned with two orthogonal arrays of stripes. Following etching, a grid of tetrahedral pits is formed, within which the quantum dots are grown. The resultant dots are lens shaped with height 5 nm, diameter 20 nm and exhibit a typical spacing between optical transitions of $\sim$45 meV.

The second class of self-assembled growth occurs on unpatterned substrates and is driven by strain effects. Hence before this technique can be described, it is necessary to briefly discuss the growth of strained quantum well structures. Quantum wells are generally based on combinations of semiconductors having identical or very similar lattice constants. If a semiconductor is grown on a substrate having a significantly different lattice constant, then initially it will grow in a strained state to allow the atoms at the substrate–epitaxial layer interface to 'fit together'. However, energy is required to strain a material and this energy builds up as the thickness of the epitaxial layer increases. Eventually sufficient energy accumulates to break the atomic bonds of the semiconductor and dislocations – discontinuities of the crystal lattice – form. The thickness of material which can be grown before dislocations start to form is known as the critical thickness. The critical thickness depends on the semiconductor being grown and also the degree of lattice mismatch between this semiconductor and the underlying semiconductor or substrate.

Dislocations provide a very efficient mechanism for non-radiative carrier recombination and a structure which contains dislocations will, in general, exhibit a very poor optical efficiency. When growing strained semiconductor layers it is therefore very important that the critical thickness is not exceeded.

A good illustration of a strained semiconductor system is provided by $In_xGa_{1-x}As$–GaAs, where the $In_xGa_{1-x}As$ provides the quantum wells and the GaAs provides the substrate and barriers. Because the substrate is much thicker than the epitaxial layers, strain is present only in the $In_xGa_{1-x}As$. As the In composition of the $In_xGa_{1-x}As$ increases, the lattice mismatch between the two semiconductors also increases (Figure 3.8), and hence the $In_xGa_{1-x}As$ becomes increasingly strained. For low In compositions, $x \leq \sim 0.2$, it is possible to grow quantum wells with thicknesses ~ 10 nm, which is below the critical thickness. However for higher x the critical thickness decreases rapidly, making the growth of quantum wells of reasonable width impossible.

The lattice constants of InAs and GaAs are very different (7%) and therefore the critical thickness for the growth of an InAs layer on GaAs is expected to be very small. Initially InAs grows on GaAs as a highly strained two-dimensional layer. However, continued two-dimensional growth quickly leads to the formation of dislocations for deposition beyond approximately two atomic layers. In contrast, for certain growth conditions it is observed that following the deposition of approximately one atomic layer, the growth mode changes to three-dimensional, leading to the formation of small disconnected islands. These islands, which sit on the original two-dimensional layer, referred to as the wetting layer, act as quantum dots.

The transition from two- to three-dimensional growth is known as the Stranski–Krastanow growth mode and is a consequence of the balance between elastic and surface energies. All surfaces have an associated energy due to their incomplete atomic bonds, and this energy is directly proportional to the area of the surface. The surface area after island formation is greater than the initial flat surface, hence the island configuration has an increased surface energy. However, within the islands, the lattice constant of the semiconductor can start to revert back to its unstrained value as the atoms are free to relax in the in-plane direction, being unconstrained by surrounding material The lattice spacing reverts smoothly back to its unperturbed value with increasing height in the dot and no dislocations are formed. This relaxation results in a reduction of the elastic energy. For systems where the reduction in elastic energy exceeds the increase in surface energy, a transition to three-dimensional growth occurs, as this represents the lowest energy and hence the most favourable state. After formation, the quantum dots are generally overgrown by a suitable barrier semiconductor. Figure 3.19 shows the main steps in the self-assembled growth of quantum dots.

The structure of self-assembled quantum dots depends on the growth conditions, most importantly the temperature of the substrate and the rate at which material is deposited. For InAs dots grown on GaAs a typical base size is between 10–30 nm, height 5–20 nm and density 1×10^9 to about 1×10^{11} cm^{-2}. However, values outside this range may be possible by careful control of the growth conditions. Because of their small size the energy separation between the confined levels is relatively large ($\sim 30-90$ meV for electrons but less for holes due to their larger effective mass) and the confinement potential is relatively deep $\sim 100-300$ meV, with both electrons and holes being confined. Self-assembled quantum dots contain no dislocations and so exhibit excellent optical properties. Their two-dimensional density is high and can be increased further by growing multiple layers. Self-assembled dots are grown by conventional epitaxial processes and can be readily incorporated in the intrinsic region of a p-i-n structure.

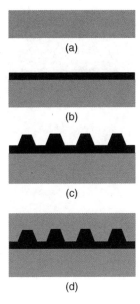

Figure 3.19 The main steps in the formation of self-assembled quantum dots: (a) initial surface, (b) deposition of the two-dimensional wetting layer, (c) formation of three-dimensional islands once the critical thickness for the 2D to 3D growth transition is reached and (d) overgrowth of the dots

Structural uniformity is reasonable (see below) but because of their small size, relatively small fluctuations in size and/or composition result in large variations in the confined energy levels. Figure 3.20 shows a cross-sectional TEM image of an uncapped InAs self-assembled quantum dot grown on a GaAs substrate. Also shown is an image, recorded along the growth direction, showing the random spatial distribution of the quantum dots. Figure 3.21 shows an AFM image of InAs self-assembled quantum dots.

Although self-assembled quantum dots have been extensively studied, there remains considerable uncertainty about their precise shape. Various shapes have been reported, including pyramids, truncated pyramids, cones and lenses (sections of a sphere). Difficulties associated with structural measurements (Section 3.7.2) may contribute to this reported range of shapes, but it also seems possible that the shape depends on the growth conditions.

Self-assembled quantum dots were first observed for InAs grown on GaAs, where the lattice mismatch is 7%. This and related InGaAs dots also grown on GaAs still form the most mature system. However, self-assembled dots appear to form for most semiconductor combinations with a sufficient degree of lattice mismatch. Examples include InP dots on InGaP, InGaN dots on GaN and Ge dots on Si.

The formation of self-assembled dots is a semi-random process. Dots at different positions start to form at slightly different times, as the amount of deposited material will vary slightly across the growth surface. Defects on the surface may also encourage or inhibit dot formation. As a result, the final shape and size (and possibly composition) will vary slightly from dot to dot, which in turn will result in an an inhomogeneous distribution of the energies of the confined states.

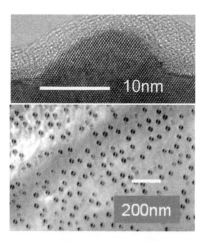

Figure 3.20 Upper panel: cross-sectional transmission electron microscope image of an uncapped InAs self-assembled quantum dot grown on a GaAs substrate. The speckled material above the dot is glue used to mount the sample. Lower panel: plan view (along the growth axis) TEM image of InAs self-assembled quantum dots grown on a GaAs substrate. In this sample the dots are overgrown with GaAs and the image results from the strain fields produced by the dots. Upper figure reproduced from P. W. Fry, I. E. Itskevich, D. J. Mowbray, M. S. Skolnick, J. J. Finley, J. A. Barker, E. P. O'Reilly, L. R. Wilson, I. A. Larkin, P. A. Maksym, M. Hopkinson, M. Al-Khafaji, J. P. R. David, A. G. Cullis, G. Hill and J. C. Clark, *Phys. Rev. Lett.* **84**, 733 (2000). Copyright 2000 by the American Physical Society

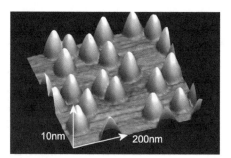

Figure 3.21 An AFM image of uncapped InAs self-assembled quantum dots grown on a GaAs substrate. Note the different horizontal and vertical scales. Image courtesy of Mark Hopkinson, University of Sheffield

The emission from a single dot has the form of a very sharp line, similar to that observed from an atom. However, many measurements simultaneously probe a large number of dots, referred to as an ensemble. Typically 10 million dots may be probed, each of which will contribute to the measured spectrum. Because each dot emits light at a slightly different energy, the sharp emission from each dot will merge into a broad emission band whose width is related to the degree of dot uniformity. The resulting emission is inhomogeneously broadened. The lower spectrum in Figure 3.22 shows the emission from an ensemble of self-assembled quantum dots, with a line width of 60 meV. The sharp emission, characteristic of a single dot, can be recovered by reducing the number of dots probed. This can be

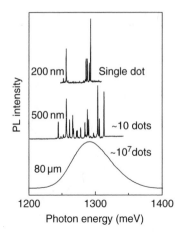

Figure 3.22 PL spectra of different numbers of self-assembled quantum dots recorded through holes of different sizes formed in an opaque metal mask. For small numbers of dots the sharp emission lines due to individual dots are observed. Data courtesy of Jonathan Finley, University of Sheffield

achieved by depositing an opaque metal mask on the sample surface in which submicron apertures are formed using electron beam lithography, followed by etching. Spectra for two aperture sizes are shown in Figure 3.22. A series of sharp lines, each arising from a different dot are visible, with the number of lines decreasing as the aperture size, and hence dot number below the aperture, decreases. The line width of each emission line is less than 30 µeV, a value limited by the resolution of the measurement system. Higher-resolution measurements have given linewidths as small as 1 µeV. Section 3.6.12 discusses mechanisms that determine this homogeneous line width, which is typically over 1000 times smaller than the ensemble inhomogeneous line width.

The non-uniformity of self-assembled quantum dots, and the resulting inhomogeneous broadening of the optical spectra, is disadvantageous for a number of device applications. For example, the absorption and emission is spread over a broad energy range instead of being concentrated at a single energy. However, there are some applications which make use of this inhomogeneous broadening. Applications of self-assembled quantum dots are discussed in detail in Section 3.8.

Once a layer of self-assembled InAs quantum dots has been deposited and overgrown with GaAs, a flat surface is regained, allowing a second layer of dots to be deposited. This procedure can be repeated, allowing the growth of multiple dot layers. In the first layer, the dot positions are essentially random. However, for multiple layers with relatively thin intermediate GaAs layers a correlation of dot positions is found, with vertical stacks of dots extending across all layers. This vertical alignment occurs as a result of a strain field formed in the GaAs immediately above the dots, produced by the InAs towards the top of the dot reverting back to its unstrained state. The size of this strain field decreases rapidly as the GaAs thickness increases but may be sufficiently large to act as a nucleation site for dots in the subsequent layer. Vertical dot alignment therefore occurs only for very thin GaAs layers (<10 nm) where the strain field at the surface remains sufficiently large. For thicker GaAs layers, the strain field is reduced essentially to zero and the dots in the next layer form at random positions. For very closely separated dot layers, dots within a vertical stack are able to interact either by a

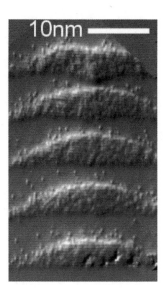

Figure 3.23 An STM cross-sectional image of the cleaved surface of a structure containing five layers of InAs self-assembled quantum dots with 10 nm thick intermediate GaAs layers. The thin GaAs layers result in a vertical alignment of the quantum dots. Reproduced by permission of the American Institute of Physics from D. M. Bruls, P. M. Koenraad, H. W. M. Salemink, J. H. Wolter, M. Hopkinson and M. S. Skolnick, *Appl. Phys. Lett.* **82**, 3758 (2003)

coupling of the carrier wavefunctions to form a one-dimensional superlattice or by carrier tunnelling between the dots. Figure 3.23 shows an STM image of a structure containing five InAs quantum dot layers, with each layer separated by 10 nm of GaAs.

3.5.13 Summary of fabrication techniques

This section has considered the different techniques that have been developed for the fabrication of inorganic semiconductor nanostructures. Each technique has relative advantages and disadvantages which, to some extent, are dependent on the precise application. Although no one technique yet satisfies all of the requirements listed in Section 3.5.1, this has not prevented the use of quantum wells, wires and dots in a number of electronic and electro-optical devices. Example applications are discussed in Section 3.8.

3.6 PHYSICAL PROCESSES IN SEMICONDUCTOR NANOSTRUCTURES

3.6.1 Modulation doping

As discussed in Section 3.2.4, the low-temperature carrier mobility of a bulk semiconductor is limited by scattering from impurities. This mechanism is particularly efficient in doped semiconductors where the scattering results from the charged dopant atoms. Consequently, the low-temperature carrier mobility in a bulk doped semiconductor is very low. However, in a semiconductor nanostructure it is possible to spatially separate the dopant atoms and the resulting free carriers. This significantly reduces impurity scattering, resulting in an extremely high carrier mobility at low temperatures. A spatial separation of the dopant atoms and free carriers is achieved through remote or modulation doping, as shown schematically for n-type doping of a quantum well structure in Figure 3.24(a). In this example the donor atoms are placed only in the wider band gap barrier material, the quantum well remains undoped. This is achieved during MBE growth by only opening the shutter in front of the cell containing the dopant atoms during growth of the barriers. In MOVPE the gas carrying the dopant atoms is similarly switched. Following thermal excitation of the electrons from the donors into the conduction band of the wider band gap semiconductor, these free electrons transfer into the lower-energy quantum well states, resulting in a spatial separation of the free electrons and the charged donor atoms. The confined electrons in the quantum well are said to form a two-dimensional electron gas (2DEG); a two-dimensional hole gas can similarly be formed by doping the barriers p-type. The non-zero charge present in both the barriers and the well (the total charge in the structure remains zero but there are equal and opposite charges in the well and barriers) results in an electrostatic bending of the conduction and valence band edges, as indicated in Figure 3.24(b). This band bending allows the formation of a modulation doping induced 2DEG at a single interface (a single heterojunction) between two different semiconductors, as shown in Figure 3.24(c). Here the combined effects of the conduction band offset and the band bending result in the formation of a triangular-shaped potential well that provides confinement of the electrons.

In a modulation-doped structure the barrier region immediately adjacent to the well is generally undoped, forming a spacer layer which further separates the charged dopant atoms and the free carriers. By optimising the width of this spacer layer and the structural uniformity of the interface, and by minimising unintentional background impurities, it is possible to achieve an extremely high carrier mobility at low temperatures. Figure 3.25 compares the temperature variation of the electron mobility of standard bulk GaAs, a very clean bulk specimen of GaAs and a series of GaAs–AlGaAs single heterojunctions. At high temperatures, where mobility is limited by phonon scattering, the mobilities of the different structures are very similar. At low temperatures the mobility in bulk GaAs is increased in the cleaner material, where a lower impurity

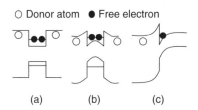

(a)   (b)   (c)

Figure 3.24   (a) n-type modulation doping of a quantum well showing the transfer of electrons from the barriers to the well; (b) the band edge profile of a modulation-doped quantum well, including the effects of band bending which results from the non-zero space charges in the well and barriers; (c) the production of a 2DEG in a modulation-doped single heterojunction

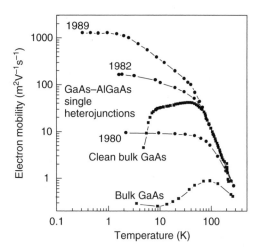

Figure 3.25   The temperature dependence of the electron mobility of two bulk GaAs samples having different purity, plus a series of n-type modulation-doped GaAs–AlGaAs single heterostructures. The labels for the heterostructures give the year of growth and demonstrate the improvement of the low-temperature mobility with time. Data reproduced by permission of the American Institute of Physics from L. Pfeiffer, K. W. West, H. L. Stormer and K. W. Baldwin, *Appl. Phys. Lett.* **55**, 1888 (1989) and C. R. Stanley, M. C. Holland, A. H. Kean, M. B. Stanaway, R. T. Grimes and J. M. Chamberlin, *Appl. Phys. Lett.* **58**, 478 (1991)

density reduces the charged impurity scattering. However, the absence of doping results in a low carrier density and, as a consequence, a low electrical conductivity. In contrast, modulation doping results in high free carrier densities and a low-temperature mobility more than two orders of magnitude higher than that of the clean bulk GaAs sample. The data for the different heterojunctions presented in Figure 3.25 demonstrates how the low-temperature mobility has increased over time, reflecting optimisation of the structure, the use of purer source materials and improved cleanliness of the MBE growth reactor. By the end of the 1980s a low-temperature mobility of $1200\,m^2V^{-1}s^{-1}$ had been achieved for a GaAs–AlGaAs single heterojunction. Since then progress has been less rapid, suggesting that the limits of material purity and interface structural perfection are being approached. The current state of the art is $3000\,m^2V^{-1}s^{-1}$ for a modulation-doped 30 nm wide GaAs–AlGaAs quantum well.

The ability to produce 2DEGs exhibiting an extremely high mobility has allowed the observation of a range of interesting and novel physical processes, a number of which are discussed in the following sections. In addition, modulation doping can be used to provide the channel of field effect transistors (FETs), where it is particularly useful for high-frequency applications. FETs that incorporate modulation doping are known as high electron mobility transistors (HEMTs) or modulation-doped field effect transistors (MODFETs). Although modulation doping provides only a minor enhancement of the room temperature carrier mobility, it produces free carriers that are confined within a two-dimensional sheet, in contrast to a layer of non-zero thickness produced by conventional doping. This precise positioning of the carriers in the channel results in FETs which exhibit improved linear characteristics and, for reasons that are still unclear, lower noise.

3.6.2 The quantum Hall effect

The Hall effect is a standard characterisation technique that can be used to determine both the density and type of majority carrier in a semiconductor. The inset to Figure 3.26 shows a schematic diagram of a Hall measurement, which can be applied to either a bulk semiconductor or a suitable nanostructure. An external voltage source causes a current I_x to flow along the bar, resulting in a current density J_x. On application of a magnetic field B_z, applied normal to the plane of the sample, a lateral electric field E_y is produced, which appears as a voltage V_y measured across the sample. The quantity E_y/B_zJ_x is known as the Hall coefficient, R_H, and for a bulk sample in which transport is dominated by one type of carrier (electrons or holes) $R_H = 1/ne$, where n is the free carrier density. The magnitude and sign of R_H allow determination of the free carrier density and the type of majority carrier, respectively.

Experimentally the electric field along the sample, E_x, can also be determined by measuring V_x as shown in Figure 3.26. This allows two resistivities to be defined and measured:

$$\rho_{xx} = \frac{E_x}{J_x}, \quad \rho_{xy} = \frac{E_y}{J_x}. \tag{3.11}$$

For a bulk semiconductor $R_H = E_y/B_zJ_x$, hence $\rho_{xy} = R_HB_z$. ρ_{xy} therefore increases linearly with increasing magnetic field, whilst ρ_{xx} remains constant. However, for a two-dimensional system, a very different behaviour is observed, as shown in Figure 3.26. In this case, although ρ_{xy} increases with increasing field, it does so in a step-like manner. In addition, ρ_{xx} oscillates between zero and non-zero values, with zeros occurring at fields where ρ_{xy} forms a plateau. This surprising behaviour is known as the quantum

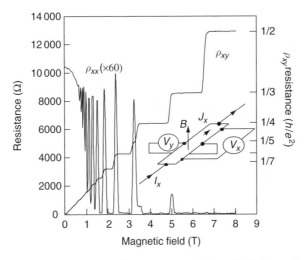

Figure 3.26 The integer quantum Hall effect as measured for a 2DEG in a GaAs–AlGaAs single heterojunction at a temperature of 50 mK. The inset shows the experimental geometry used for Hall effect measurements. Data reproduced from M. A. Paalanen, D. C. Tsui and A. C. Gossard, *Phys. Rev. B* **25**, 5566 (1982). Copyright 1982 by the American Physical Society

Hall effect and was discovered in 1980 by Klaus von Klitzing, a discovery for which he was awarded the 1985 Nobel physics prize. The quantum Hall effect arises as a result of the form of the density of states in a two-dimensional system in a magnetic field. This corresponds to that of a fully quantised system, with quantisation in one direction resulting from the physical structure of the sample and in the remaining two directions by the magnetic field. A full discussion of the physics underlying the quantum Hall effect, including the importance of disorder, is beyond the scope of this book. Suggestions for further reading are given at the end of this chapter.

An important practical application of the quantum Hall effect arises from the plateau values of ρ_{xy}. It can be shown that these are given by

$$\rho_{xy} = \frac{1}{j}\frac{h}{e^2} = \frac{25813}{j}\Omega, \tag{3.12}$$

where j is an integer whose value decreases with increasing magnetic field. ρ_{xy}, which is independent of the sample, can be measured to very high accuracy and is now used as the basis for the resistance standard and also to calculate the fine structure constant $\alpha = \mu_0 c e^2/2h$, where the permeability of free space, μ_0, and the speed of light, c, are defined quantities.

The parameter j in Equation (3.12) is known as the filling factor and denotes the number of different states occupied by the free carriers. The degeneracy (the maximum number of electrons or holes a given state can contain) of the states formed in a magnetic field increases with increasing field; consequently, for a constant total electron number, the number of states occupied, and hence j, decreases with increasing field. The quantum Hall effect discussed so far, and shown in Figure 3.26, occurs for integer values of j and is therefore known as the integer quantum Hall effect. However, in samples with very high carrier mobilities, plateaus in ρ_{xy} and minima in ρ_{xx} are also observed for fractional values of j, giving rise to the fractional quantum Hall effect. The discovery and theoretical interpretation of the fractional quantum Hall effect, which results from the free carriers behaving collectively rather than as single particles, led to the award of the 1998 Nobel physics prize to Stormer, Tsui and Laughlin.

3.6.3 Resonant tunnelling

Quantum mechanical tunnelling, in which a particle passes through a classically forbidden region, is the mechanism by which α-particles escape from the nucleus during α-decay and electrons escape from a solid in thermionic emission. In both cases the tunnelling probability is a very sensitive function of the energy of the particle and the thickness and height of the potential barrier. Carrier tunnelling can also be observed in nanostructures where a tunnelling barrier is formed by sandwiching a thin layer of a wide band gap semiconductor between layers of a smaller band gap semiconductor. Incorporated into a suitable device, this allows the behaviour of electrons incident on the barrier to be studied. The energy of the electrons is determined by the voltage applied to the device, and the probability of tunnelling through the barrier is reflected by the magnitude of the current that flows. Of greater interest, however, is the case of

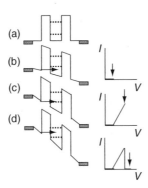

Figure 3.27 Schematic conduction band diagram of a double-barrier resonant tunnelling structure. The band structure is shown for different applied voltages and with the corresponding current: (a) no applied voltage, (b) voltage below the first resonance, (c) in resonance with the lowest energy state in the quantum well, and (d) above the first resonance

two barriers separated by a thin quantum well, a double-barrier resonant tunnelling structure (DBRTS). A schematic diagram of a DBRTS is shown in Figure 3.27. Quantised energy levels are formed in the quantum well, as described in Section 3.3.1. A DBRTS is generally grown between two doped layers (n-type in Figure 3.27) which provide reservoirs of carriers.

Figure 3.27 shows a DBRTS for various applied voltages. For the sign of voltage shown, electrons travel from left to right. Electrons are first incident on the leftmost barrier, through which they attempt to tunnel into the well, followed by tunnelling out of the well via the second barrier. At low voltages, condition (b), the electron energy following tunnelling into the well is below that of the lowest confined state. Hence there are no available states in the well and it acts as a further barrier. For this condition (b) the two barriers plus the well act as one effective thick barrier, as a consequence the tunnelling probability, and hence the current, is very low. As the voltage is increased to condition (c), the energy of the electrons tunnelling through the first barrier comes into resonance with the lowest state in the well. The effective barrier width is now reduced and it becomes much easier for the electrons to pass through the structure. As a result, the current increases significantly. For a further increase in voltage, condition (d), the resonance condition is lost and the current decreases. Although for condition (d) the energy of the tunnelling electrons coincides with higher energy states in the quantum well subband, these states correspond to non-zero in-plane motion (Section 3.3.1). The tunnelling electrons are moving parallel to the growth direction only, so tunnelling into these subband states is not possible. However, for higher applied voltages, additional resonances may be observed, corresponding to higher confined states. Figure 3.27 also shows the expected current–voltage characteristic of a DBRTS, indicating the relationship between specific points on the characteristic and the different bias conditions of the structure. Figure 3.28 shows experimental results obtained for a DBRTS consisting of a 20 nm GaAs quantum well confined between 8.5 nm AlGaAs barriers. Resonances with five confined quantum well states are observed. Beyond each resonance a DBRTS exhibits a negative differential resistance, a region where the current decreases as the applied voltage is increased. This characteristic has a number of applications, including the generation and mixing of microwave signals.

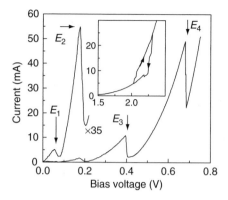

Figure 3.28 Current–voltage relationship for a double-barrier resonant tunnelling structure consisting of a 20 nm GaAs quantum well with 8.5 nm $Al_{0.4}Ga_{0.6}As$ barriers. Resonances with four confined well states are observed. The inset shows the hysteresis observed for an asymmetrical device with 8.5 and 13 nm $Al_{0.33}Ga_{0.67}As$ barriers and a 7.5 nm $In_{0.11}Ga_{0.89}As$ quantum well. Data courtesy of Philip Buckle and Wendy Tagg, University of Sheffield

DBRTS devices may also exhibit hysteresis in their current–voltage characteristics, particularly when the two barriers have unequal thicknesses. A thinner first barrier allows carriers to tunnel easily into the well whilst a thicker second barrier impedes escape, resulting in charge build-up in the well. This charge build-up modifies the voltage dropped across the initial part of the structure and maintains the resonance condition to higher voltages than would occur in an empty well. This broadened resonance is only observed as the voltage is increased, allowing charge to accumulate in the well. When the voltage is finally taken above the resonance condition, the well empties. If the voltage is now decreased, a narrower resonance is observed as there is now no charge accumulation. For such a structure the current follows a path that is dependent upon the direction in which the voltage is swept; the current–voltage characteristics exhibit hysteresis. The inset to Figure 3.28 shows the characteristics of an asymmetrical DBRTS with 8.5 and 13 nm thick $Al_{0.33}Ga_{0.67}As$ barriers and a 7.5 nm $In_{0.11}Ga_{0.89}As$ quantum well.

3.6.4 Charging effects

A charge carrier in a quantum dot is highly spatially localised. If a quantum dot already contains one or more carriers, then significant energy is required to add an additional carrier, as a result of the work that must be done against the repulsive electrostatic force between like charges. This charging energy modifies the energies of the dot states relative to their energies in the uncharged system.

The inset to Figure 3.29 shows the conduction band profile of a structure consisting of a quantum dot placed close to a reservoir of free electrons. Applying a voltage to a metal contact on the surface of the structure allows the energy of the dot to be varied with respect to the reservoir. If a given energy level in the dot is below the energy of the reservoir, then electrons will tunnel from the reservoir into the dot level. Alternatively, if the energy level is above the reservoir, then the level will be unoccupied. Hence by

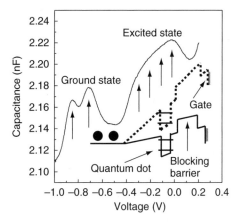

Figure 3.29 Capacitance–voltage profile of approximately 2 million self-assembled quantum dots showing features as successive electrons are added to the dots. The vertical arrows indicate the positions of the features corresponding to the loading of electrons into the ground state (two features) and the excited state (four unresolved features). The inset shows the conduction band profile for two bias voltages. Data reproduced by permission of EDP Sciences from M. Fricke, A. Lorke, J. P. Kotthaus, G. Medeiros-Ribeiro and P. M. Petroff, *Europhys. Lett.* **36**, 197 (1996)

varying the gate voltage, the dot states can be sequentially filled with electrons. This filling can be monitored by measuring the capacitance of the device, which will exhibit a characteristic feature each time an additional electron is added to the dot.

Figure 3.29 shows the capacitance of a device containing approximately 2 million self-assembled quantum dots. These dots have two confined electron levels: the lowest level (ground state) is able to hold two electrons (degeneracy 2) with the excited level able to hold four electrons (degeneracy 4). In the absence of charging effects, only two features would be observed in the capacitance trace, one at the voltage corresponding to the filling of the ground state, the other when the voltage reaches the value required for electrons to transfer into the excited state. However, once one electron has been loaded into the ground state, charging effects result in an additional energy, hence a higher voltage, being required to add the second electron. This leads to two distinct capacitance features corresponding to the filling of the ground state. Similarly, four distinct features are expected as electrons are loaded into the excited state, although in the present case inhomogeneous broadening prevents them being individually resolved. This charging behaviour is known as Coulomb blockade and is observed experimentally when the charging energy exceeds the thermal energy k_BT.

Coulomb blockade effects may also be observed in transport processes where carriers tunnel through a quantum dot. Suitable dots may be formed electrostatically using split gates to define the dot and to provide tunnelling barriers between the dot and two 2DEGs which act as carrier reservoirs. The device is analogous to the resonant tunnelling structures considered in Section 3.6.3 but with the quantum well replaced by a quantum dot. An additional gate electrode allows the energy of the dot to be varied with respect to the carrier reservoirs. The relatively large dot size (>100 nm) results in Coulomb charging energies – given by $e^2/2C$ where C is the dot capacitance – much larger than the confinement energies. The Coulomb charging energies therefore

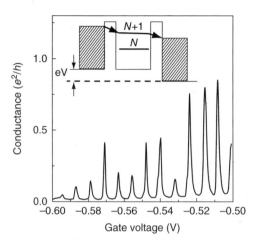

Figure 3.30 Coulomb blockade oscillations for an electrostatically defined quantum dot measured at a temperature of 10 mK. The inset shows the transport of a single electron through the structure. Data reproduced from L. P. Kouwenhoven, N. C. van der Vaart, A. T. Johnson, W. Kool, C. J. P. M. Harmans, J. G. Williamson, A. A M. Staring and C. T. Foxon, *Zeitschrift für Physik B* **85**, 367 (1991). Copyright 1991 Springer-Verlag

dominate the energetics of the system. The inset to Figure 3.30 shows a schematic diagram of the structure with a small bias voltage applied between the left and right reservoirs. The dot initially contains N electrons, resulting in a dot energy indicated by the lower horizontal line. An additional electron can tunnel into the dot from the left-hand reservoir but this increases the dot energy by an amount equal to the charging energy. This process is therefore only energetically possible if the energy of the dot with $N + 1$ electrons lies below the maximum energy of the electrons in the left-hand reservoir. Tunnelling of this additional electron into the right-hand reservoir may subsequently occur, but only if the $N + 1$ dot energy lies above the maximum energy of this reservoir, as the electron can only tunnel into an empty state. If these two conditions are satisfied, requiring that the dot energy for $N + 1$ electrons lies between the energy maxima of the two reservoirs, a sequential flow of single electrons through the structure occurs. For this condition the system exhibits a non-zero conductance. As the gate voltage varies the dot energy, rigidly shifting the different confined states up or down, the condition for sequential tunnelling will be successively satisfied for different values of N, and a series of conductance peaks will be observed. An example is shown in Figure 3.30 for a dot of radius 300 nm. This large dot size results in a large capacitance and a correspondingly small charging energy (0.6 meV for the present example). Hence measurements must be performed at very low temperatures in order to satisfy the condition $e^2/2C \gg k_BT$. Practical applications of Coulomb blockade are described in Section 8.7.

3.6.5 Ballistic carrier transport

The carrier transport considered in Section 3.2.4 is controlled by a series of random scattering events. However the high carrier mobilities which can be obtained by the use of modulation doping correspond to very long distances between successive scattering events, distances that can significantly exceed the dimensions of a nanostructure. Under these conditions a carrier can pass through the structure without experiencing a scattering event, a process known as ballistic transport. Ballistic transport conserves the phase of the charge carriers and leads to a number of novel phenomena, two of which will now be described.

When carriers travel ballistically along a quantum wire there is no dependence of the resultant current on the energy of the carriers. This behaviour results from a cancellation between the energy dependence of their velocity ($v = \sqrt{2E/m^*}$) and the density of states, which in one dimension varies as $1/\sqrt{E}$ (Section 3.4). For each occupied subband a conductance equal to $2e^2/h$ is obtained, a behaviour known as quantised conductance. If the number of occupied subbands is varied then the conductance of the wire will exhibit a step-like behaviour, with each step corresponding to a conductance change of $2e^2/h$. Quantum conductance is readily observable in electrostatically induced quantum wires (Section 3.5.7). The gate voltage determines the width of the wire, which in turn controls the energy spacing between the subbands. For a given carrier density, reducing the subband spacing results in the population of a greater number of subbands, and hence increased conductance. Figure 3.31 shows quantum conductance in a 400 nm long, electrostatically induced quantum wire. The structure of the device is shown in the inset. Such measurements are generally performed at very low temperatures to obtain the very high mobilities required for ballistic transport conditions. In contrast to the highly accurate values observed for ρ_{xy} in the quantum Hall effect, which are independent of the structure and quality of the device, the quantised conductance values of a quantum wire are very sensitive to any potential fluctuations, which may result in scattering events. This sensitivity prevents the use of quantum conductance as a resistance standard.

The inset to Figure 3.32 shows a structure where a quantum wire splits into two wires, which subsequently rejoin after having enclosed an area A. Under ballistic

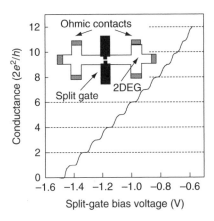

Figure 3.31 Quantised conductance steps in a 400 nm long electrostatically defined quantum wire measured at a temperature of 17 mK. The wire is produced by a split gate formed on the surface of a modulation-doped GaAs–AlGaAs heterostructure. The inset shows the form of the electrical contacts. Data reproduced by permission of the American Institute of Physics from A. R. Hamilton, J. E. F. Frost, C. G. Smith, M. J. Kelly, E. H. Linfield, C. J. B. Ford, D. A. Ritchie, G. A. C. Jones, M. Pepper, D. G. Hasko and H. Ahmed, *Appl. Phys. Lett.* **60**, 2782 (1992)

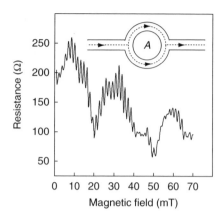

Figure 3.32   The Aharonov–Bohm effect in a 1.8 μm diameter ring, measured at a temperature of 280 mK. The inset shows the geometry of the structure. Data and figure reproduced from G. Timp, P. M. Mankiewich, P. deVegvar, R. Behringer, J. E. Cunningham, R. E. Howard, H. U. Baranger and J. K. Jain, *Phys. Rev. B* **39**, 6227 (1989). Copyright 1989 by the American Physical Society

transport conditions the wavefunction of an electron incident on the loop will split into two components which, upon recombining at the far side of the loop, will interfere. This process requires that the phase of the electron wavefunction is conserved as it transits the structure. If a magnetic field is now applied normal to the plane of the loop, the wavefunctions acquire or lose an additional phase, depending on the sense in which they traverse the loop. The phase difference between the two paths increases by 2π when the magnetic flux through the loop, given by the area multiplied by the field ($=BA$) changes by h/e. As the magnetic field increases, the system oscillates between conditions of constructive interference (corresponding to a high conductance) and destructive interference (corresponding to low conductance). The change in field, ΔB, between two successive maxima (or minima) is given by the condition $\Delta BA = h/e$, resulting in the conductance of the system oscillating periodically with the field. An example of this behaviour, known as the Aharonov–Bohm effect, is shown in Figure 3.32 for a loop of diameter 1.8 μm, formed from the 2DEG of a GaAs–AlGaAs single heterostructure by electron beam lithography.

3.6.6 Interband absorption in semiconductor nanostructures

As discussed in Section 3.2.5, a semiconductor can absorb a photon in a process where an electron is promoted between the valence and conduction bands. The strength of this absorption is proportional to the density of states in both bands – the joint density of states. The joint density of states has a form similar to the individual density of states and is therefore a strong function of the dimensionality of the system. In addition, the absorption will be modified by the quantised energy levels of a nanostructure, resulting in a number of different energy transitions occurring between the confined hole and electron states.

A further modification arises from the influence of excitonic effects. In a nanostructure the electron and hole are prevented from moving apart in one or more directions and, as a result, their average separation is decreased and the exciton binding energy is increased. For an ideal two-dimensional system, the exciton binding energy is increased by a factor of four compared to the bulk value. However, in a real quantum well the electron and hole wavefunctions penetrate into the barriers, and the ideal two-dimensional case is never achieved, although up to an approximately twofold enhancement is possible. A larger binding energy decreases the probability of exciton ionisation at high temperatures and, as a consequence, stronger excitonic effects are observed in a nanostructure at room temperature than in a comparable bulk semiconductor.

Figure 3.33 shows the absorption spectrum of a 40-period GaAs multiple quantum well structure with wells of width 7.6 nm and AlAs barriers. The gross form of the spectrum consists of a series of steps, representing the density of states of a two-dimensional system (Section 3.4), with an excitonic enhancement at the onset of each step. A number of transitions are observed between the mth confined hole state and nth confined electron state. These transitions are subject to selection rules that result in transitions between identical electron and hole index states ($m = n$) being the most intense. Three orders of such transitions are observed in the spectrum of Figure 3.33. Weaker transitions occur when the index changes by an even number ($|n - m| = 2, 4$, etc.). Transitions where the index changes by an odd number ($|n - m| = 1, 3$, etc.) are forbidden by parity considerations. The spectrum is further complicated by the presence of two different valence bands, known as the heavy and light hole bands, which result in two distinct series of confined valence band states. Transitions are possible from both series to the conduction band.

Figure 3.34 compares the room temperature absorption of bulk GaAs, and a GaAs multiple quantum well structure with 10 nm wide wells. The enhancement of the excitonic strength in the quantum well, due to the increased exciton binding energy, is

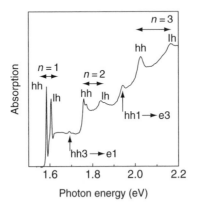

Figure 3.33 Low-temperature absorption spectrum of a 40-period GaAs–AlAs multiple quantum well structure with 7.6 nm wide wells. The most intense features result from transitions between the nth ($n = 1, 2, 3$) confined light hole (lh) and heavy (hh) hole states and identical index electron states. In addition, two weaker transitions are observed between the first and third heavy hole and electron states (hh3 → e1 and hh1 → e3). Data reproduced by permission of Taylor and Francis Ltd from A. M. Fox, *Contemp. Phys.* **37**, 111 (1996)

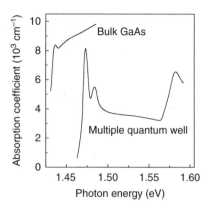

Figure 3.34 Comparison of the room temperature absorption spectra of bulk GaAs and a 77-period GaAs$-$Al$_{0.28}$Ga$_{0.72}$As multiple quantum well with 10 nm wide wells. Data reproduced by permission of the American Institute of Physics from D. A. B. Miller, D. S. Chemla, D. J. Eilenberger, P. W. Smith, A. C. Gossard and W. T. Tsang, *Appl. Phys. Lett.* **41**, 679 (1982)

clearly visible. The presence of an exciton provides a sharp onset of the band edge absorption, and this has a number of practical applications, for example in optical modulators. Nanostructures allow excitonic effects to be exploited more readily as these effects persist to considerably higher temperatures than in bulk semiconductors.

The absorption spectrum of a quantum wire should be similar to that of a quantum well, but with a further enhancement of the exciton binding energy provided by the additional quantum confinement. There will also be a modification due to the $1/\sqrt{E}$ form of the density of states. However, to date the quality of available quantum wires is not sufficient to observe these effects clearly, with spatial fluctuations of the wire cross section leading to significant inhomogeneous broadening of the absorption spectrum.

The energy levels of a quantum dot are already discrete, hence excitonic effects are less obvious in zero-dimensional systems. In this case they reduce the energies of the optical transitions compared to those that would occur between single-particle, non-interacting states. The absorption spectrum of a quantum dot resembles that of an atom, consisting of a number of discrete absorption lines between which there is zero absorption.

3.6.7 Intraband absorption in semiconductor nanostructures

The interband absorption discussed in the previous section occurs between the valence and conduction bands. However, in nanostructures, absorption between confined electron (or hole) levels can also occur, a process known as intraband absorption. The inset to Figure 3.35 shows intraband absorption for the conduction band of a quantum well. Electrons are required in the initial state and these are generally provided by doping. Photon absorption involves the excitation of electrons between the $n = 1$ and higher confined electron states. In some structures excitation between confined well states and the unconfined states of the barrier is also possible. The energy separation between the confined states is typically $\sim 10-200$ meV, corresponding to wavelengths $\sim 6-120$ μm

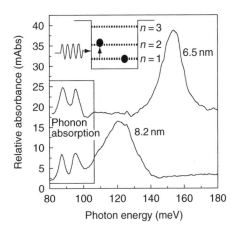

Figure 3.35   Intraband absorption spectra of two 50-period GaAs multiple quantum well structures with wells of width 6.5 and 8.2 nm. Transitions occur between the $n = 1$ and $n = 2$ confined electron levels. The inset shows a schematic diagram of the absorption process. Data reproduced by permission of the American Institute of Physics from L. C. West and S. J. Eglash, *Appl. Phys. Lett.* **46**, 1156 (1985)

and resulting in absorption in the infra-red region of the electromagnetic spectrum. The energies of the transitions can be varied over this wide range by altering the well width, allowing the absorption to be tuned to a specific wavelength. Intraband absorption spectra are shown in Figure 3.35 for two quantum wells of widths 6.5 and 8.2 nm. With decreasing well width, the separation between the $n = 1$ and $n = 2$ states increases and the absorption shifts to higher energy. There are no excitonic effects associated with intraband absorption, because only one type of carrier is involved; excitons require both an electron and a hole, which are only present in interband processes.

Intraband absorption is also observed between the confined states of quantum wires and dots. For quantum wells, one important selection rule for intraband absorption requires that the incident radiation has an electric field component normal to the plane containing the wells. As a consequence, for light incident along the growth direction, which is the most convenient experimental geometry, intraband absorption does not occur. Intraband absorption in a quantum well therefore requires the use of less convenient geometries, including light incident on the edge of the structure or normally incident light that is bent into the structure by a diffraction grating deposited on the surface. In contrast, for quantum dots this selection rule is modified and intraband absorption is possible for normally incident light. This difference is particularly advantageous for the practical application of intraband absorption in infra-red detectors.

3.6.8 Light emission processes in nanostructures

Electrons and holes can be created in a semiconductor either optically (Section 3.2.5), with incident photons of energy greater than the band gap, or by electrical injection in a pn junction (Section 3.2.7). The electrons and holes are typically created with excess energies above their respective band edges. However, the time required to lose this excess energy is generally much shorter than the electron–hole recombination time, consequently the electron and holes relax to their respective band edges before recombining to emit a photon. Emission therefore occurs at an energy corresponding to the band gap of the structure, with a small distribution due to the thermal energies of the electrons and holes. The influence of rapid carrier relaxation is demonstrated in the emission spectrum of a structure containing quantum wells of five different widths (Figure 3.11). Only emission corresponding to the lowest-energy transition of each well is observed, even though the wider wells contain a number of confined states.

Higher-energy transitions in a nanostructure can be observed in emission if the density of electrons and holes is sufficiently large that the underlying electron and hole states are populated. This can occur under high-excitation conditions where the lower energy states become fully occupied and carriers are prevented from relaxing into these states by the Pauli exclusion principle. Figure 3.36 shows emission spectra of an ensemble of self-assembled quantum dots for different optical excitation powers. At low powers the average number of electrons and holes in each dot is very small, and consequently only the lowest-energy, ground state transition is observed. However with increasing power the ground state, which has a degeneracy of two, is fully occupied and emission from higher-energy, excited states is observed.

As discussed in Section 3.5.12, fluctuations in the size, shape and composition of quantum dots result in the significant inhomogeneous broadening of optical spectra recorded for large numbers of dots. Only by probing a small number of dots can the predicted very sharp emission be observed. Figure 3.37 shows emission spectra obtained from a single self-assembled InAs quantum dot as a function of the incident laser power.

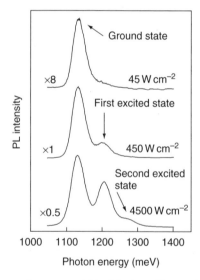

Figure 3.36 Emission spectra of an ensemble of InAs self-assembled quantum dots for three different laser power densities. At the highest power, emission from three different transitions is observed. The numbers by each spectra indicate the relative intensity scale factors. Data reproduced from M. J. Steer, D. J. Mowbray, W. R. Tribe, M. S. Skolnick, M. D. Sturge, M. Hopkinson, A. G. Cullis, C. R. Whitehouse and R. Murray, *Phys. Rev. B* **54**, 17738 (1996). Copyright 1996 by the American Physical Society

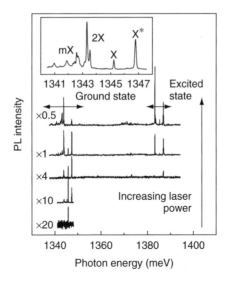

Figure 3.37 Emission spectra of a single InAs self-assembled quantum dot as a function of laser power. The inset shows emission from the ground state in greater detail. Emission lines are observed corresponding to an exciton recombining in an otherwise empty dot (X), the dot occupied by one additional exciton (the biexciton 2X), the dot occupied by > 1 additional excitons (the mX lines) and one additional hole (a charged exciton X^*). Data reproduced from J. J. Finley, A. D. Ashmore, A. Lemaître, D. J. Mowbray, M. S. Skolnick, I. E. Itskevich, P. A. Maksym, M. Hopkinson and T. F. Krauss, *Phys. Rev. B*, **63**, 073307 (2001). Copyright 2001 by the American Physical Society

At low powers a single, very sharp line is observed, arising from the recombination of an electron and hole from their respective ground states. At high powers these states are fully occupied and carriers are forced to occupy higher-energy excited states from which emission is then observed.

An added complication in the spectra of Figure 3.37 is that, at high excitation powers, multiple emission lines are observed for the ground state and excited state transitions. This is shown more clearly in the inset to Figure 3.37, which depicts the ground state emission for high laser power. These multiple emission lines result from interactions between the carriers confined within the dot. If the dot is occupied by a single electron and hole, one exciton, then emission will occur at a particular energy. If, however, the dot is occupied by two electrons and two holes, two excitons, then the energy of the first recombining electron and hole will be perturbed by the Coulomb interactions between the two excitons, and the emission will be slightly shifted in energy. For a dot initially occupied by three excitons, the recombination of the first electron and hole will be further perturbed. Lines corresponding to different recombination processes are observed in the same spectrum because the carrier population of the dot fluctuates during the time required to record the emission spectra, typically a few seconds. The recombination of an electron and hole in an otherwise empty dot is known as exciton recombination (X), that of an electron and hole in a dot occupied by an additional electron and hole as a biexciton (2X). The inset to Figure 3.37 also shows a line labelled X^*, arising from exciton recombination in the presence of just an additional hole. This process is possible if the carrier capture probability of the dot is different for electrons and holes.

It is also possible to observe emission when carriers make an intraband transition between the quantised electron or hole states of a nanostructure. However, this emission is generally relatively weak as there are other competing processes, including the emission of one or more phonons. An important application of intraband emission is in a new and increasingly important class of laser, the quantum cascade laser, which is discussed in detail in Section 3.8.2.

3.6.9 The phonon bottleneck in quantum dots

In both bulk semiconductors and nanostructures, electrons and holes may lose any excess energy by emitting a series of phonons. Figure 3.38(a) shows a typical case for electrons in a quantum well. The electrons initially have zero energy as they are at the conduction band edge of the barrier, but on transferring into the well they are left with an excess energy. Associated with each confined well state is a continuum of states, resulting from the in-plane motion (Section 3.3.1), and this allows the electron to lose energy by emitting a sequence of phonons, as shown in the figure. For III–V semiconductors the carriers interact most strongly with longitudinal optical (LO) phonons and it is these phonons which are emitted as the carriers lose energy. A typical LO phonon energy is $\sim$30 meV and a typical time to emit a single LO phonon is $\sim$150 fs. Once the carriers reach an energy less than one LO phonon energy from the band edge, further LO phonon emission becomes impossible and the final energy is lost by the emission of low-energy acoustic phonons, a slower process compared to LO phonon emission.

Figure 3.38(a) also applies to bulk semiconductors and quantum wires, both of which have a continuum of states. However, the situation is very different for a quantum dot, where the energy levels are discrete. Here emission by a series of LO phonons is only possible if the spacing between the energy levels equals the LO phonon energy, an unlikely coincidence. Hence carrier relaxation in a quantum dot must occur by an alternative, slower process. Possibilities include the emission of multiple phonons, for example an LO plus an acoustic phonon, or an Auger process where the energy released by one carrier as it relaxes is transferred to a second carrier, which is excited to a higher

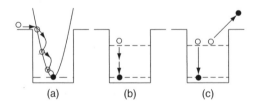

(a)   (b)   (c)

Figure 3.38   (a) Electron relaxation in a quantum well. The electron is injected into the well with excess energy but the continuum of well states allows this energy to be lost by the emission of a sequence of LO phonons. In a quantum dot, the separation between the discrete, confined states does not generally match the LO phonon energy and the electron relaxes by (b) simultaneously emitting two different phonons or (c) by transferring energy to a second electron that is excited into the continuum states of the barrier

energy state, for example the continuum states associated with the barrier. These processes are shown schematically in Figure 3.38(b) and (c). The slow carrier relaxation predicted to result from the discrete energy levels of a quantum dot is known as the phonon bottleneck. In severe cases this slow relaxation may affect the performance of quantum dot devices, particularly for high-speed applications. Experimental studies of carrier relaxation mechanisms are discussed in Section 3.7.1.

3.6.10 The quantum confined Stark effect

Applying an electric field to a semiconductor causes the band edges to tilt along the field direction (inset of Figure 3.39(a)). Although spatially separated, states in the conduction band are now closer in energy to states in the valence band, resulting in absorption occurring below the band gap energy. This process is known as the Franz–Keldysh effect and can be used as the basis for an electro-optical modulator for light tuned to an energy slightly below the band gap. A practical modulator requires a high on/off contrast ratio, which is possible if there is a steep absorption onset at the band gap. In a bulk semiconductor the $\sqrt{E}$ variation of the absorption provides only a weak onset, although this is enhanced by the presence of an exciton. However, the application of only a relatively weak electric field pulls apart the

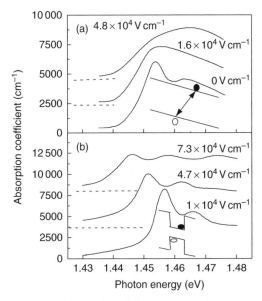

Figure 3.39 Quantum well absorption spectra for (a) electric fields applied in the plane of the well, equivalent to a bulk semiconductor, and (b) along the growth axis. For spectra along the growth axis, the potential wells prevent ionisation of the excitons, so they are observed in the spectra to considerably higher fields. The insets show the band edge profiles for the two cases. Data reproduced from D. A. B. Miller, D. S. Chemla, T. C. Damen, A. C. Gossard, W. Wiegmann, T. H. Wood and C. A. Burrus, *Phys. Rev. B* **32**, 1043 (1985). Copyright 1985 by the American Physical Society

electron and hole, ionising the exciton. With increasing field, excitonic effects are therefore quickly lost from the absorption spectra of a bulk semiconductor, as demonstrated in Figure 3.39(a), which shows absorption spectra as a function of electric field applied in the plane of a quantum well, equivalent to applying a field to a bulk semiconductor. In contrast, when an electric field is applied normal to the plane of a quantum well, the potential wells prevent the separation of the electron and hole, inhibiting the ionisation of the exciton (inset of Figure 3.39(b)). Excitonic effects are observed to considerably higher fields, as shown in Figure 3.39(b). The field-induced shift of the transition energy in a low-dimensional structure is known as the quantum-confined Stark effect and has practical applications in electro-optical modulators. Although initially observed in quantum wells, the quantum-confined Stark effect is also observed in quantum wires and dots.

3.6.11 Non-linear effects

The presence of a high density of additional carriers destroys an exciton. This results from a number of physical effects, including the screening of the Coulomb interaction between the electron and hole by the additional carriers and, more importantly, that these carriers occupy the states from which the exciton is formed – Pauli exclusion blocking. Figure 3.40 shows quantum well absorption spectra for the unperturbed state and soon after a very short laser pulse has created a large density of electrons and holes. The excitonic features are quenched by the optically created carriers. This behaviour provides the possibility for an all-optical modulator, where the carriers created by an intense laser beam switch a second weaker beam that is tuned to one of the exciton features. Although similar effects are observed in bulk semiconductors, the incident power required to 'switch off' an exciton in a quantum well is found to be significantly

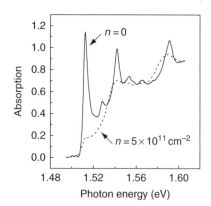

Figure 3.40   Absorption spectra of a 156-period GaAs–Al$_{0.3}$Ga$_{0.7}$As multiple quantum well with 20.5 nm wide wells. Spectra are shown for the unperturbed system ($n = 0$) and 100 ps after a pulsed laser has created a carrier density of $n \approx 5 \times 10^{11}$ cm^{-2} in each well, quenching the exciton features. Data reproduced by permission of Elsevier from C. V. Shank, R. L. Fork, R. Yen, J. Shah, B. I. Greene, A. C. Gossard and C. Weisbuch, *Solid Stat. Commun.* **47**, 981 (1983). Copyright 1983

less than in a bulk semiconductor. In addition, excitons are also present in quantum wells at room temperature, an essential requirement for practical devices. A further manifestation of the ability to control the strength of an exciton is that the fractional absorption experienced by light tuned to one of the exciton energies is dependent on the intensity of that light. With increasing power, larger numbers of carriers are excited and the exciton strength is reduced, decreasing the absorption. The resultant variable absorption is a non-linear effect and has a number of practical applications in optical systems.

3.6.12 Coherence and dephasing processes

When an exciton is optically excited in a semiconductor there is a well-defined phase relationship between the exciton wavefunction and the optical field. The field and exciton are coherent. Over time, various scattering processes will randomly alter the exciton phase, and the phase relationship with the field will be lost. This is known as dephasing. Alternatively, if a collection of excitons are created, they will initially have the same phase but this coherence will decay at a rate determined by the dephasing time. The most important dephasing mechanisms in semiconductors are the scattering of the exciton by other excitons, free carriers or phonons. These mechanisms are very efficient, hence dephasing times are generally very short. Typical values for a quantum well are 10 ps at 4 K, decreasing to 0.1 ps at room temperature. Recently there has been increased interest in dephasing processes as a result of the possible use of excitons for the basic elements of quantum computers; so-called qubits. For this application a system is required that retains its phase over a time interval sufficiently long to allow a number of logical operations to be performed. Quantum dots are of particular interest for this application as the spatial localisation of excitons within a dot should prevent interaction with other excitons, and the limited number of confined energy levels may reduce phonon scattering. Measurements on self-assembled quantum dots at low temperatures reveal very long dephasing times, with a value of $\sim$1 ns as the temperature approaches absolute zero. However, the dephasing time decreases with increasing temperature, and by room temperature it has a value similar to that found for a quantum well. The dephasing time of the exciton in a quantum dot also determines the homogeneous line width of the emission via the uncertainty relationship $\tau\Delta E = 2\hbar$, where τ is the dephasing time and ΔE is the linewidth. A dephasing time of 1 ns equates to a very narrow linewidth of $\sim$1 μeV.

3.7 THE CHARACTERISATION OF SEMICONDUCTOR NANOSTRUCTURES

In this section the main experimental techniques that can be used to probe the structural, electronic and optical properties of inorganic semiconductor nanostructures are reviewed. Emphasis is placed on the information provided by the techniques, and their relative advantages and disadvantages. Additional background material, relevant to many of the techniques, is given in Chapter 2.

3.7.1 Optical and electrical characterisation

The most commonly applied optical characterisation technique is photoluminescence (PL). PL involves creating electrons and holes by illuminating the structure with photons of sufficient energy, generally using light from a laser as described in Section 2.7.1.2. Typically photon energies are used such that absorption occurs in the barriers, as this produces a relatively high density of electrons and holes and therefore results in a strong PL signal. After creation the carriers diffuse spatially and are captured and localised in different parts of the structure. This is followed by relaxation as the carriers lose any excess kinetic energy, generally reaching the lowest possible energy states before recombining to emit a photon. The energies of the emitted photons are determined using a spectrometer and suitable detector.

The carrier transport efficiency between different parts of a nanostructure affects which regions contribute to the PL. In quantum wells and self-assembled quantum dots there is a very rapid transfer of carriers from the barriers into the lower energy states provided by the wells or dots. As a result, emission from the barriers, and from the wetting layer in quantum dot structures, is generally not observed. However, in nanostructures where the different regions are spatially well separated, efficient carrier transport may not be possible, and a number of regions may contribute to the PL. For example, in V-groove quantum wire structures (Section 3.5.12), the quantum wires, the quantum wells formed on the sides of the grooves and between the grooves, and the bulk GaAs may all give distinct emission, as shown in Figure 3.41. This behaviour complicates the interpretation of the emission spectrum and is a serious disadvantage for optical devices where emission at a single energy is generally required.

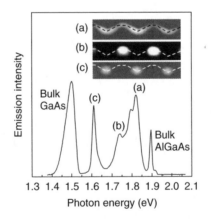

Figure 3.41 Cathodoluminescence (CL) spectrum of a V-groove quantum wire structure showing emission from the different regions of the structure. Scanning CL images are shown recorded for the detection of emission lines (a), (b) and (c). The forms of these images are consistent with line (a) arising from carrier recombination in the quantum wells formed on the side walls of the grooves, (b) quantum wells formed on the planar regions between the grooves, and (c) the quantum wires at the bottom of the grooves. The dashed lines in the images indicate the approximate position of the pre-growth surface. Data and images courtesy of Dr Gerald Williams, QinetiQ, Malvern

In conventional PL measurements the exciting laser beam is focused to a spot of diameter $\sim100\,\mu$m. Because of the high density of dots or wires in a typical structure, this results in the simultaneous measurement of many nanostructures. For example, V-groove quantum wires are typically spaced by $\sim1\,\mu$m and self-assembled quantum dots may have a density $\sim10^{11}\,$cm^{-2}. For these densities a spot of diameter $100\,\mu$m excites 100 wires or 10 million dots. Consequently, the PL spectra are inhomogeneously broadened due to unavoidable fluctuations in the wire or dot shape, size and composition. Although the magnitude of this broadening provides information on the homogeneity of the nanostructures, it prevents the study of processes that occur on comparable or smaller energy scales, such as the perturbation of the emission energy of a quantum dot as additional excitons are added (Section 3.6.8).

In order to study physical processes occurring on an energy scale smaller than the inhomogeneous broadening, it is necessary to probe individual dots or wires. This can be achieved through reducing the size of the focused laser beam, either by using a large-aperture microscope objective, for which a diffraction-limited spot size of $\sim1\,\mu$m is possible, or by using a scanning near-field optical microscope (SNOM) that allows the diffraction limit to be circumvented. The principle and operation of a SNOM are discussed in Sections 2.4.1 and 9.1.1. Both techniques allow single quantum wires to be studied. However, the tendency of carriers to diffuse away from the illuminated area, coupled with the typically large quantum dot densities, necessitates the use of additional steps to probe single quantum dots. This generally involves the reduction of the dot density, achieved by modifying the growth conditions, followed by the physical isolation of a single dot, either by etching submicron mesas or forming small holes in an otherwise opaque metal surface mask. Examples of single quantum dot spectroscopy are shown in Figure 3.22, where the spectra are recorded through different sized holes formed in a metal mask, and Figure 3.37, which shows spectra of a single quantum dot isolated in a 200 nm diameter mesa.

Carriers may also be created electrically by placing a nanostructure in a p-i-n device. In this case the process of light emission is known as electroluminescence (EL). EL has the advantage that the rate of carrier injection is uniform across the area studied, in contrast to PL, where the incident laser beam has a non-uniform, Gaussian profile. EL from a single quantum dot may be observed by exciting a relatively large number of dots and selecting the emission from a single dot by using a small aperture in a metal mask, which also forms one of the contacts. EL can also be excited using current injection from a scanning tunnelling microscope (STM). This method offers high spatial resolution and is therefore particularly suitable for the study of single quantum dots.

A third mechanism for the excitation of luminescence is a beam of high-energy electrons, such as found in an electron microscope: this is the cathodoluminescence (CL) technique described in Section 2.7.3.2. Although the nominal spatial resolution provided by the electron beam is degraded by carrier diffusion, a careful choice of beam voltage and beam current makes it possible to observe the emission from a single quantum wire and a few quantum dots. One powerful application of CL is in identifying the origin of the different features present in the emission spectra of complex structures; for example V-groove quantum wires. Initially the emission spectrum for excitation of a large area is recorded. The system is then set to detect photons corresponding to one of the emission features, and the electron beam is raster scanned over the sample. Regions of the structure responsible for the selected emission appear bright in the resultant image, allowing their position and shape, and hence their origin, to be determined. This procedure is then repeated for the different emission features. Scanning CL images from the cleaved edge of a V-groove quantum wire structure are shown in Figure 3.41. Detecting photons corresponding to line (c) results in emission located at the bottom of the grooves, and therefore originating from the quantum wires. Lines (a) and (b) result in emission from the side walls of the grooves and the flat surfaces between the grooves, consistent with emission from the side and top quantum wells, respectively. These wells have different thicknesses, hence they emit at different energies, a result of the dependence of the GaAs growth rate on surface orientation.

Carrier relaxation is generally much faster than radiative recombination (Section 3.6.9); hence only the lowest-energy, ground state transition is observed in emission. Emission from higher energy states can be observed by increasing the carrier excitation or injection rate so that the population of carriers in the ground state is sufficient to block relaxation from the excited states. An example is shown for quantum dots in Figure 3.36. Although high carrier injection allows the excited states to be observed, the system is highly occupied and the interaction between the carriers may significantly perturb the energies of the transitions. In addition, emission techniques do not allow the true relative strengths of the transitions to be determined, as the emission intensity is dependent not only on the intrinsic transition strength but also on the carrier populations in the initial and final states.

A determination of the unperturbed energies and the strengths of optical transitions therefore requires the use of an absorption technique. However, it is very difficult to measure the direct absorption of a single nanostructure because only a very small fraction of the light is absorbed in comparison to the majority of the light that simply passes through the sample. Measuring this small change in transmitted light against the large background is technically very difficult. For quantum wells it is possible to use multiple-well structures to increase the absorption to a measurable level, and for colloidal and nanocrystal dots it is possible to produce films or solutions containing the dots of sufficient thickness to allow direct absorption measurements. However, absorption measurements of epitaxially grown wires and dots are more difficult because, as they occupy only a relatively small fraction of the cross-sectional area, their intrinsic absorption is very low. The absorption spectrum of a single layer of self-assembled quantum dots can be measured, but this requires the use of very sensitive and expensive commercial systems. Even with such systems the very small absorption signal requires the use of extremely long integration times, typically a few hours. By focusing a laser beam to a spot size <1 μm, using a large-aperture microscope objective, it is possible to measure the absorption spectrum of a single self-assembled quantum dot, whose physical cross section is now a reasonable fraction of the laser beam area. However, the absorption is still too low to allow the direct measurement of the absorption. Consequently, a modulation technique consisting of an oscillating electric field that varies the transition energies via the quantum-confined Stark effect is used. The resulting spectra correspond to the first derivative of the absorption.

Because of these experimental difficulties, absorption studies of quantum wires and dots generally make use of an absorption-related technique; photocurrent (PC) spectroscopy and photoluminescence excitation (PLE) are the main examples. In PC, incident photons create electrons and holes in a nanostructure, which is placed in the intrinsic region of a p-i-n device. Under suitable conditions the carriers are able to escape from the nanostructure before recombining, giving rise to a current that can be measured by an external circuit. As the energy of the incident photons is varied, a change in absorption will alter the number of carriers created, and hence the magnitude of the photocurrent. In PLE the intensity of the photoluminescence is monitored as the energy of the exciting photons is varied. A change in the absorption alters the number of carriers created, and hence the intensity of the photoluminescence. Both PC and PLE measure a small signal against a zero background; although the majority of incident light still passes through the structure, this does not contribute to the measured signal. PC and PLE are therefore referred to as background-less techniques and are particularly suited to the study of nanostructures that absorb only very weakly. However, both techniques involve an additional step beyond photon absorption. In PC the photo-created carriers must escape from the nanostructure and in PLE they must relax to the transition being monitored. If the probability of these processes is not constant, but depends on the initial energy of the carriers, then the form of the resultant spectra will not reflect the true absorption.

A further important class of optical characterisation techniques is time-resolved spectroscopy. Here a structure is excited by a very short pulse of light, and the subsequent temporal changes in its properties are determined as carriers recombine or relax in energy. The pulse length and time resolution of the measurement system are typically in the range 1 ns to 0.1 ps; the precise value is dependent on the physical processes being studied. Time-resolved spectroscopy has been used to study a range of carrier processes in semiconductor nanostructures, including the rate at which carriers are captured from the barriers into the nanostructure, the rate at which carriers relax between confined levels, dephasing times and recombination times. Figure 3.42 shows results from a study of carrier relaxation mechanisms in InGaAs self-assembled quantum dots. Electrons and holes are initially created in the GaAs barriers by 1.5 ps pulses

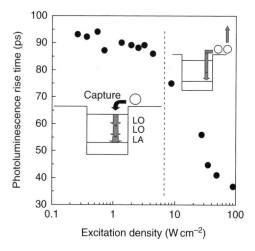

Figure 3.42   The rise time of the PL from InGaAs self-assembled QDs measured as a function of laser power density. The rise time indicates the time required for the carriers to be captured by the dots followed by energy relaxation to the ground state. Two possible relaxation processes are shown schematically. The vertical dashed line indicates the laser power corresponding to the creation of an average of one electron and hole per quantum dot. Data from B. Ohnesorge, M. Albrecht, J. Oshinowo, A. Forchel and Y. Arakawa, *Phys. Rev. B* **54**, 11532 (1996). Copyright 1996 by the American Physical Society

of light. The PL from the ground state of the quantum dots is detected and the rise time of this emission provides an indication of the time taken for carriers to be captured into the dots, followed by relaxation to their ground state. In Figure 3.42 the photoluminescence rise time is plotted as a function of the laser power, which is used to vary the density of carriers created in the structure. For low powers, equivalent to less than one electron and hole per dot, the rise time is relatively long, and reflects the slow relaxation of carriers between the discrete states of the dots by the emission of multiple phonons. This is a manifestation of the phonon bottleneck (Section 3.6.9). This slow rise time is retained until the laser power becomes sufficient to excite an average of one electron and hole per dot, indicated by the vertical dashed line in the figure. Above this power the rise time starts to decrease, reflecting carrier capture and relaxation by a much faster process in which the energy lost by one carrier is transferred to a second carrier. This mechanism, known as an Auger process, requires an electron or hole in addition to the relaxing carriers, hence it only becomes possible above an average electron–hole number of one per dot. Further increase in the number of carriers per dot increases the efficiency of this process, resulting in the continuous decrease of the rise time observed at high laser powers. The insets to Figure 3.42 show the carrier relaxation processes relevant to the low and high carrier density regimes.

Electrical measurements in their basic form involve the determination of currents and voltages from which sample resistances and electrical conductivity can be obtained. With the addition of a magnetic field, the integer and fractional quantum Hall effects can be studied, and measurements at very low fields allow a determination of carrier mobility. Two-dimensional carrier densities are determined from the period of the ρ_{xx} oscillations in the integer quantum Hall regime; this is also known as the Shubnikov–de Haas effect. Information on the dominant carrier scattering mechanism may be obtained by studying the temperature variation of the carrier mobility.

The capacitance of a quantum dot structure is affected by the charge state of the dots and this provides a method for probing the number of electrons or holes which have been loaded into a dot (Figure 3.29 and discussion in Section 3.6.4). A refinement of this technique is deep-level transient spectroscopy (DLTS), which measures the temporal evolution of the capacitance following the application of a voltage pulse. If the size and sign of the pulse are chosen such that excess carriers are loaded onto the dot, then a transient change in the capacitance will occur. Following removal of the pulse, the capacitance will revert back to its original value as the excess carriers escape from the dot. By measuring the rate at which the capacitance recovers it is therefore possible to determine the carrier escape rate, and measurements as a function of temperature allow the height of the potential barrier confining the carriers to be determined.

3.7.2 Structural characterisation

A number of techniques are available for the direct study of the physical structure of quantum wells and superlattices. These include: X-ray diffraction, which can determine periodicity, layer uniformity and in some cases composition; and transmission electron microscopy, which can determine layer thicknesses and compositions and which enables the nature of interfaces to be directly imaged. X-ray studies of quantum wires and dots are more limited, although shallow incidence X-ray diffraction has been used to determine the distribution of indium in self-assembled quantum dots. Transmission electron microscopy has been applied extensively to the study of quantum dots and wires, and examples of cross-sectional images are presented in Figure 3.18 and Figure 3.20. Planview images recorded along the growth axis, and for a geometry sensitive to strain, reveal the spatial distribution of self-assembled quantum dots (Figure 3.20), although because the strain field extends beyond the dots, their size and shape is not directly imaged. A complication arising with cross-sectional images of quantum dots is that for dot shapes other than cuboidal the apparent shape depends on where the dot is sectioned during sample preparation. This complication also applies to compositional studies.

Compositional information is provided by electron energy loss spectroscopy (EELS) and energy-dispersive X-ray spectroscopy (EDX), techniques which are described in Section 2.7.3. For quantum wells and wires, which have a uniform cross section through the thinned specimen, it is possible to obtain absolute compositions from both EELS and EDX by first calibrating the system with samples of a known composition. However, for quantum dots the uncertainty in how the dot has been sectioned makes it difficult to achieve absolute compositional determination. Figure 3.43 shows an EELS image of the gallium distribution in an InAs self-assembled quantum dot grown within a GaAs matrix. In addition, the two traces show EDX line scans of the indium distribution along directions parallel to the growth axis and passing through either just the wetting layer or both the dot and wetting layer. The profile of the latter trace is broader, reflecting the additional indium-containing region of the quantum dot.

Atomic force microscopy (AFM) is routinely used to study nanostructures as no complicated sample preparation is required. AFM allows the shape, size and distribution of self-assembled quantum dots to be determined (Figure 3.21) but these dots must

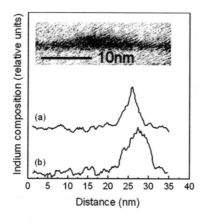

Figure 3.43 EDX profiles of a self-assembled InAs quantum dot structure recorded for line scans (a) passing through the wetting layer and (b) passing through the wetting layer and a quantum dot. Also shown is an EELS image of the Ga composition in an InAs self-assembled quantum dot structure with GaAs barriers. Data reproduced by permission of the Institute of Physics from M. A. Al-Khafaji, A. G. Cullis, M. Hopkinson, L. R. Wilson, S. R. Parnell, D. J. Mowbray and M. S. Skolnick, *Inst. Phys. Conf. Ser.* **161**, 585 (1999)

be situated on the sample surface, requiring the growth to be terminated immediately after the dots have been formed. In contrast, dots for optical and other applications must be covered by a relatively thick 'protective' layer to prevent the strong non-radiative carrier recombination which otherwise occurs at a free surface. As there is some evidence that material diffuses into and out of the dots during the growth of the capping layer, altering their shape, size and composition, it is possible that dot parameters determined by AFM may be different to those of covered dots.

AFM may also be used to determine alloy compositions in aluminium-containing nanostructures. This requires the nanostructure to be cleaved and exposed to air, causing the aluminium-containing materials to oxidise and bow outwards slightly from the surface. The degree of bowing is proportional to the local aluminium content and can be measured by an AFM. By calibrating the system with samples of known composition it is possible to produce a two-dimensional image of the aluminium composition.

Structural information may also be obtained from scanning tunnelling microscopy (STM). Figure 3.23 shows an STM image of a cleaved sample containing a stack of five InAs self-assembled quantum dots. High tunnelling current corresponds to regions of high indium composition, a result of a smaller local band gap and the increased strain which causes the cleaved surface to bow out slightly. A simulation of these two contributions to the tunnelling current allows a determination of the indium distribution across the cleaved surface.

3.8 APPLICATIONS OF SEMICONDUCTOR NANOSTRUCTURES

A number of practical applications of semiconductor nanostructures have been briefly discussed previously and in this section some of these are considered in greater detail. There is insufficient space to consider all current or potential future applications, and the aim is to provide examples of the more important ones and to demonstrate the breadth of possibilities provided by inorganic semiconductor nanostructures.

3.8.1 Injection lasers

Semiconductor injection lasers are physically small, convert electrical energy into light with high efficiency, have very long operating lifetimes, and can be switched on and off (i.e., modulated), extremely rapidly. This makes them very suitable for many applications, including data transmission and storage, printing and medical uses. However, initial devices, which were based on bulk, three-dimensional semiconductors, were far from ideal. Their threshold current, the current that must be applied for the laser to operate, was relatively high and increased rapidly with device temperature. This section shows that the use of nanostructures in the emitting region of a laser can result in significant performance improvements.

A laser is an optical oscillator and therefore contains a gain mechanism by which light is amplified. Gain is the inverse of absorption and in a semiconductor it occurs when there are a large number of electrons in the conduction band and a large number of holes in the valence band, a condition known as a population inversion. A population inversion is achieved in a forward-biased p-i-n structure where large numbers of electrons and holes are injected into the intrinsic region. For the system to oscillate or lase, the total gain must balance the total loss. This requires the creation of sufficient densities of electrons and holes, hence the injection of a corresponding current, to achieve the required gain. The gain is proportional to the density of occupied states: hence, because the density of states in a bulk semiconductor increases with energy (Section 3.4), the gain will increase with increasing carrier density as higher energy states are successively filled. In a bulk system it is therefore necessary to 'fill' the density of states to a point where sufficient gain is reached, with the carriers at lower energies effectively wasted. This process is shown schematically in Figure 3.44(a). These wasted carriers result in a relatively large threshold current.

The situation depicted in Figure 3.44(a), with only the lowest energy states occupied, corresponds to absolute zero temperature. At non-zero temperatures, carriers will be thermally excited to higher states, as shown in Figure 3.44(b). This further wastes carriers by exciting them into states not involved with the lasing process, a particularly serious problem in a bulk laser where the form of the density of states results in a large number of states at higher energies. As the temperature is increased, an increasing fraction of the carriers are thermally excited out of the lasing states and in Figure 3.44(b) the maximum gain is now below that required for lasing action. Consequently, a higher current must be applied to achieve the required gain. This mechanism is the reason why the threshold current of a laser increases with increasing temperature.

The modification of the density of states in a nanostructure overcomes, to various extents, the limitations of bulk semiconductor lasers. In particular, the density of states is increased at low energies with respect to higher energies, which results in both a decrease in the absolute threshold current and also its sensitivity to temperature. In the ultimate limit of a quantum dot with only one confined electron state and one confined hole state, all of the carriers must have the same energy at all temperatures – there are no states available for thermal excitation – and the laser should exhibit a very low and temperature-insensitive threshold current.

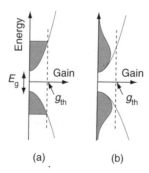

Figure 3.44 Electron and hole distributions in the conduction and valence bands of a bulk semiconductor laser at (a) absolute zero temperature and (b) non-zero temperature. In (b) thermal excitation of carriers to higher energy states has reduced the maximum gain below the value, g_{th}, required for laser action

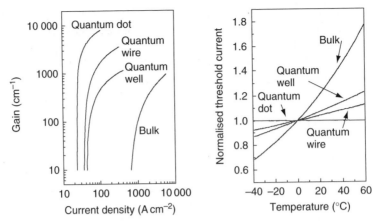

Figure 3.45 Left-hand panel: calculated gain variation with current density for lasers having different dimensionalities. For a given current density, the gain increases as the dimensionality decreases. Data reproduced by permission of the American Institute of Physics from Y. Arakawa and H. Sakaki, *Appl. Phys. Lett.* **40**, 939 (1982). Right-hand panel: calculated temperature variation of the threshold current density for bulk, quantum well, quantum wire and quantum dot lasers. Decreasing the dimensionality increases the temperature stability. Data reproduced from M. Asada, Y. Miyamoto and Y. Suematsu, *IEEE J. Quant. Electron.* **QE-22**, 1915 (1986). Copyright 1986 IEEE

The potential advantages arising from the use of nanostructures in a laser were first predicted from calculations performed for idealised systems. Figure 3.45 shows how the gain of systems having different dimensionalities develops with increasing current; it also shows the dependence of the threshold current on temperature. A given gain is reached at successively lower currents as the dimensionality is decreased, with threshold current densities of 1050, 380, 140 and 45 A cm^{-2} calculated for bulk, quantum well, quantum wire and quantum dot lasers, respectively.

The temperature dependence of the threshold current density of a semiconductor laser can be described by an equation of the form

$$J_{th} = J_{th}^0 \exp(T/T_0), \qquad (3.13)$$

where J_{th}^0 is the threshold current density at 0 °C and T_0 is a parameter that determines the temperature sensitivity; a larger T_0 corresponds to a lower sensitivity. For the idealised lasers considered in Figure 3.45, T_0 values of 104, 285, 481 °C and infinity are calculated, demonstrating a reduced sensitivity with the progression bulk→well→wire→dot.

Semiconductor lasers based on quantum wells are now used extensively for many applications, including CD and DVD data storage and fibre-optic data transmission. Quantum wire lasers have been fabricated, although their performance has been limited by the lack of systems with suitable energy spacings between the confined subbands. Greater progress has been made with quantum dot lasers based on self-assembled dots. At room temperature, threshold current densities are approximately one-third the value of comparable quantum well lasers, and the stability of the threshold current is improved by approximately a factor of two. That these improvements are not as great as predicted by the calculations summarised in Figure 3.45 is due to the departure of self-assembled dots

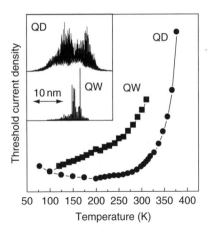

Figure 3.46 Comparison of the temperature dependence of the threshold current density of a self-assembled quantum dot laser (circles) and a quantum well laser (squares). Because of the different device designs, the threshold current densities are normalised to their 100 K values. The inset compares emission spectra of a quantum well (QW) laser and a quantum dot (QD) laser. Quantum well laser data reproduced from T. Higashi, S. J. Sweeney, A. F. Phillips, A. R. Adams, E. P. O'Reilly, T. Uchida and T. Fujii, *IEEE J. Select. Topics Quant. Electron.* **5**, 413–419 (1999). Copyright 1999 IEEE. Quantum dot laser data courtesy of Ian Sellers, University of Sheffield

from idealised systems. In particular, self-assembled dots have a number of confined levels into which carriers may be thermally excited, as well as the surrounding barrier material, resulting in a finite T_0. In addition, there may be non-radiative processes and the dot ensemble is inhomogeneously broadened. Figure 3.46 shows a comparison of the temperature dependence of the threshold current densities of a self-assembled quantum dot laser and a quantum well laser. At low temperatures the expected temperature stability of the quantum dot laser is achieved and below 200 K the quantum dot laser is significantly more stable than the quantum well laser. However, above 200 K the threshold current density of the quantum dot laser increases due to the mechanisms previously discussed, and by 300 K the quantum well and dot lasers exhibit very similar temperature sensitivity. One of the key goals in quantum dot laser development is to extend the temperature-insensitive regime to room temperature and beyond.

In addition to their low threshold current density and high temperature stability, quantum dot lasers offer a number of additional advantages over lasers of higher dimensionality. Their maximum modulation frequency should be higher, although this may be compromised if carriers are unable to relax sufficiently rapidly to the states from which lasing occurs – the phonon bottleneck. Once carriers have been captured into a quantum dot, their subsequent motion is restricted unlike the case, for example, of a quantum well laser where carriers captured into the well are still free to move within the plane of the well. This carrier localisation prevents the diffusion of carriers to non-radiative centres which may be formed on the surface of the device or within the device by, for example, radiation damage. Quantum dots are therefore particularly suitable for the fabrication of very small lasers, with a high area-to-volume ratio, and lasers for use in harsh radiation environments. Carrier localisation also prevents carriers in different dots from directly interacting and subsets of dots with similar emission energies may act as independent lasers. The dot ensemble may therefore behave as a collection of sub-lasers, with lasing occurring over a significant fraction of the inhomogeneous line width. Although it is a disadvantage for many applications where a single lasing energy is required, a broad emission allows the fabrication of a tunable laser by placing the system in an external cavity, allowing selection of one particular frequency. Figure 3.46 compares quantum dot and quantum well laser spectra; the increased width of the quantum dot spectra is clearly visible. Finally, quantum dots allow emission at new wavelengths to be obtained. GaInAs quantum well lasers grown on GaAs substrates are limited to <1.2 µm by the critical thickness of this strained system (Section 3.5.12). However, InAs quantum dots allow the fabrication of lasers operating in the important 1.3 µm telecommunications band, with some prospects for devices operating in the related 1.55 µm band. Current 1.3 and 1.55 µm lasers require the use of quantum well lasers grown on less technologically convenient InP substrates.

The most common type of semiconductor laser is based on a transverse geometry where light propagates in the plane defined by the quantum well or dots. In this geometry the mirrors at the ends of the optical cavity are formed by cleaving the semiconductor along a particular crystal direction to give atomically flat surfaces. The refractive index difference between air and the semiconductor produces mirrors with a reflectivity of typically ~30%. Although this reflectivity is relatively low, and represents a large loss for photons within the cavity, the gain of the system can be increased to compensate for this loss by increasing the length of the cavity. Typical cavity lengths are ~1 mm. Although used for many applications, transverse lasers have a number of disadvantages: the cleaving step is difficult and adds significantly to fabrication costs, the different dimensions normal to and in the plane of the laser result in an output beam with an elliptical cross section, and it is not possible to fabricate the two-dimensional arrays which may be required for optical interconnects in future generations of microprocessors. These problems may be overcome by using a geometry in which the light propagates normal to the plane, to give a vertical cavity surface-emitting laser (VCSEL). In this geometry the available gain is much smaller than for a transverse laser as the light passes only once through a layer of quantum dots or a quantum well, instead of along the layer. Consequently, the losses must be reduced significantly and a mirror reflectivity of ~99.9% is required, a value that cannot be achieved with a simple cleaved semiconductor–air interface. Instead the mirrors are formed by depositing alternating layers of two semiconductors having different refractive indexes. The resultant Bragg stack (Section 3.8.8) has a reflectivity that increases with increasing layer number, allowing a reflectivity >99.9% to be obtained with typically ~20–30 layer pairs. Although the growth sequence of a VCSEL is more complex than that of a transverse laser, the subsequent fabrication is greatly simplified, with the devices formed by lithography and etching to produce circular mesas; there is no cleaving step. Dense two-dimensional arrays can be formed, with the circular cross section of the devices resulting in symmetrical output beams.

3.8.2 Quantum cascade lasers

The lasers discussed in the previous section are based on interband transitions, with lasing occurring between states in the valence and conduction bands, and are classed as bipolar because their operation requires both electrons and holes. Interband lasers based on various semiconductor combinations are able to cover a spectral range extending from the near infrared through to the violet, $\sim 2.0-0.4\,\mu m$. However, lasers operating in the infra-red region between $\sim 2-100\,\mu m$ are more difficult to obtain, due to a lack of semiconductors with sufficiently small band gaps. This spectral region is technologically important as it contains many molecular absorption bands. Quantitative detection of a specific gas is possible if a laser can be tuned to one of the gas absorption bands.

Recently a new type of semiconductor laser has been developed, the quantum cascade laser, which provides emission in the infrared spectral region. This is an intraband, unipolar device that relies only on electrons, which make transitions between confined quantum well conduction band states. Figure 3.47 shows a typical band structure of a quantum cascade laser. The lasing transition occurs between quantum well levels 3 and 2, the separation of which is typically in the range of $\sim 12-350\,meV$, corresponding to wavelengths $\sim 100-3.5\,\mu m$. The lasing wavelength can be varied by selecting a suitable well width. Electrons are injected into the upper lasing level 3 by tunnelling through the left-hand barrier, B, before relaxing to the lower level 2 by emitting a photon. Level 1 is required to efficiently remove electrons from the lower laser level, as a build-up of electrons in this level can prevent the attainment of a population inversion, which requires a greater number of electrons in level 3 than level 2. The gain provided by a single stage is generally not sufficient to overcome the losses of the system, and structures based on typically 25 coupled stages are used, although devices with as many as 100 stages have been reported. Electrons are transported between stages via the miniband of a superlattice, which is designed to inject electrons having the correct energy into the next stage. In this scheme the electrons cascade down through the multiple stages of the structure.

Quantum cascade lasers based on a number of different designs have been operated successfully, with emission wavelengths from 3.5 to 106 μm. However operation at room temperature has only been achieved for the more limited range of 4.5 to 16 μm, and only at 6 and 9.1 μm has continuous room temperature operation been demonstrated; all other devices operate in pulsed mode. Room temperature continuous operation at only two wavelengths reflects efforts to optimise the relevant devices, and similar operation over approximately the entire range 6 to 9.1 μm should be possible. In general, continuous

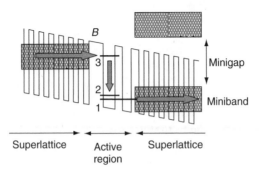

Figure 3.47 The band structure of a quantum cascade laser. Electrons are transported between stages via the miniband of a superlattice, before tunnelling through barrier B into the upper lasing state, state 3. The minigap formed between the minibands of the superlattice prevents electrons tunnelling directly out of state 3. Courtesy of John Cockburn, University of Sheffield

operation at room temperature is made difficult by significant heat generation, a result of the large threshold current densities, which are typically up to two orders of magnitude higher than those of interband lasers. This inefficiency arises because electrons are able to relax between the lasing levels by emitting a phonon instead of a photon, resulting in a large fraction of the electrons being wasted. In addition, electrons may be thermally excited out of the quantum well from the upper lasing state. However, despite their high threshold current densities, quantum cascade lasers are being investigated for a number of applications, including gas monitoring, free space communication systems and medical imaging.

3.8.3 Single-photon sources

It is possible to transmit information using the polarisation state of single photons. Because performing a measurement will disturb the polarisation, attempts to break into the system are readily detectable. This property forms the basis of an intrinsically secure quantum cryptography system, the security being guaranteed by the laws of quantum mechanics. The key component on the transmission side of a quantum cryptography system is the production of a stream of regularly spaced single photons. This can be achieving by using a highly attenuated pulsed laser. However, because of the statistical nature of the number of photons in each pulse, it is necessary to attenuate the laser considerably below an average of one photon per pulse. At higher levels, a significant number of pulses will contain two or more photons, which may allow the integrity of the system to be broken. A maximum of one photon per 100 pulses is typically used, which significantly reduces the data transmission rate.

An alternative method for producing a regular sequence of single photons is based on the emission from a single quantum dot. As discussed in Section 3.6.8, the energy of a photon emitted from a dot containing two excitons is slightly different from that containing a single exciton, a result of the interaction between the charged carriers which perturbs the energy of the system. Hence a dot initially loaded with, for example, five excitons will emit five photons, each having a slightly different energy, as the excitons recombine. By filtering the emission so that only the photon due to the recombination of the final exciton is selected, only one photon will be produced each time the dot is loaded with excitons, irrespective of the initial number of excitons. Single-photon sources have been demonstrated based on the filtered emission from a single self-assembled InAs quantum dot that is periodically loaded with excitons, either optically with a pulsed laser or electrically by placing the dot within a p-i-n structure and applying a pulsed bias voltage. Although the dot produces one photon per cycle, the efficiency with which these photons can escape from the structure is low, due to reflection losses at the air–semiconductor interface. However, the number of external photons per pulse is comparable to that achieved using an attenuated laser and may be increased by the use of a photonic structure (Section 3.8.8). An additional problem is that the selection of one of the emission lines is only possible if it is separated from the neighbouring lines by an energy greater than its line width. At helium temperatures this condition is satisfied but at higher temperatures the emission line width is homogeneously broadened by interaction with phonons, and this broadening limits the use of InAs-based dots to below $\sim$80 K. Higher-temperature operation may be possible with the use of II–VI dots, which have a larger separation between the different emission lines, although ideally the emission energy should be matched to the 1.3 or 1.55 μm transmission maxima of current optical fibres.

3.8.4 Biological tagging

Free-standing colloidal quantum dots (Section 3.5.11) have been developed as a means of tagging biological materials. Different sized dots emit light at different energies or colours when excited with a suitable laser. By preparing a number of dot sizes, various combinations of these sizes can be placed in protective latex beads. When illuminated the beads produce a combination of colours that is characteristic of the dots contained. This produces an optical bar code which can be read with a suitable detection system, allowing the identification of beads containing different dot combinations. The latex beads can be attached to biological materials, allowing the reactions between different materials to be studied or the transport of a particular material through an organism to be monitored. In comparison to traditional tagging that uses organic dyes, colloidal quantum dots offer a greater number of different colours, all of which can be excited with the same laser, in contrast to organic dyes where different dyes require different wavelength excitation. In addition, the emission intensity of colloidal quantum dots is more stable than that of organic dyes following prolonged excitation.

3.8.5 Optical memories

Charge storage in a quantum dot has the potential to provide high-density memory systems, with the possibility of both optical and electrical reading and writing. These systems may provide an alternative to the purely electronic random access memory used in present-day computers. Self-assembled quantum dots may have area densities as high as 1×10^{11} cm^{-2}, a figure that can be increased by at least a factor of ten by using multiple dot layers. A storage density of 1×10^{12} bits/cm^2 considerably exceeds that of conventional electronic memories. For memory applications the inhomogeneous broadening of the dots is an advantage, as potentially it allows individual dots to be addressed by the use of a laser tuned to their specific emission energy.

If both an electron and a hole are trapped in a quantum dot, they will recombine radiatively on a timescale of ∼1 ns, far too short for a practical memory system. A realistic quantum dot memory device therefore requires the storage of only electrons or holes. Figure 3.48 shows the band structure of a prototype quantum dot memory device based on InAs self-assembled quantum dots. Electrons and holes are created by the direct optical excitation of the dots. An applied electric field causes electrons to rapidly escape from the dots, but hole escape is prevented by the large barrier placed to the left of the dot layer. An alternative scheme is to use type II quantum dots; e.g., GaSb dots grown on GaAs, which result in the confinement of one carrier type only. The resultant net positive charge on the dots of Figure 3.48 repels a fraction of the holes from a two-dimensional hole gas (2DHG) placed parallel to the quantum dot layer. This reduced 2DHG density increases its resistance, a change which is monitored by an external

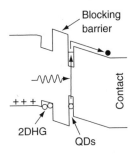

Figure 3.48   Band structure of a quantum dot memory device

electrical circuit. The device is reset by applying a voltage pulse to a surface gate, causing electrons to flow into the dots, where they recombine with the stored holes. At a temperature of 145 K it is possible to store charge in the dots for in excess of 8 h. This storage time decreases at higher temperatures where the holes may be thermally excited out of the dots, although charge storage up to ~200 K has been observed. In current devices a significant number of dots must be excited as the influence of a single charged dot on the conductivity of the 2DHG is too small to measure. This indicates a significant problem with these devices in that although writing to a single dot appears possible, methods for reading and resetting single dots are unclear. At high temperatures, in addition to the loss of carriers from the dots by thermal excitation, the homogeneous broadening of the dot transitions appears to prevent optical writing to individual dots.

3.8.6 Impact of nanotechnology on conventional electronics

The information technology revolution has been driven by the rapid increase in computing power, exemplified by Moore's law which predicts that the number of transistors in microprocessors increases exponentially with time. The initial form of Moore's law, based on integrated circuit development in the mid 1960s, predicted a doubling of transistor number every year. A decade later this was redefined to a doubling every 24 months to account for increased circuit complexity. Figure 3.49 shows a plot of transistor number versus year, demonstrating that Moore's law has held for the previous 30 years or so, with a doubling of transistor number every 26 months. However, increasing the number of transistors on a chip necessitates a commensurate reduction in component size, which also allows the operating frequency to be increased. Over the period covered by Figure 3.49, operating frequencies have increased from ~1 MHz to ~3 GHz. Transistor dimensions will soon enter the nanoscale regime, where their fabrication and operation will cease to follow behaviour scaled down from larger sizes. This section discusses the problems that will be encountered in attempting to continue Moore's law over the next two decades and considers possible solutions, including the application of nanotechnology concepts discussed earlier in this chapter.

The transistors in a microprocessor are of the metal oxide semiconductor field effect transistor (MOSFET) type, the generic structure of which is shown in Figure 3.50. Current is carried between the drain and source contacts by free carriers in the channel:

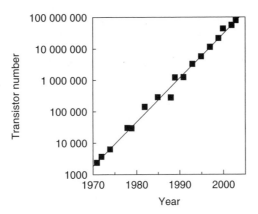

Figure 3.49 Increase in microprocessor transistor number since 1970. The solid line is a fit to the data and indicates a doubling in number every 26 months

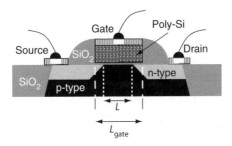

Figure 3.50 Schematic structure of an n-type metal oxide semiconductor field effect transistor (MOSFET)

either electrons in an n-type device (NMOS) or holes in a p-type device (PMOS). A third contact, the gate, is placed above the channel, with a thin SiO_2 insulating layer preventing the flow of current between the gate and the channel. The doping type in the channel immediately below the gate is opposite to the doping type of the source and drain regions. However, application of a suitable gate voltage repels the majority carriers and attracts minority carriers to the channel–insulator interface region, allowing current conduction to occur between the source and the drain. By varying the gate voltage, this conduction path can be switched on or off, giving the basic switching action of the transistor. Neighbouring devices are electrically isolated from each other by insulating SiO_2 regions. Si provides the ideal semiconductor for MOSFET construction as its natural oxide (SiO_2) is highly insulating, it can form very thin layers and the Si–SiO_2 interface has a high structural quality, ensuring that carrier mobilities are not significantly degraded by interface roughness. In practice the metal gate of a MOSFET is replaced with polysilicon to reduce the size of the switching voltage.

The size of a MOSFET is defined by the gate length, and it is this parameter that has been continuously reduced to allow increasing numbers of faster transistors to be fabricated within a given circuit area. Table 3.1 shows the progress required in gate length reduction if Moore's law is to hold until 2018.

Table 3.1 The reduction in microprocessor transistor sized required to continue Moore's law until 2018. Each generation is classified in terms of the technology node number, which specifies the half-pitch distance of related dynamic random access memories (DRAMs), defined as the smallest separation between lines in the first metal layer. The major nodes are in bold. The data is from the International Roadmap for Semiconductors, 2003

	2004	2005	2006	2007	2008	2009	2010	2013	2016	2018
Node number (nm)	**90**	80	70	**65**	57	50	**45**	**32**	**22**	18
Transistor gate length in resist (nm)	53	45	40	35	32	28	25	18	13	10

The first problem encountered in fabricating ever smaller devices is their lithographic definition. As a result of diffraction effects, the minimum feature size that can be imaged in the resist is

$$k\frac{\lambda}{\text{NA}}, \tag{3.14}$$

where λ is the wavelength of the radiation, NA is the numerical aperture of the optical system and k (typically 0.5–0.9) is the technology constant that accounts for non-ideal behaviour of the system. It is difficult to increase the value of the numerical aperture significantly above unity, so for conventional lithography the minimum feature size is typically the same order as the radiation wavelength. Some reduction of the minimum feature size for a given wavelength is possible via refinements of the lithographic process. Examples include the use of phase-shifting techniques, off-axis illumination and immersion lithography. In the first modification a phase difference of π is introduced for light passing through adjacent features in the mask. Consequently, light diffracted into the region between the features interferes destructively, increasing the contrast and hence resolution of the features. In immersion lithography a liquid in contact with the photo-resist decreases the effective wavelength of the incident light, thereby decreasing the minimum feature size that can be formed. For a combination of phase shifting and off-axis illumination it is possible to obtain a k value of 0.3. The use of immersion lithography provides a further decrease equal to the refractive index of the liquid (1.43 for water).

Current commercial high-volume lithography is based on deep ultraviolet (DUV) light produced by excimer lasers, with shorter wavelengths used for successive nodes; examples are 248 and 193 nm for the 130 and 90 nm nodes, respectively. DUV light at 157 nm may be required for the 65 nm node, although recent advances in immersion lithography suggest that 193 nm immersion lithography will provide the dominant technology for the 65 nm node and may be extendable to the 45 nm node. 157 nm immersion lithography may be applicable down to almost the 32 nm node but this seems likely to form the practical limit of DUV lithography; new technologies will be required for future nodes. One possibility is extreme ultraviolet (EUV) lithography which employs wavelengths as short as 13 nm; EUV lithography is predicted to be implemented for the 32 nm node and should be extendable down to the 22 nm node (and possibly the 16 nm node). However, there are many difficult problems with this technology, including generation of the EUV radiation (possibilities include synchrotrons, which are very large and expensive, or laser plasma sources, which are not yet a mature technology for reflective optics), the requirement that EUV is absorbed by all materials, preventing the use of conventional transmissive optics, and that the optics must be flat to within ~0.1 nm to reduce aberrations to an acceptable level. The ultimate limit of photon-based lithography is X-ray proximity lithography which uses wavelengths of ~1 nm. In contrast to DUV and EUV lithography, where the final image size is obtained from a larger mask size by focusing optics, the absence of suitable X-ray optics requires the use of a mask:image ratio of 1:1. Here the mask is placed slightly above the surface of the semiconductor wafer, projecting an image directly on to the photoresist. Because of the separation between the mask and the wafer, the minimum defined feature size is significantly greater than the radiation wavelength, although features of order 10 nm may be possible. Problems that need to be overcome with X-ray lithography include the generation of X-rays having sufficient intensity, the production of the 1:1 scale masks and the prevention of mask damage by the high-energy photons.

An alternative to photon lithography is electron beam (e-beam) lithography. The electron wavelength is controlled by the accelerating potential, resulting in the possibility of defining feature sizes below 10 nm. Conventional e-beam systems use a serial approach in which the pattern is written by a single electron beam scanned over the surface. However, this approach is too slow for mass volume production. Instead systems consisting of multiple beams or the projection of a broad beam through a suitable mask are being developed, although they are still a long way from commercial systems.

As the feature size becomes ever smaller, fluctuations due to the polymer unit size of the resist become significant. Such fluctuations, which are transferred to the fabricated devices, may impair carrier mobilities, requiring the development of new resists. The ultimate fabrication limit appears to be provided by AFM and related techniques, which allow individual atoms to be positioned on a surface. However, these techniques are relatively slow and it is difficult to see how they can be scaled up to allow mass volume production.

Even if increasingly smaller size MOSFETs can be fabricated, their electrical properties will eventually deviate significantly from those of larger devices. MOSFET technology has generally followed a constant field scaling, in which the physical dimensions and operating voltage are reduced by the same factor, with the substrate doping increased by a similar factor to maintain an unchanged electric field pattern within the channel. This scaling also requires the gate oxide thickness d_{ox} to be reduced, and the relationship $L_{gate} \approx 45 d_{ox}$ is typically used. Although SiO_2 thicknesses of 1.5 nm or less can be formed (1.2 nm is used for the present 90 nm node), below ~1.5 nm significant quantum mechanical tunnelling of carriers through the gate oxide occurs, and this leakage current adds to the power consumption of the system. A solution to this problem requires materials with a higher dielectric permittivity than SiO_2, allowing thicker gate insulating layers to be used. A short-term solution is provided by the use of SiON, which should provide oxide thicknesses equivalent to slightly less than 1 nm of SiO_2. In the longer term, more exotic materials are needed, such as ZrO_2 and HfO_2, although more development is required before commercial production is possible. Similar high-permittivity materials will also be required for future generations of dynamic random access memories (DRAMs). These devices are based on the storage of charge by arrays of tiny capacitors. As the capacitor size decreases, the use of a higher-permittivity material allows a given capacitance, and hence stored charge, to be maintained.

As the channel length is reduced, the channel potential distribution becomes strongly distorted in the vicinity of the drain region, due to the drain voltage. In addition, charge may leak between the drain and the source in the region away from the gate–channel interface, increasing the current in the off state of the transistor and greatly adding to the power consumption of the device. Modified structures are required to overcome these effects. Possibilities include devices with an underlying SiO_2 layer – Si on insulator (SOI) – or with material directly under the channel removed – silicon on nothing (SON). In addition, devices with dual gates and with either a horizontal or vertical channel are being investigated. The vertical channel structure has the advantage that the channel length is defined by an epitaxial growth step, rather than lithography, although the subsequent positioning of the dual gates is very difficult. Higher operating frequencies may also be obtained by increasing the mobility of carriers in the channel. This can be achieved by the use of strained Si or SiGe, both of which have a higher carrier mobility than unstrained Si. However, the use of these materials requires an additional, non-standard step in the fabrication process.

Other problems that will eventually arise as transistor sizes are reduced include limits to doping densities, resulting from the solubility limits of the dopant atoms, and the quantisation of energy levels and charge. Power consumption, which increases with increasing frequency but is also affected by current leakage, is a serious problem to be overcome. Current microprocessors generate heat with a power density equivalent to that of a hotplate. By 2010, at current trends, the power density will be comparable to that produced by a rocket nozzle! Power dissipation problems are worse still for devices fabricated on SiO_2 due to this material's poor thermal conductivity. The ultimate size limit of inorganic semiconductor devices may be single-electron transistors, discussed in the next section. Further size reductions are likely to require radically different approaches, such as transistors based on single molecules (Section 8.8.1.2).

Finally, as transistor switching frequencies increase, the speed with which signals propagate between different parts of a microprocessor must also increase. The present generation of microprocessors use copper interconnects, which give reduced propagation delays compared to aluminium which has been traditionally used. Surrounding the interconnect wires with insulators having a lower relative permittivity than the currently used fluorine-doped SiO_2 will reduce propagation delays further. In the longer term, critical direct electrical connections are likely to be replaced with optical or radio frequency interconnects.

At the time of writing (early 2004) state-of-the-art production is switching to the 90 nm node, based on a combination of 248 and 193 nm DUV lithography. This technology, which includes copper interconnects with low relative permittivity dielectrics, strained silicon and 1.2 nm thick gate oxide layers, is used for the production of microprocessors with 77 million transistors of gate lengths $\sim$50 nm and operating frequencies of 3.4 GHz. The same technology has been used to produce 512 Mbit static random access memory (SRAM) chips with 330 million transistors. At a research level, transistors with gate lengths as small as 10 nm have been fabricated, although these are of a conventional structure, so for gate lengths below $\sim$30 nm their performance is degraded due to the reasons discussed earlier. MOSFET operating frequencies in excess of 1 THz (1000 GHz) have been reported. Intel has recently announced a new transistor design that incorporates a ZrO_2 high-permittivity gate oxide, a thin Si channel on SiO_2 to reduce drain–source leakage, and modified drain and source contact regions to minimise series resistances. It is envisaged that this structure will eventually be suitable for mass production of devices operating above 1 THz. Although these devices are still some way from mass production, they demonstrate that Si-based electronics will provide the dominant technology for the foreseeable future. If the trends described in the International Roadmap for Semiconductors 2003 are followed, then by 2016 microprocessors will contain approximately 9 billion transistors with 10 nm gate lengths and operating at a frequency of 28 GHz.

3.8.7 Coulomb blockade devices

In section 3.6.4 we described how the discrete nature of electronic charge, coupled with the small capacitance of a quantum dot, results in the ability to add and remove carriers being a function of the charged state of the dot. For the electrostatically defined Coulomb blockade device shown in Figure 3.30, the current between the reservoirs is controlled by the gate voltage in a manner similar to transistor action. However, in this device, transistor action occurs for the transport of only one electron at a time, although the number of electrons held on the dot may be significantly greater than one. The device can therefore be thought of as a single-electron transistor and, as such, represents the ultimate limit in transistor scaling given that a single electron represents the smallest possible unit of charge. In addition, Coulomb blockade also provides the possibility for single-electron memory cells. A number of schemes have been proposed but, in the simplest form, writing involves the transfer of an additional electron to the dot, with reading relying on the modification of the gate voltage by the presence of the additional electron.

Coulomb blockade effects may be observed in any conducting system of suitably small size, although silicon-based devices are required for compatibility with existing electronics. The observation of Coulomb blockade requires a charging energy considerably greater than the thermal energy, which for silicon-based devices operating at room temperature equates to a dot size of less than 10 nm. This size is well below what is achievable with present lithographic techniques, so devices fabricated in this way can only operate at low temperatures. Alternative fabrication possibilities include dots formed by thickness fluctuations in a thin silicon layer after treatment with an alkali-based solution, small silicon crystallites in a polysilicon layer and germanium self-assembled quantum dots. The first two techniques have both demonstrated room temperature memory operation, although the inherent randomness of the dot formation may make scaling up to large arrays difficult.

For the electrostatically defined dot shown schematically in the inset to Figure 3.30 it is possible to separately control the heights of the two tunnelling barriers via their defining gate voltages. If the height of the left-hand barrier is initially set low, with the height of the right-hand barrier set high, a single electron may tunnel on to the dot but is prevented from leaving the dot. In addition, a second electron is prevented from tunnelling on to the dot because the first electron raises the potential of the dot. If the heights of the barriers are now reversed, the electron can tunnel out of the dot, with the overall effect of these two steps being the transfer of one electron between the reservoirs. This process can be repeated continuously by applying AC voltages, with a suitable phase shift, to the gates. If the applied frequency is f then f electrons per second will move between the reservoirs, giving a current $I = fe$. Such a device has potential as a current standard.

3.8.8 Photonic structures

The nanostructures discussed so far modify the electronic properties of the underlying semiconductor. However, in structures or devices that produce or detect light there is also interest in modifying the properties of the photons by creating a photonic structure. The simplest photonic structure uses a one-dimensional optical cavity to confine photons, and this can be combined with a quantum well so that both photons and electrons are confined. An optical cavity is created with two high-reflectivity parallel mirrors, separated by a multiple of the wavelength of the photons to be confined. In a semiconductor system the mirrors are formed by growing a Bragg stack, which consists of a repeated sequence of alternating semiconductor layers. The refractive index change between the semiconductors results in only a small reflectivity, however this is enhanced by the repeated nature of the stack if the thickness of each layer equals a quarter of the wavelength of the photons. By using a Bragg stack consisting of $\gtrsim 20$ layers, a reflectivity $\gtrsim 99\%$ is possible. A similar structure is used to form vertical cavity surface-emitting laser (VCSEL) devices (Section 3.8.1).

The inset to Figure 3.51 shows a schematic diagram of a one-dimensional optical cavity, known as a microcavity. Two Bragg stacks and a GaAs layer of thickness equal to the photon wavelength form the optical cavity, which confines photons travelling along the growth direction. A quantum well placed at the centre of the cavity provides confinement of the excitons. The curves in Figure 3.51 show the unperturbed energies of the confined exciton and photon as a function of temperature, which is used to vary their relative energy. The energy of the exciton, which is related to the band gap of the semiconductor, is relatively temperature dependent; the energy of the photon, given by the thickness of the cavity, is less temperature dependent. At resonance the exciton and photon couple together to form a state known as a polariton. This coupling is demonstrated by the experimental data points, which deviate from the calculated, unperturbed

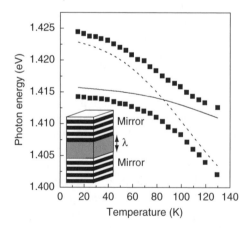

Figure 3.51 Solid symbols show the temperature dependence of the energies of the quantum well exciton and cavity mode (photon) of a GaAs–AlGaAs microcavity, which exhibit an anti-crossing around 90 K. The lines show the calculated exciton (dashed line) and cavity (solid line) energies in the absence of coupling. The inset shows the physical structure of the microcavity. Data courtesy of Adam Armitage, University of Sheffield

energies to produce an anti-crossing of the exciton and photon states. This mixed exciton–photon polariton state has a number of novel physical properties that result from an in-plane wave-vector energy dependence which is very different from that of the uncoupled photons or excitons. Possible applications include extremely low threshold lasers. Additional applications result from the use of microcavities with very small diameters ($\lesssim 1\,\mu\text{m}$). These allow the radiative lifetime of the exciton to be decreased and also increase the fraction of photons that are able to escape from the device. Microcavities containing single quantum dots may provide the efficiency required for practical single-photon sources, as discussed in Section 3.8.3.

Confinement of photons in more than one dimension may be achieved by the use of a periodic refractive index array. A two-dimensional structure can be created by the use of electron beam or DUV lithography to define a periodic pattern on a semiconductor wafer, which is subsequently etched to form a series of pillars and holes. The periodic modulation of the refractive index produces a photonic band structure, with bands of allowed photon states separated by band gaps, in a similar manner to the electronic band structure of a solid which arises from the periodic arrangement of the atoms. Figure 3.52 shows images of a two-dimensional photonic structure that consists of periodic arrays of holes and a series of unpatterned regions, which act as ultrasmall cavities within which photons are confined by the surrounding periodic structure. Photonic structures allow the direction of light propagation to be controlled (bending around corners is possible), permit the control of spontaneously emitted light (e.g.,

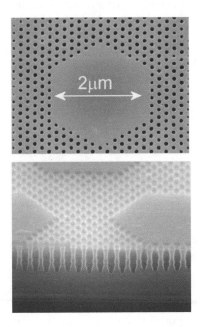

Figure 3.52 Two-dimensional photonic structure consisting of periodic arrays of holes and a series of unpatterned regions which act as ultra-small cavities within which photons are confined by the surrounding photonic structure. The upper image is recorded via the upper surface of the structure, the lower image is of a cleaved edge. Data courtesy of Alan Bristow, University of Sheffield

suppressing it if it occurs within a band gap), provide strong non-linear effects, modify the dispersion of light (including reducing the propagation speed), and allow the fabrication of very small lasers.

The structure shown in Figure 3.52, an example of a micro laser, provides photon confinement along the growth direction by using a layer of a large refractive index semiconductor sandwiched between lower refractive index layers, forming a planar waveguide. The extension to true three-dimensional periodic structures is desirable, but is difficult to achieve using lithographic techniques. Although photonic structures may be fabricated in a range of materials, semiconductor-based structures can be directly integrated with conventional electro-optical devices, and quantum dots may be incorporated directly into the structure to provide efficient light-emitting centres.

3.9 SUMMARY AND OUTLOOK

The importance of inorganic semiconductor nanostructures arises from the opportunities they provide for the controllable modification of the electronic and optical properties of the underlying semiconductors. Consequently, they are of interest for both the study of novel physical phenomena in reduced dimensionality systems, and for a wide range of electronic and electro-optical device applications. Quantum wells are now used in the majority of semiconductor lasers and modulation doping is used in specialist transistors, particularly those required for high-frequency or low-noise applications. The quantum Hall effect provides a resistance standard. It is now possible to purchase lasers based on self-assembled quantum dots and also colloidal quantum dots for biological tagging. Quantum cascade lasers are likely to be used in commercial systems within the next few years.

However, many applications require further developments, particularly in fabrication techniques. For example, it is desirable to increase energy level separations for high-temperature applications, particularly for holes which generally have a larger effective mass. This requires a reduction in the size of nanostructures. Greater control of the self-assembly technique is required to obtain improved uniformity and to be able to position quantum dots at specific positions, possibly by growing on prepatterned surfaces. For some applications the presence of the wetting layer is undesirable and techniques for its removal need to be developed. The ability to fabricate nanostructures emitting or absorbing light at a range of specific wavelengths, from the infrared through the visible to ultraviolet spectral regions, will open up a number of applications. Over the past few years, increasing progress has been made in the study of single nanostructures, and this should lead to a number of applications based on individual quantum dots, including single-electron transistors and memories, and single-photon sources. There is considerable interest in using the spin properties of electrons rather than their charge, a field referred to as spintronics. Nanostructures allow the spins of electrons and holes to be manipulated, and mechanisms which relax the spin may be inhibited in quantum dots, resulting in extremely long spin coherence times. Integration with other areas of nanotechnology is likely to occur, such as the inclusion of a small number of magnetic ions in a quantum dot to give a magneto-optical zero-dimensional structure, and the combination of Si-based electronics with polymer or biological systems. Finally, a convergence will occur between traditional technologies used for the manufacture of microprocessors, memories and related electronic systems and recently developed nano-technologies, driven by the continued reduction of device dimensions in state-of-the-art electronic circuits.

BIBLIOGRAPHY

Basic semiconductor properties are covered in many solid-state physics textbooks, for example *Introduction to Solid State Physics* (7th edn) by C. Kittel (Wiley, Chichester, 1996) and *Solid State Physics* by J. R. Hook and H. E. Hall (Wiley, Chichester, 1991). Textbooks dealing specifically with semiconductors include *Fundamentals of Semicon-ductors* by P. Y. Yu and M. Cardona (Springer, Berlin, 2001) and *The Physics of Semiconductors with Applications to Optoelectronic Devices* by K. F. Brennan (CUP, Cambridge, 1999). Brennan's book also covers the operation of a number of conven-tional semiconductor electronic and electro-optical devices, as do *Semiconductor Devices: Basic Principles* by J. Singh (Wiley, Chichester, 2001) and *Semiconductor Optoelectronic Devices* (2nd edn) by P. Bhattacharya (Prentice Hall, New York, 1997). Two recent textbooks, *Band Theory and Electronic Properties of Solids* by J. Singleton (OUP, Oxford, 2001) and *Optical Properties of Solids* by A. M. Fox (OUP, Oxford, 2001), provide good coverage of the electronic and optical properties of bulk semiconductors and semiconductor nanostructures. Singleton's book contains a clear discussion of the quantum Hall effect. *Semiconductor Optics* by C. F. Klingshirn (Springer, Berlin, 1997) provides a comprehensive discussion of the optical properties of semiconductors with a more limited discussion of quantum well structures.

A number of books deal specifically with various aspects of inorganic semiconductor nanostructures. General texts include *Physics of Semiconductors and their Heterostruc-tures* by J. Singh (McGraw-Hill, New York, 1993), *Quantum Semiconductor Structures: Fundamentals and Applications* by C. Weisbuch and B. Vinter (Academic Press, London, 1991) and *Low-Dimensional Semiconductors: Materials, Physics, Technology, Devices* by M. J. Kelly (OUP, Oxford, 1995). Kelly's book contains a good discussion of a range of device applications. The self-assembly technique for the fabrication of quantum dots is covered in considerable depth in *Quantum Dot Heterostructures* by D. Bimberg, M. Grundmann and N. N. Ledentsov (Wiley, Chichester, 1999) and *Heterojunction Band Discontinuities: Physics and Device Applications* edited by F. Capasso and G. Margaritondo (North-Holland, Amsterdam, 1987) covers the calculation and mea-surement of band offsets, in addition to the general properties and applications of quantum well systems. The *Physics of Low-Dimensional Semiconductors: An Introduction* by J. H. Davies (CUP, Cambridge, 1998), *Wave Mechanics Applied to Semiconductor Heterostructures* by G. Bastard (Halsted Press, Paris, 1988) and *Quantum Wells, Wires and Dots: Theoretical and Computational Physics* by P Harrison (Wiley, Chichester, 1999) all provide a mathematically based treatment of inorganic semiconductor nanostructures.

Nanoelectronics and Information Technology: Advanced Electronic Materials and Devices edited by R. Waser (Wiley VCH, Weinheim, 2003) covers the present status of silicon MOSFETs, memory devices and lithographic techniques, and discusses possible future advances and alternative, non-silicon-based technologies. Information concerning future trends in microprocessor development can be obtained from the International Technology Roadmap for Semiconductors. This is updated yearly and the current edition can be found at the Semiconductor Manufacturing Technology (SEMATECH) website, www.sematech.org/. Finally, the silicon research section of the Intel website, www.intel.com/research/silicon/, contains many articles relevant to current and future microprocessor development.

4

Nanomagnetic materials and devices

4.1 MAGNETISM

Nanotechnology as a term applied in the field of magnetism may be regarded as either relatively mature, or as a subject at a nascent stage in its development. Many of the permanent magnets in common use today, in devices ranging from high-efficiency motors to fridge magnets, have properties dictated by the physical nanostructure of the material, and the subtle and complex magnetic interactions that this produces. The use of nanostructured magnets has already enabled very significant savings in energy consumption and weight for motors, which contributes strongly to the green economy. Data storage density has increased with a compound growth rate of 60%, giving (in 2004) disk drives for the mass PC market in excess of 400 Gbyte. The internet, image handling in cameras and data storage drive the demand for increases in data storage capacity. Fujitsu recently demonstrated (spring 2002) 100 Gbit/in^2 data storage capacity in a form of conventional longitudinal recording media. This requires an effective 'bit size' of 80 nm square. Figure 4.1 is a reproduction of an IBM road map (dated 2000) for the projected development of storage capacity. There are fundamental physical barriers to be overcome (see the superparamagnetic limit discussed in Section 4.1.4) in order to keep to this road map for increasing capacity with current technologies, and a paradigm shift will be needed which will be nanotechnology led.

We begin this chapter with a brief introduction to the *language* and *technical terms* involved in the study of nanomagnetism. There are many suitable texts which may be consulted to provide a more extensive coverage of the background. These are listed in the bibliography at the end of the chapter.

4.1.1 Magnetostatics

In common usage, magnetic materials are those that exist in a state of permanent magnetization without the need to apply a field. For all permanent magnets there is a

Nanoscale Science and Technology   Edited by R. W. Kelsall, I. W. Hamley and M. Geoghegan
© 2005 John Wiley & Sons, Ltd

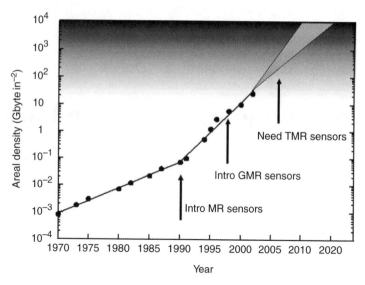

Figure 4.1 Schematic data storage road map drawn up from various internet sources. The introduction of new recording head technologies (magnetoresistance plus giant and tunnelling magnetoresistance) is marked by the arrows

north pole (N) and a south pole (S), and lines of magnetic field pass from N to S. If we begin with the simple premise that the net properties of a magnet are some form of sum over a large number of atomic or electronic magnets (referred to as magnetic dipole moments), we can separate magnetic materials into three categories.

4.1.2 Diamagnetism, paramagnetism and ferromagnetism

There are three categories of magnetism that we need to consider: diamagnetism, paramagnetism and ferromagnetism. Diamagnetism is a fundamental property of all atoms (molecules), and the magnetization is very small and opposed to the applied magnetic field direction. Many materials exhibit paramagnetism, where a magnetization develops parallel to the applied magnetic field as the field is increased from zero, but again the strength of the magnetization is small. In the language of the physicist, ferromagnetism is the property of those materials which are intrinsically magnetically ordered and which develop spontaneous magnetization without the need to apply a field. The ordering mechanism is the quantum mechanical exchange interaction. It is this final category of magnetic material that will concern us in this chapter. A variation on ferromagnetism is ferrimagnetism, where different atoms possess different moment strengths but there is still an ordered state below a certain critical temperature.

We will now define some key terms. The magnetic induction **B** has the units of tesla (T). To give a scale, the horizontal component of the earth's magnetic flux density is approximately $20\,\mu T$ in London; stray fields from the cerebral cortex, are about $10\,fT$. The magnetic field strength **H** can be defined by

$$\mathbf{B} = \mu_0 \mathbf{H}, \tag{4.1}$$

where $\mu_0 = 4\pi \times 10^{-7} \mathrm{H\,m^{-1}}$ is the permeability of free space. This gives the horizontal component of the earth's magnetic field strength as approximately $16\,\mathrm{A\,m^{-1}}$ in London. The flux $\Phi = BA$ can be defined, where A is a cross-sectional area. Flux has the units of weber (W).

If a ferromagnetic material is now placed in a field $\mathbf{H}$, Equation (4.1) becomes

$$\mathbf{B} = \mu_0(\mathbf{H} + \mathbf{M}) = \mu_0(\mathbf{H} + \chi\mathbf{H}) = \mu_r\mu_0\mathbf{H}, \qquad (4.2)$$

where $\mathbf{M}$ is the magnetization of the sample (the magnetic dipole moment per unit volume). We define χ as the susceptibility of the magnetic material, and $\mu_r = 1 + \chi$ as the relative permeability of the material (both χ and μ are dimensionless). Table 4.1 summarizes this classification scheme, and gives values for typical suscep- tibilities in each category.

If we take *any* ferromagnet and increase the field from zero to some peak value, decrease the field back through zero to an equal and opposite value and then return again to the original peak value, we trace out the hysteresis loop (Figure 4.2). The name 'hysteresis' comes from the Greek for loss. Energy is dissipated in traversing the loop. This energy loss is primarily manifested as heat. There are several key points on the loop. The saturation magnetization $\mathbf{M}_s$ is the maximum value that the magnetic dipole moment per unit volume can take in the direction of an applied magnetic field. In this state all the contributing atomic (electronic) moments are aligned in the field direction.

Table 4.1   Classification of magnetic materials by susceptibility

Diamagnet ($\chi < 0$)	Paramagnet ($\chi > 0$)	Ferromagnet ($\chi \gg 0$)
Cu: -0.11×10^{-5}	Al: 0.82×10^{-5}	Fe: $> 10^2$
Au: -0.19×10^{-5}	Ca: 1.40×10^{-5}	
Pb: -0.18×10^{-5}	Ta: 1.10×10^{-5}	

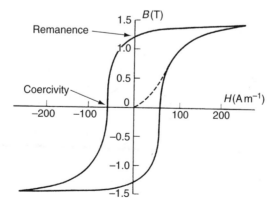

Figure 4.2   A schematic diagram of a magnetization (or hysteresis) loop. Remanence is the magnetization remaining when the field is reduced to zero from that required to saturate the sample. Coercivity is the field required to bring the magnetization to zero from remanence

The remanent magnetization $\mathbf{M}_r$ is the magnetization remaining in the sample when the applied field is reduced to zero from the value which saturated the sample. The coercivity $\mathbf{H}_c$ is the field required to reduce the magnetization to zero from saturation (Figure 4.2), and is applied in the opposite direction to the original saturating field. If the magnetization is reduced to zero from less than its saturating value then $\mathbf{H}_c$ is termed the coercive field (always less than the coercivity). The ratio M_r/M_s and the area enclosed in the loop are important in determining the applicability of a given ferromagnet (Section 4.2).

4.1.3 Magnetic anisotropy

If we take a *single crystal* of a ferromagnet, and apply a magnetic field along different crystallographic directions, the magnetization varies with field depending on the chosen direction for the applied field (Figure 4.3). This phenomenon goes under the generic title of magnetic anisotropy. For Fe, as shown in Figure 4.3, $\langle 100 \rangle$ are the easy directions (low field required to saturate), $\langle 110 \rangle$ and $\langle 111 \rangle$ directions are hard (high field to saturate). This form of anisotropy is termed magnetocrystalline anisotropy and is closely related to the detailed electronic structure of the crystal.

Fe is a body-centred cubic solid; that is, the Fe atoms arrange on the cube corners and at each cube centre in a regular three-dimensional pattern. The $\langle 100 \rangle$ directions are along a cube edge, the $\langle 110 \rangle$ directions along a face diagonal and the $\langle 111 \rangle$ directions along a cube body diagonal.

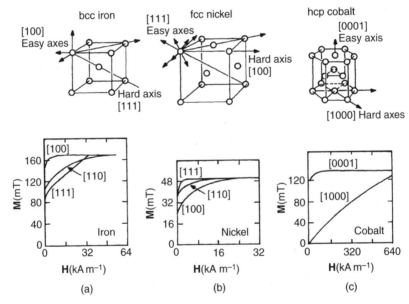

Figure 4.3 The crystal structure and magnetization loops For Fe, Ni and Co, demonstrating the anisotropic nature of the magnetization process. After R. C. O'Handley, *Modern Magnetic Materials: Principles and Applications*, Wiley, New York, 2000, p. 180. Copyright 2000 John Wiley & Sons Inc.

In certain crystals such as cobalt, which has a hexagonal closed-packed (hcp) crystal structure, there is only one easy axis (in this case perpendicular to the close-packed plane; see Figure 4.10 for another example), leading to uniaxial anisotropy. The magnetocrystalline anisotropy energy is a minimum when the magnetization lies along this easy axis. The magnetocrystalline anisotropy energy is a maximum when the magnetization is at 90° to the easy axis. We can express this anisotropy energy E_a by expanding in terms of material-dependent anisotropy constants K_i using a series of powers of $\sin^2\theta$ (to reflect the symmetry), where θ is the angle between the easy axis and the magnetization:

$$E_a = K_0 + K_1 \sin^2\theta + K_2 \sin^4\theta + \ldots \tag{4.3}$$

In all expressions for anisotropy energy there is a constant term K_0. This term is ignored, as in practice it is changes in energy due to moment rotations that are of interest. For a given crystal class (e.g., cubic) direction cosines (α_i) related to the cube edges are used to define the direction of magnetization. The anisotropy energy is then expanded in a polynomial series involving the α_i. By symmetry arguments (for details see books listed in the bibliography)

$$E_a = K_1\left(\alpha_1^2\alpha_2^2 + \alpha_2^2\alpha_3^2 + \alpha_1^2\alpha_3^2\right) + K_2\left(\alpha_1^2\alpha_2^2\alpha_3^2\right) + \ldots \tag{4.4}$$

There can also be a magnetic anisotropy associated with the shape of a magnetic material; it has its origins in the demagnetizing field H_d within the material. Magnetic field lines always run from north poles to south poles, and so *inside* a magnet there is a field in the opposite direction to the magnetization induced by an external field. This reduces the net field experienced by the magnet.

We now rewrite Equation (4.2) to take account of this field internal to the magnet and opposite in direction to the magnetization

$$\mathbf{B} = \mu_0\mathbf{H}_a + \mu_0\mathbf{M} - \mu_0\mathbf{H}_d. \tag{4.5}$$

In the absence of an applied field, $\mathbf{H}_a = \mathbf{0}$, the magnetostatic energy per unit volume, E_s, is given by

$$E_s = \frac{1}{2}\mathbf{B}\cdot\mathbf{M}, \qquad \mathbf{B}_i = \mu_0\mathbf{H}_d = \mu_0 N_d\mathbf{M}, \tag{4.6}$$

where $0 \leq N_d \leq 1$ is the geometrically defined demagnetizing factor. The factor $1/2$ is to ensure that each dipole interaction with the field is counted only once. For a prolate spheroid, taking the major axis (c) as the easy axis, we have

$$E_s = \frac{1}{2}\left((M\cos\theta)^2 N_c + (M\sin\theta)^2 N_a\right)\mu_0 = \frac{1}{2}\mu_0 N_c M^2 + \frac{1}{2}\mu_0(N_a - N_c)M^2\sin^2\theta. \tag{4.7}$$

This should be compared with Equation (4.3), and allows us to define K_s as the shape anisotropy constant. In this case:

$$K_s = \frac{1}{2}\mu_0(N_a - N_c)M^2. \tag{4.8}$$

For a sphere $N_a = N_c$ and $K_s = 0$ and there are no easy directions due to shape anisotropy.

It is now obvious that as physical dimensions are decreased, and we start to produce magnetic materials with one or more dimensions on the nanometre scale, the possibility of significant anisotropy due to shape increases. It is very important to remember that when more than one type of anisotropy is present they do not add vectorially, rather one type of anisotropy will be dominant, determining the local direction of magnetization in the absence of a magnetic field.

When a ferromagnet undergoes a change in the state of magnetization there may be an accompanying change in physical dimensions (the Joule effect). A fundamental material constant can be defined, the saturation magnetostriction constant λ_s, which decreases with temperature, approaching zero at $T = T_c$. T_c is the Curie temperature, and is the temperature above which thermal energy is sufficient to overcome the moment alignment produced by the exchange interaction. The measured strain depends on the crystallographic direction for the measurement. When the distance, r, between two atomic magnetic moments can vary, the interaction energy must in general be a function of r and ϕ, where ϕ is the angle between the magnetic moments (assumed parallel due to exchange) and the bond axis. The crystal lattice will therefore be deformed when ferromagnetism is established ($T < T_c$); this is the spontaneous magnetostriction. As the angle ϕ is varied (say by applying a field at an angle to an easy direction) the interaction energy may also change (this is magnetic anisotropy) and there may be an accompanying change in r to keep the total energy a minimum. This is termed field-induced magnetostriction. There is a fundamental link between anisotropy and magnetostriction; magnetostriction is related to the strain derivative of the anisotropy energy.

A phenomenological model can be produced which relates the field-induced strain to fundamental magnetostriction constants and angles in a crystal. It can be shown (see bibliography) that for a cubic crystal:

$$\lambda_{si} = \frac{3}{2}\lambda_{100}\left(\alpha_1^2\beta_1^2 + \alpha_2^2\beta_2^2 + \alpha_3^2\beta_3^2 - \frac{1}{3}\right) + 3\lambda_{111}(\alpha_1\alpha_2\beta_1\beta_2 + \alpha_2\alpha_3\beta_2\beta_3 + \alpha_3\alpha_1\beta_3\beta_1).$$

$$(4.9)$$

λ_{si} is the strain from the ideal demagnetized state to saturation, α_i are the direction cosines defining the applied field direction, and β_i are the direction cosines defining the strain direction. To determine the strain in the same direction as the magnetization, put $\alpha_1\alpha_2\alpha_3 = \beta_1\beta_2\beta_3$ in Equation (4.9) to give:

$$\lambda_{si} = \lambda_{100} + 3(\lambda_{111} - \lambda_{100})\left(\alpha_1^2\alpha_2^2 + \alpha_2^2\alpha_3^2 + \alpha_3^2\alpha_1^2\right). \tag{4.10}$$

λ_{si} is uniquely defined and is a fundamental property (however the demagnetized state is hard to uniquely define in theory or in practice).

Just as the presence of magnetostriction may cause a dimensional change as the magnetization varies in an applied field, so an applied strain may cause a change in magnetization (inverse Joule or Villari effect). An easy direction may be defined associated with the applied strain (or stress).

When one or more dimensions of a magnetic material is on the nanometre scale, the surface area to volume ratio must increase. The atomic relaxations at free surfaces and the lattice strains at interfaces may thus have a profound effect on anisotropy and magnetostriction in nanoscale magnetic materials. More will be said about this later.

4.1.4 Domains and domain walls

If elements such as iron(Fe) are ferromagnetic, that is all the moments are aligned as a result of the exchange force, why is it that not all pieces of iron behave like permanent magnets? Consider a block of ferromagnet magnetized to saturation; from earlier we have that the magnetostatic energy E_s for a sample volume V is

$$E_s = \frac{1}{2}\mu_0 N_d M_s^2 V. \tag{4.11}$$

For Fe, $M_s = 1.73 \times 10^6\,\mathrm{A\,m^{-1}}$; taking $N_d = 1/3$ and $V = 10^{-6}\,\mathrm{m^3}$, $E_s = 0.627\,\mathrm{J}$ in this saturated state. If the sample is allowed to divide into two regions (domains) of equal volume but opposite directions of magnetization, then

$$E \approx \frac{1}{2}E_s + \gamma_w A_w, \tag{4.12}$$

where γ_w is the energy per unit area of the 'wall' separating the domains and A_w is the wall area. The factor 1/2 comes in as before, to avoid counting interactions twice. For Fe: $\gamma_w = 1.24 \times 10^{-3}\,\mathrm{J\,m^{-2}}$, and $E = 0.314\,\mathrm{J}$. Division into further domains to reduce E will continue until it costs more energy to create another wall than is gained by division:

$$E \approx \frac{1}{n}E_s + (n-1)\gamma_w A_w, \tag{4.13}$$

where n is the number of domains in the sample. Figure 4.4 is a schematic diagram of what is happening. E can be reduced further by the formation of closure domains,

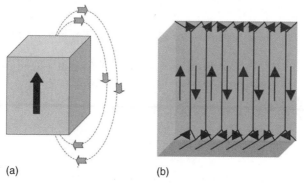

      (a)   (b)

Figure 4.4 (a) A single-domain ferromagnetic particle, (b) a multidomain ferromagnetic particle with closure domains on the top and bottom surfaces which remove the presence of free poles from the sample surface. In a domain, all the moments are aligned (i.e., locally the sample is magnetized to saturation by the exchange interaction). A neighbouring domain may be magnetized in a different direction (but still locally to saturation, and usually with the moments in an easy direction).

We need to consider how the direction of magnetization varies across the boundary between two domains. This reduces the question to how local anisotropies and exchange interactions compete to determine the direction of the local magnetic moment. We obtain a basic picture from energy considerations. If two neighbouring magnetic moments are not aligned with each other, some exchange energy is stored:

$$E_{ex} = -2JS^2 \cos \phi, \tag{4.14}$$

where J is the exchange integral and S is the spin quantum number. If ϕ is small (justified later) then:

$$\cos \phi = 1 - \frac{\phi^2}{2} + \frac{\phi^4}{24} - \cdots \quad \text{and} \quad E_{ex} = 2JS^2 \frac{\phi^2}{2} - 2JS^2. \tag{4.15}$$

If the domain wall is abrupt (i.e., the moments rotate from one domain direction to another from one lattice site to the next) then a large exchange energy is stored. Exchange energy considerations taken on their own argue in favour of a wide domain wall with the moments rotating slowly from site to site. If the rate of moment rotation across the wall is slow, the *extra* energy stored due to the presence of the domain wall is

$$E_{ex} = JS^2 \phi^2 \text{ per spin pair.} \tag{4.16}$$

It is more useful to know the exchange energy per unit wall area, and to estimate this energy we consider a simple cubic lattice of side a and a plane domain wall parallel to a cube face of thickness N atoms. There are $1/a^2$ rows of N atoms per unit wall area. If γ_{ex} is the energy per unit wall area:

$$\gamma_{ex} = \frac{JS^2 N\phi^2}{a^2} = \frac{JS^2 \pi^2}{Na^2} \tag{4.17}$$

for a 180° domain wall (i.e., $\phi_{total} = \pi$ radians) and equal increments in ϕ from site to site (i.e., $\phi = \pi/N$).

In zero applied field, the moments within a domain usually lie in easy directions. In the domain wall, the moments are usually directed away from easy directions. Hence anisotropy energy is stored in the region of the domain wall. Also, the presence of anisotropy energy favours a narrow domain wall (few moments out of easy directions):

$$E_{an} = K_1 \times \text{volume of the wall.} \tag{4.18}$$

Let γ_{an} be the anisotropy energy stored per unit wall area:

$$\gamma_{an} = K_1 Na. \tag{4.19}$$

The total wall energy per unit area, γ, can then be written as:

$$\gamma = \gamma_{ex} + \gamma_{an} = \frac{JS^2\pi^2}{Na^2} + K_1 Na. \tag{4.20}$$

The domain wall thickness may be de defined by $\delta = Na$, and then minimization of γ with respect to δ yields

$$\frac{d\gamma}{d\delta} = -\frac{JS^2\pi^2}{\delta^2 a} + K_1 = 0, \quad \text{where} \quad \delta = \left(\frac{JS^2\pi^2}{Ka}\right)^{1/2} = \sqrt{\frac{A}{K_1}} \quad \text{and} \quad \gamma = 2K\delta \propto \sqrt{AK_1}. \tag{4.21}$$

A is called the exchange stiffness; typical values are $\delta = 120$ atoms, $\gamma = 3\,\text{mJ}\,\text{m}^{-2}$. The energy balances can change when one or more dimension is reduced in to the nanometre range so that shape anisotropy becomes significant.

The domain wall in Figure 4.5 is approximated by an elliptical cylinder of width W and height D. If $W > D$ the demagnetising energy flips the wall into Néel mode. Here the magnetization is constrained to rotate in the plane of the film. This reduces the magnetostatic contribution to the total energy. A 90° Néel wall has less energy than a 180° Bloch wall. Further energy minimization can be achieved by forming cross-tie walls which further reduce magnetostatic terms in the energy.

By further examining the total energy, it is possible to see that there are cases where it is energetically unfavourable to divide into domains. Take a uniformly magnetized sphere of radius r and divide it into two domains:

$$E_{s-2\text{domains}} \approx \frac{1}{2} E_{s-1\text{domain}}. \tag{4.22}$$

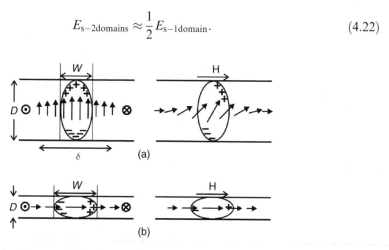

Figure 4.5 Schematic diagrams of Bloch and Néel domain walls represented by regions enclosed by the ellipses. Redrawn after A. Hubert and R. Schäfer, *Magnetic Domains*, Springer, 1998, p. 239 with permission Copyright 1998 Springer-Verlag. (a) In the Bloch wall, the magnetic moments (represented by the arrows) rotate in a plane perpendicular to the sample surface, resulting in N (+) and S (−) poles on the surfaces. (b) In the Néel wall, the moments in the wall rotate in the sample plane, giving rise to no surface poles. The direction of the magnetization in the domains is out of the plane of the paper but in the sample plane, as marked by the circles dots and crosses

To create two domains, the domain wall energy $\pi r^2 \gamma$ is added. If

$$E_s \leq \frac{E_s}{2} + \pi r^2 \gamma \tag{4.23}$$

there is no division into domains as there is no net energy reduction. For a sphere:

$$\left(\frac{1}{4}\mu_0 N_d M^2\right)\left(\frac{4}{3}\pi r^3\right) \leq \pi r^2 \gamma, \quad \text{implying} \quad r \leq \frac{9\gamma}{\mu_0 M^2} \text{ for } N_d = 1/3. \tag{4.24}$$

For Fe_3O_4 the critical radius below which there is only a single domain is $r \leq 6\,nm$. We can therefore conclude that making nanoscale magnetic materials may result in changes to the bulk domain structure, or even the inability to create domain walls at all.

In the regime of single-domain physics, as the particle size reduces further, the total free energy contribution from the magnetization approaches $k_B T$ (0.025 eV at room temperature). In this case the magnetization may change direction under thermal excitation in the absence of a field. This is the so-called superparamagnetic limit. This limit presents a barrier to the continuous decrease in magnetic particle size whilst retaining stability in the magnetization.

4.1.5 The magnetization process

Magnetic materials often experience applied fields that change both magnitude and direction with time. As can be seen from Figure 4.2, the behaviour is very complex; plotting B or M against H yields curves of this generic form for all magnetic materials. Hard magnetic materials require high remanence and coercivity but soft magnetic materials require low coercivity and low anisotropy. At the fundamental level it is the direction and magnitude of anisotropies, and the ease or difficulty of moving domain walls through the sample, that determine whether a particular sample class is magnetically hard or soft. As a rule of thumb, soft magnetic materials have coercivities below $1\,kA\,m^{-1}$ and anisotropy energy densities less than $1\,kJ\,m^{-3}$. Hard magnetic materials would have coercivities and anisotropy energy densities higher than these values.

Over the past twenty years, the nanostructure of alloys has come under sufficient control by process developments and subtle variations of alloy chemistry, that record hard ($H_c > 2 \times 10^6\,A\,m^{-1}$) and soft ($H_c < 1\,A\,m^{-1}$) magnetic materials can be produced. We now demonstrate this by considering some specific examples of nanomagnetic materials.

4.2 NANOMAGNETIC MATERIALS

A broad classification of nanomagnetic materials may be made as follows:

- *Particulate nanomagnets*: granular solids where one or more phases are magnetic; nanograined layers grown on columnar films; quasi-granular films made by heat-treating multilayers of immiscible solids.

- *Geometrical nanomagnets*: columnar films or nanowires; needle-shaped particles in a matrix.

- *Layered nanomagnets*: multilayers with metallic and/or other nanometre-thick films on a supporting substrate (Section 4.2.2).

4.2.1 Particulate nanomagnets

Rapid solidification processing can be regarded in essence as an exercise in basic calorimetry. If a hot body (small melt puddle) and a cold body (large chill block) are brought together, the resultant temperature will lie between the two and will tend towards that of the larger thermal mass (chill block). Cooling rates of order $10^6 \, \text{K s}^{-1}$ are easily achieved in the laboratory.

There are two routes to producing nanophase material in this way: (1) form an amorphous phase then heat-treat to develop the nanostructure; (2) cool more slowly so that the nanostructure is developed without further treatment. The alloy chemistry can be varied to go from the softest (lowest coercivity and anisotropy) to the hardest (highest coercivity and anisotropy) magnetic materials known:

$$\text{soft} = \text{FeCuNbSiB} \Rightarrow \text{NdFeB} = \text{hard}.$$

For soft ferromagnets $\text{Fe}_{73.5}\text{B}_9\text{Si}_{13.5}\text{Nb}_3\text{Cu}$ is the archetype, going under the trade name FINEMET. The nanostructure comes about as Cu enhances crystal nucleation, producing many centres for crystals to form. The presence of Nb retards crystal growth, stopping the grains growing on the nucleation centres. The result is bcc-Fe_3Si grains of size 10–15 nm in an amorphous ferromagnetic matrix. A more detailed discussion of the nanostructure formation of FINEMET is included in Section 5.3.2.

It is the physical nanostructure and the resultant interplay between exchange and anisotropy energies that is significant here. In very small grains that are magnetically coupled, exchange forces cause the moments in neighbouring grains to lie parallel, overcoming the intrinsic easy-axis properties of the individual grains. The coupling dominates over length scales shorter than the exchange length, L_{ex}, for given exchange stiffness, A. This length is of the order of the domain wall width defined earlier, and is an expression for the length scale over which the direction of magnetization can vary:

$$L_{ex} = \sqrt{\frac{A}{K_1}} = 35 \, \text{nm for } \text{Fe}_3\text{Si}. \tag{4.25}$$

We can make a simple model of this coupling using the *random anisotropy model*. Figure 4.6 is a schematic diagram of square nanograins, showing the magnetization direction wandering from grain to grain. Starting with a grain size D, and each grain having anisotropy constant K_1, it is assumed that the anisotropy directions are random from grain to grain. The effective anisotropy $\langle K \rangle$ results from averaging over $N = (L_{ex}/D)^3$ grains in a volume $V = L_{ex}^3$. For a finite number of grains there will always be some net

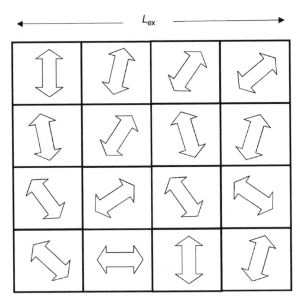

Figure 4.6 A schematic illustration of a nanostructured soft ferromagnet. Each square represents one grain, and the block arrows represent the direction of magnetization within each grain. L_{ex} is the exchange length

easy direction defined by an effective anisotropy constant $\langle K \rangle$. $\langle K \rangle$ is the mean anisotropy fluctuation amplitude of the grains and is given by:

$$\langle K \rangle \approx \frac{K_1}{\sqrt{N}} = K_1 \left(\frac{D}{L_{ex}} \right)^{3/2}. \tag{4.26}$$

From Equation (4.25) with K_1 replaced by $\langle K \rangle$ we have:

$$\langle K \rangle \approx \frac{K_1^4}{A^3} D^6. \tag{4.27}$$

Figure 4.7 is a schematic diagram of how the strength of local anisotropy energy interacts with the exchange energy, demonstrating the extremes of spin alignment for weak and strong local anisotropy energy. If the local anisotropy is strong, the magnetization is well aligned whereas for weak local anisotropy, the magnetization changes direction gradually, effectively overcoming the direction of the local anisotropy. The weak anisotropy limit is relevant for soft nanophase magnets. Figure 4.8 shows the variation of coercivity with grain size, the coercivity being directly proportional to $\langle K \rangle$. The D^6 law (4.27) is clearly obeyed over a range of nanometric grain sizes. Figure 4.9 illustrates how the annealing temperature must be carefully chosen to ensure that the correct microstructure (that required for low coercivity) is produced. The optimum conditions produce iron–silicon grains a few tens of nanometres in diameter surrounded by a magnetically soft amorphous matrix. This produces the behaviour seen in the lower part of Figure 4.7.

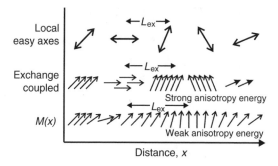

Figure 4.7 A schematic diagram of how the exchange and anisotropy strengths interplay to determine the net exchange length in a nanomagnet. Redrawn after R. C. O'Handley, *Modern Magnetic Materials: Principles and Applications*, Wiley, New York, 2000, p. 337. Copyright 2000, with permission of John Wiley & Sons Inc.

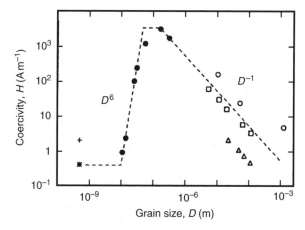

Figure 4.8 The variation of coercivity with grain size in five soft magnetic alloys. Reproduced from G. Herzer, *IEEE Trans. Mag.* **26**, 1397 (1990). Copyright 1990, with permission of IEEE

For *permanent nanomagnets* the archetype comes from a nanocomposite of $Nd_2Fe_{14}B$ grains of size 20–30 nm in a paramagnetic Nd- and B-rich matrix. Figure 4.10 illustrates the unit cell of $Nd_2Fe_{14}B$ which has the *c*-axis as the uniaxial anisotropy axis (Section 4.1.3). To be useful, a permanent magnet material requires a high remanence (M_r/M_s close to 1), a high coercive field and a high Curie temperature.

Nanocomposites are usually used to make working magnets. The nanocomposite is formed by taking rapidly solidified flake, or particles produced from flake, and then bonding with a filler (e.g., a polymer) or by forming diffusion bonds between particles or flakes by heat treatment under high pressure at elevated temperature. The route used to produce the final product may significantly affect the overall magnetic properties, as illustrated by Figure 4.11.

In the nanocomposite material there can be exchange coupling between grains, especially if the alloy is Fe-rich when bcc-Fe nanoparticles accumulate between the NdFeB grains, leading to remanence enhancement effects whereby the remanence can be increased significantly (Figure 4.12).

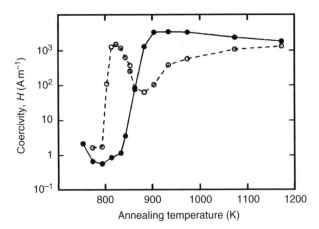

Figure 4.9 The variation of coercivity with annealing temperature in the FeCuNbSiB system. The open circles show the material without the crystal growth-retarding Nb present, whereas the closed circles correspond to the presence of a small amount (1 at%). The coercivity is significantly lower when the grain size is constrained by the Nb additive. Reproduced from G. Herzer, *IEEE Trans Mag.* **26**, 1397 (1990). Copyright 1990, with permission of IEEE

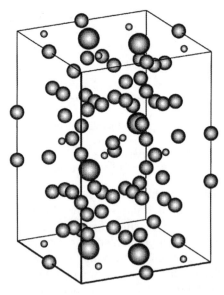

Figure 4.10 Tetragonal unit cell of $Nd_2Fe_{14}B$. The c axis (direction of uniaxial magnetic anisotropy) is vertical in this image

Figure 4.13 illustrates the importance of controlling the intergranular Fe thickness to ensure good exchange coupling between the $Nd_2Fe_{14}B$ grains. If the thickness of the Fe layer exceeds the exchange length, then coupling between NdFeB grains will be lost as the direction of magnetization in the Fe grain will rotate towards the intrinsic easy direction in the Fe crystal.

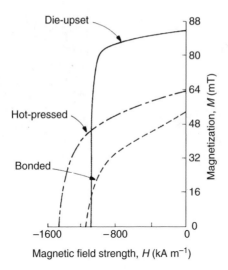

Figure 4.11 The effects of final magnet production on hysteresis loop shape in the second quadrant for NdFeB magnets. The die-upset nanocomposite involves forcing liquid metal into the die to minimize porosity. A hot-pressed nanocomposite magnet is a powder or flake consolidated under high temperature and pressure and the bonded nanocomposite is also a powder or flake treated with a binder, such as an epoxy resin. After R. C. O'Handley, *Modern Magnetic Materials: Principles and Applications*, Wiley, New York, 2000, p. 508

It is the continual development of alloy composition and control of physical nanostructure which has taken us very close to the theoretical limit for performance with the NdFeB family of magnets. Cordless power tools and miniature earphones on portable audio devices are just two current applications of such advanced nanomagnetic materials. Work is also in progress to produce high-efficiency electric motors for use in transportation. The energy product $(BH)_{max}$, is the parameter used to define the energy

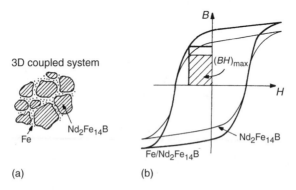

Figure 4.12 (a) A schematic diagram of the exchange-coupled alloy microstructure of $Nd_2Fe_{14}B$. (b) The corresponding hysteresis loop After R. C. O'Handley, *Modern Magnetic Materials: Principles and Applications*, Wiley, New York, 2000, p. 464. Copyright 2000, with permission of John Wiley & Sons, Inc.

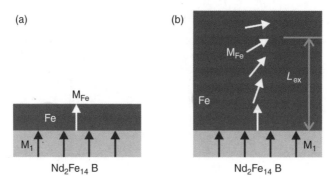

Figure 4.13 A schematic illustration of the effect of soft layer thickness on exchange coupling between grains in a permanent magnet nanocomposite. Black arrows represent the magnetization direction in the $Nd_2Fe_{14}B$ phase and white arrows represent the magnetization direction in the Fe layer. Redrawn after R. C. O'Handley, *Modern Magnetic Materials: Principles and Applications*, Wiley, New York, 2000, p. 465, Copyright 2000, with permission of John Wiley & Sons, Inc.

stored in an air gap in the magnet. This point is marked on Figure 4.12 for illustration. The highest value reported so far in an NdFeB type magnet is around $450\,kJ\,m^{-3}$.

Another system in this category is granular solids, produced when two immiscible species are co-deposited; e.g., FeCu or CoCu with 20–30% Fe or Co. A composite may be formed of nanomagnetic particles in a non-magnetic matrix. These materials may demonstrate giant magnetoresistance (Section 4.3.2) but high fields are usually needed to see maximum effect, and thus they are not best suited to devices.

The final system in this category is where a nanoparticulate magnetic film is grown on a suitable seeding layer. If the seed layer can be grown with a high crystallographic texture (close-packed direction) out of the plane, and with small grain size, then the functional overlayer can be both textured and nanostructured. Such films are candidates for recording media (Figure 4.14).

Modern recording media require small grain size to ensure low noise with many grains per bit. High coercivity is also essential so that the written information remains stable over time. If the grains can be magnetically decoupled, using non-magnetic grain

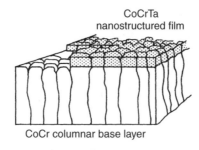

Figure 4.14 A schematic picture of a granular magnetic recording medium. In this example the crystallographic texture of the CoCr underlayer ensures that the functional CoCrTa layer has the necessary nanostructure. After R. C. O'Handley, *Modern Magnetic Materials: Principles and Applications*, Wiley, New York, 2000, p. 463. Copyright 2000, with permission of John Wiley & Sons Inc.

boundaries, stability against thermal erasure of information may also be increased. The growth of a suitably doped nanomagnetic layer on a columnar seed layer ensures these criteria are met. The doped grains are about 30 nm thick.

4.2.2 Geometrical nanomagnets

Self-assembly is developing as a very important technique in this area. One route to the production of nanomagnets comes from templating. Using lithographic techniques, a mask is formed which can be stamped in to a polymer resist. Then using lift-off or etching, a pattern of magnetic islands can remain. Alternatively, layered films may be directly lithographically patterned.

Block copolymer techniques (Chapter 8), or porous alumina, may be used to give a template in to which nanowires may be grown by electrodeposition. This whole subject area is still at the very early stages of exploration, but biological and data storage applications may come to rely on this new paradigm for the construction of smart magnetic materials.

If nanoparticles can be produced, then self-assembly by surfactant and ligand structuring can be used. This offers a number of advantages in that large areas might be covered by self-assembly, and the spacing between nanoparticles might be controlled by variation of the ligand lengths. There has been a recent report[1] of a self-assembled nanocomposite formed using FePt and Fe_3O_4 nanoparticles as precursors, giving an energy product of $160 \, kJ \, m^{-3}$. This is already within a factor of three of the best NdFeB-type magnets. Thin film self-assembled permanent magnets have great potential for application in micro- and nanoelectromechanical systems (MEMS and NEMS) and also data storage.

Biomimetics can be employed, where the magnetic particles are carried in large molecules in an analogous way to haemoglobin in red blood corpuscles. The encapsulation of a magnetic nanoparticle in a protein, or other biocompatible shell, offers significant benefits for drug delivery and cancer therapy, as well as smart diagnostic systems.

It is worth noting that there have been geometrical permanent nanomagnet materials available for a number of years that rely on nanoscale geometrical effects for their properties. Alnico, an alloy of Fe with Ni, Co and Al additives, was first developed in 1938. There is an energetic preference for particle elongation along the $\langle 100 \rangle$ direction during production. This is driven by minimization of the magnetic charge density at the interface between particle and matrix. This can be practically realized by applying a field in a chosen direction during spinodal decomposition (a form of precipitation discussed briefly in Section 8.6.1). This leads to periodic (not sharp) composition fluctuations throughout the magnet volume.

It is shape anisotropy of the particles that produces the necessary anisotropy to produce hard magnetic properties in this instance. Figure 4.15 is a TEM micrograph showing the elongated particles in the matrix. The magnitude of shape anisotropy is not great, so the performance of the Alnico magnets is not as impressive as of the NdFeB magnets discussed above.

[1] H. Zeng, J. Li, J. P. Liu, Z. L. Wang and S. Sun, *Nature*, **420**, 395 (2002).

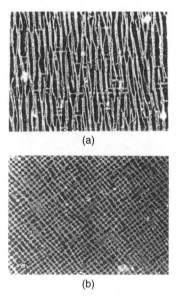

(a)

(b)

Figure 4.15 The nanostructure of Alnico 8 after isothermal heat treatment in a magnetic field for 9 h at 800 °C: (a) view perpendicular to the direction of the applied field, (b) view along the direction of the applied field. The microstructure has a periodicity of approximately 32 nm. After R. C. O'Handley, *Modern Magnetic Materials: Principles and Applications*, Wiley, New York, 2000, p. 481

The final category is layered magnets, where one dimension is in the nanometre range. It is in this category where effects at surfaces and interfaces have been observed. For example, there is growing evidence that in chemically homogeneous films with thickness less than 20 nm, the magnetoelastic coefficients are markedly different to the bulk. This is most clearly illustrated in data from our own work on $Ni_{81}Fe_{19}$ plotted in Figure 4.16, where the composition has been chosen for zero magnetostriction *in the*

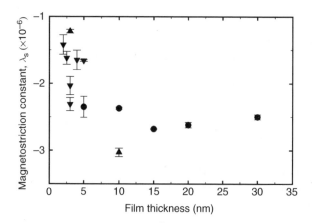

Figure 4.16 Plot of magnetostriction constant λ_S against film thickness for NiFe on SiO_2. The different symbols refer to different data sets. Data courtesy of M. P. Hollingworth

bulk. There are major implications for such devices as spin valves, where the soft magnetic layers are 'buried' in a stack of many layers. This is inevitably going to place the layers under strain.

The reduced symmetry at a surface or interface (atoms only coordinated over a hemisphere) can lead to changes in magnetic anisotropy. As films (or particles) decrease in dimension, surface effects may dominate volume effects.

4.3 MAGNETORESISTANCE

4.3.1 Contributions to resistivity in metals

The basic electronic properties of solids are covered in Chapters 1 and 3 and the reader may need to refer to these for background. We outline here the basis of Drude theory, which assumes that the mean free path λ between electron scattering events is very much longer than the lattice spacing. The incremental velocity δv acquired by an electron in the direction of an electric field E in time τ is given by

$$\delta v = \frac{eE\tau}{m^*}. \qquad (4.28)$$

This gives a conductivity σ of

$$\sigma = \frac{ne^2\tau}{m^*} = \frac{ne^2\lambda}{\hbar k_F}, \qquad (4.29)$$

where n is the electron density and k_F is the wavevector at the Fermi level. The finite mean free path arises from the presence of defects, phonons, magnetic fluctuations and occasionally electron–electron scattering.

In the effective mass approximation $E_k = \hbar^2 k^2/2m^*$ where m^* is close to the electron mass except when k is close to the edge of the Brillouin zone where the electron is Bragg reflected. The Brillouin zone of a crystal with N unit cells contains Nk points, so a full zone can hold $2N$ electrons (counting electron spin). Hence if the number of electrons per atom is less than or equal to order unity then the electrons behave like free electrons.

The s electrons behave as almost free electrons, so an s electron carries the current efficiently; any scattering off impurities is weak. The overall effect of all the atom cores is very weak (the electrons are free) and if an atom is displaced as in a phonon, it produces very little scattering. The band and density of states pictures make it clear that the *only* electron states that matter are those at the Fermi level.

Electron bands for d orbitals are much narrower than those for s electrons, so m^* will be higher. Also if the band is narrow, then the electrons will interact much more strongly with the lattice than for a wide free electron band. This means that they are much more affected by any imperfections, so that we expect that τ will be much shorter for d electrons than for s electrons. Hence we should expect d electrons to contribute less than s electrons to the conductivity σ, even though there are more of them. In fact, d electrons seriously impede the s electrons. However, there is an alternative view, where hybridization occurs (mixing of s and d states – see Section 6.1.2.1 for more details) and we should think of conduction by the hybridized bands. These have mostly d character, hence they have high effective mass and so conduct badly.

From Fermi's golden rule we find that the probability of scattering depends on the number of available states, which is the density of states at the Fermi surface. Thus we see that the s electrons have a greatly reduced lifetime, τ, when there are available d states for scattering.

In copper the s electrons carry the current efficiently and any scattering off impurities is weak; the s electrons are behaving as almost free electrons. There are also very few states for an electron to scatter into because the density of states is so low. If it is scattered, it stays an s electron and so continues to contribute to the conductivity.

In a transition metal the s electrons scatter readily into d electron states because there are so many available states. The d electrons have a very short scattering time and a large effective mass and so they do not contribute to the conductivity. The result of this is surprising; some electron states at the Fermi level are needed for conductivity.

In order to see significant spin-dependent transport effects in ferromagnets, two criteria must be satisfied. There must be different conductivities for electrons in the majority (spin-up) and minority (spin-down) spin bands, and little spin mixing so that the two channels act independently.

Spin flip (sf) scattering occurs because of the spin–orbit interaction and also because of the scattering of the electrons by the spin waves, which makes spin flip scattering temperature dependent. An electron must change its spin and its wave vector (spatial wavefunction) simultaneously. This is a weak effect because the spin–orbit interaction energy is much less than the Coulomb interaction energy. We expect that an electron will undergo N spin-conserving scatterings before it flips its spin. N is a large number and temperature dependent. We can write $\tau_{sf} = N\tau_0$.

In order to make a device, we need to know how far an electron travels before a spin flip occurs. Because of scattering, the electron is travelling in a random walk as it is scattering N times. So we need to consider the distance travelled in a random walk, $\langle R^2 \rangle^{1/2}$:

$$\langle R^2 \rangle^{1/2} = N^{1/2}\lambda = \tau_0 v_F \sqrt{\frac{\tau_{sf}}{\tau_0}} = v_F \sqrt{\tau_0 \tau_{sf}}. \tag{4.30}$$

Spin flip scattering is usually ignored, which means that we consider the conductivity in a ferromagnet in terms of two independent spin channels conducting in parallel. The total current density is given by $I = j_\uparrow + j_\downarrow$, and the voltage drop across the devices is $V = \sigma_\uparrow j_\uparrow = \sigma_\downarrow j_\downarrow$. Here j is the current density in either the spin-up or spin-down channel. We are now able to consider the effects of magnetic fields on electron transport.

4.3.2 Giant magnetoresistance

With the advent of nanomagnetic materials in the form of granular solids or multilayers (Section 4.2) there are now systems that demonstrate very significant spin flip scattering. Giant magnetoresistance (GMR) is a generic term applied to a range of phenomena where there is a significant (>5%) change in electrical resistivity as the sample magnetization magnitude or direction is varied.

We can consider a number of features in a study of GMR. Firstly, the structure demonstrating GMR is usually a combination of normal metal and ferromagnetic layers. The electrical conductivity depends on the relative orientation of magnetization in successive ferromagnetic layers in the stack. Switching the relative magnetization of the layers from parallel to antiparallel defines two states, high and low resistivity, respectively. In terms of sensor technology, these two states are easily discriminated and may be ascribed to the binary states 0 and 1. This has opened up the possibility of applications in data storage technology (see below).

Two geometries are commonly used in GMR studies. The current may flow in the plane of the layers (CIP) or current perpendicular to the plane of the layers (CPP). These are shown schematically in Figure 4.17. As the layers are only a few nanometres thick in each case, the CIP geometry offers high resistance from the small cross-sectional area, and the attendant voltages are easy to measure. To make CPP measurements tractable, lithography must be used to create pillars of small cross-sectional area to achieve voltages suitable for detection.

To satisfy the criteria developed above for spin-dependent scattering, the lateral dimensions must be small compared with the electron mean free path. Hence each electron samples different layers and/or directions of magnetization between scattering events.

A simple model may be developed based on resistor networks. In a ferromagnet there are two different resistances for the two spin channels: $\rho_\downarrow$ for the minority spin electrons

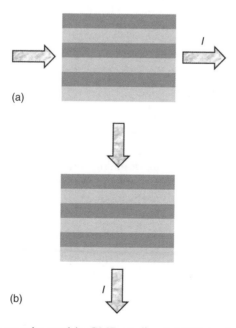

Figure 4.17 The two geometries used in GMR studies: (a) CIP in plane and (b) CPP current perpendicular to plane

and $\rho_\uparrow$ for the majority spin electrons. The resistivity of the non-magnetic metal will be independent of the spin and is designated ρ_0.

We take the thicknesses of the ferromagnetic layers to be t_F and the thickness of the spacer layer to be t_s. Consider the resistance in CPP for two ferromagnetic layers aligned with magnetizations parallel separated by a non-magnetic spacer. We take A as the planar area of the multilayer stack. For two ferromagnetic layers separated by a single non-magnetic conductor, we can use a simple resistor model again.

The total resistivity for the spin-up electrons is $R_\uparrow = 2t_F\rho_\uparrow + t_s\rho_0$. The total resistivity for the spin-down electrons is $R_\downarrow = 2t_F\rho_\downarrow + t_s\rho_0$. The net resistance for the two conduction paths is obtained by adding the two spin contributions in parallel:

$$\frac{1}{R_{\uparrow\uparrow}} = \frac{1}{R_\uparrow} + \frac{1}{R_\downarrow} = \frac{2t_F(\rho_\uparrow + \rho_\downarrow) + 2t_s\rho_0}{(2t_F\rho_\uparrow + t_s\rho_0)(2t_F\rho_\downarrow + t_s\rho_0)}. \tag{4.31}$$

The same considerations apply for two ferromagnets aligned antiparallel. In the parallel case each spin is a minority spin in one ferromagnetic region and a majority spin in the other. The resistances are equal, so

$$R_\uparrow = R_\downarrow = t_F(\rho_\uparrow + \rho_\downarrow) + t_s\rho_0, \tag{4.32}$$

giving $R_{\uparrow\downarrow} = R_\uparrow/2$ (again by conductors in parallel). The GMR ratio is commonly defined as $(R_{\uparrow\downarrow} - R_{\uparrow\uparrow})/R_{\uparrow\uparrow}$.

The resistance is always larger for the antiparallel alignment (both species of electron must pass through the region of high resistance) and lower for parallel alignment as one species of electron 'shorts out' the other. Hence GMR is always positive and can be very large.

Writing $\alpha^+ = \rho_\uparrow/\rho_0$ and $\alpha^- = \rho_\downarrow/\rho_0$, we have:

$$\text{GMR} = \frac{(\alpha^+ - \alpha^-)^2}{4\left(\alpha^+ + \dfrac{t_F}{2t_s}\right)\left(\alpha^- + \dfrac{t_F}{2t_s}\right)}. \tag{4.33}$$

The effect is bigger if the contribution to the resistance from the spacer is small.

It is straightforward to include any boundary resistance by redefining the thickness of the ferromagnetic layer to take account of it. We write $t_F\rho_\uparrow + r_j = t_F^j\rho_\uparrow$ and similarly for the minority electron, which gives the modified result

$$\text{GMR} = \frac{(\alpha^+ - \alpha^-)^2}{4\left(\alpha^+ + \dfrac{t_F^j}{2t_s}\right)\left(\alpha^- + \dfrac{t_F^n}{2t_s}\right)}, \tag{4.34}$$

where r_j is the junction resistance. The model is not exact. There is still much dispute concerning the details, and it is believed that electron interference effects give small deviations from the simple model.

There are two basic reasons why there is GMR. Firstly, the resistivities for the majority and minority electrons are different, as discussed above. There is also likely to be scattering and reflection at the interfaces. If the interfaces are rough, this will increase the amount of scattering. As the electrons enter the ferromagnetic layer, the scattering depends on the density of states for the spin-up and spin-down electrons.

The second reason is due to interface resistance. The probability of reflection of a wave from one medium to another (R) where the wavenumbers are k and k' is

$$R = \left| \frac{k - k'}{k + k'} \right|^2 .$$

This means that the probability of reflection vanishes if the wavenumbers are very similar in the layers at either side of the boundary, and is large if they are very different. It has been shown that there is a nearly perfect match between the wavevector for spin-up electrons in Co, and the wavevector of free electrons in Cu. There is a poor match between the wave vector of spin-down electrons in Co and the wave vector of free electrons in Cu. Hence most of the spin-down electrons will be reflected while the spin-up electrons will pass straight through the boundary. The reverse effect is seen for Fe/Cr, so it is the minority electrons which pass through.

We shall now consider the constraints that are placed on the dimensions of the layer thicknesses for GMR to occur. If the spin flip scattering may be neglected, the spin currents are conserved. In CIP geometry the mean free path in the normal metal must be less than the spin flip scattering length. At any boundary, in equilibrium, $\mathbf{j} \cdot \hat{\mathbf{n}} = 0$ ($\hat{\mathbf{n}}$ is the unit vector normal to the boundary) is continuous (because $\nabla \cdot \mathbf{j} = 0$ from Maxwell's equations), hence in a CPP structure the important requirement is that the total thickness of the device is less than λ_{sf}, the spin diffusion length.

For CIP we consider two cases: layer thickness large or small compared with the scattering length between collisions, λ_0. It is the shortest such length that matters. For small spacings, we take an average resistance over the whole sample; that is, no Ohm's law behaviour for each component. For large spacings, we take separate Ohm's law behaviour for each component. The current flow is in the material with the low resistance (the spacer). Electrons do not travel much between different components, and the GMR vanishes.

The resistance of a few monolayers of transition metal is extremely small. Almost all GMR experiments are done using a CIP geometry. A superconducting circuit is needed to detect CPP unless lithographic patterning is used to reduce the effective area.

Consider a multilayer made of layers of Co (n Å) and of Cu (m Å) with N repeats. ($1 \, \text{Å} = 10^{-10} \, \text{m}$). The interatomic spacings in Co and Cu are both $\sim 2.5 \, \text{Å}$, so the multilayers are made with only a few atomic planes. This is written as $[\mathrm{Co}(n\,\text{Å})/\mathrm{Cu}(m\,\text{Å})]_N$. The exchange interaction between Co layers oscillates with the thickness of the spacer layer. It may be assumed that the spacing has been chosen so that the interaction is antiferromagnetic, and the magnetizations of the two layers are taken as M_A and M_C.

Now assume that we have a magnetic field B acting. The energy per unit area will be

$$U = -J M_A M_C - B t_{\mathrm{Co}} (M_A - M_C). \tag{4.35}$$

The net field on the C layer changes sign when $B = J M_A / t_{\mathrm{Co}}$ and the magnetization of the C layer flips round. The resistivity changes from that appropriate for an antiferromagnetic structure to that for a ferromagnetic structure. Figure 4.18 shows resistance with magnetic field for an Fe/Cr multilayer. The saturation field, H_s, can be used to deduce the exchange from the above theory. In this diagram $R_{\uparrow\downarrow}$ is the same as $R(H = 0)$ and $R_{\uparrow\uparrow}$ is equal to $R(H > H_s)$.

Figure 4.19 shows the variation of saturation field H_s for an NiFe/Ru multilayer as a function of the ruthenium layer thickness. The regions where the exchange is ferromagnetic correspond to the minima in the saturation field.

As described in Section 4.2.1, we can consider a granular alloy for GMR containing small, usually irregular grains of a ferromagnet in a non-magnetic matrix. In the absence

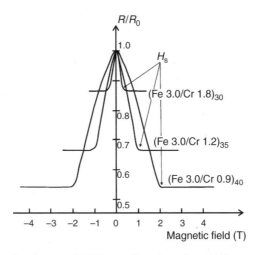

Figure 4.18   Normalized resistance (R/R_0) as a function of applied magnetic field for an Fe/Cr multilayer. The notation (Fe x/Cr y)$_z$ denotes a stack of z bilayers of x nm of iron and y nm of chromium. H_S is the saturation field as defined in the text. Redrawn with permission after B. Heinrich and J. A. C. Bland, *Ultrathin Magnetic Structures*, Vol. I, Springer-Verlag, Berlin, 1994, p. 98. Copyright 1994 Springer-Verlag

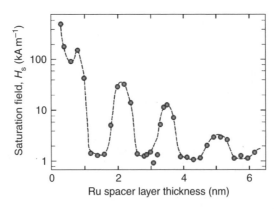

Figure 4.19   Dependence of saturation field on Ru spacer layer thickness in an NiFe/Ru multilayer at room temperature. Redrawn with permission after B. Heinrich and J. A. C. Bland, *Ultrathin Magnetic Structures*, Vol. I, Springer-Verlag, Berlin, 1994, p. 161. Copyright 1994 Springer-Verlag

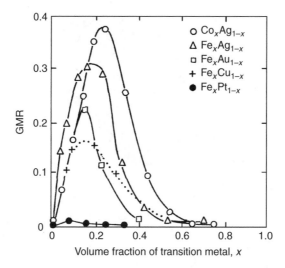

Figure 4.20 Variation of GMR with volume fraction of magnetic components. For several transition metal–noble metal granular systems. After R. C. O'Handley, *Modern Magnetic Materials: Principles and Applications*, Wiley, New York, 2000, p. 594. Copyright 2000 with permission of John Wiley & Sons Inc.

of an applied field, the grains are randomly arranged due to their own shape anisotropy and a long-range exchange (RKKY) interaction between them.[2] An applied field lines up the grains.

Provided that the electron mean free path is larger than the spacing between the ferromagnetic granules, the resistance will drop in the aligned state. Transition metal (e.g., iron) and noble metal (e.g., silver) systems show this GMR. Figure 4.20 shows some data for transition and noble metal composites.

4.3.3 Spin valves

In applications such as data storage there is a requirement for a sensor that converts a changing magnetic field into a changing voltage. The changing field is derived from the stray field from a recording medium such as a hard disk or tape. The voltage, which needs to be linearly related to the stray field direction and/or magnitude, is used as the read signal and is transposed into audio or digital signals for interpretation.

To create a hard disk drive reader, one must have a device that responds at very high frequencies to very low magnetic fields. The high frequencies are necessary in order to have high rates of data transfer. One such device, derived from the basic GMR physics set out above, is the spin valve. This is a device with only two ferromagnetic layers. One layer is pinned; its direction of magnetization being unperturbed by the changes in field that are to be detected. The other layer, called the free layer, is sensitive to the stray

---

[2] See R. C. O'Handley, *Modern Magnetic Materials: principles and applications* (Wiley, New York 2000) for a full discussion.

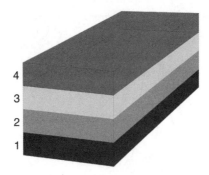

Figure 4.21 A schematic diagram of the structure of a spin valve. Layer 1 is the antiferromagnetic layer (e.g. FeMn) and layer 2 is the pinned soft magnetic layer (e.g., NiFe). A copper spacer layer, layer 3, separates the lower two layers from a free soft magnetic layer, layer 4 (e.g., NiFe).

fields; that is, the direction of magnetization in the free layer may be reversed by the stray fields. Permalloy, an alloy of Ni and Fe, is usually chosen for both layers. The chosen alloy has a very low coercive field and anisotropy field (but note the discussion in Section 4.2.2) and is a strong ferromagnet, so it has a good GMR response. The pinned layer is also NiFe and has a direction of magnetization defined by exchange or anisotropy bias (Section 4.3.4) from an adjacent antiferromagnetic layer; e.g., IrMn or FeMn.

The pinned layer is magnetized in-plane as it is prepared in such a way as to have uniaxial in-plane anisotropy. The structure is cooled down through the ordering temperature of the antiferromagnet in an external applied field (Section 4.3.4).

The spacer, usually Cu or Ru, is chosen so as to give the smallest possible magnetic coupling between the two ferromagnetic layers.

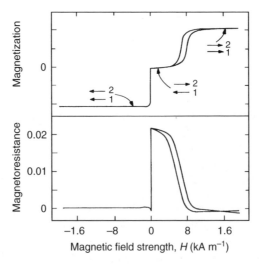

Figure 4.22 The magnetization and magnetoresistance curves for a spin valve. The arrows indicate the magnetization direction in the two ferromagnetic layers. After R. C. O'Handley, *Modern Magnetic Materials: Principles and Applications*, Wiley, New York, 2000, p. 596. Copyright 2000, with permission of John Wiley & Sons Inc.

Figures 4.21 and 4.22 illustrate the principles of the spin valve. The magnetic layers are NiFe or CoFe alloys. For all applied fields that are negative, the pinned and free layer magnetizations represented by the arrows in Figure 4.22 are parallel and the device is in a low-resistance state. As the applied stray field becomes sufficiently positive, the free layer may rotate, reducing the net magnetization to zero and causing an almost stepwise increase in the resistance. Further increases of stray field in the same sense will eventually cause the pinned layer to rotate, returning the device to a low-resistance state. A practical device would only operate for small field excursions around $H = 0$, leaving the pinned layer unchanged. It is simple to see that the two resistance states may be ascribed the logical states 0 and 1. The changes in state are detected in terms of the voltage change across the spin valve for a constant current flowing through the device.

A full understanding of the microscopic origins of exchange bias remains to be found, but many current disk drive systems are now using spin-valve technology for the read head.

4.3.4 Tunnelling magnetoresistance

Whilst spin valves are already developed sufficiently to be in production, research work continues on another form of magnetoresistive device that relies on electron tunnelling through an oxide barrier between two ferromagnets. This results in tunnelling magneto-resistance (TMR).

Consider two ferromagnets separated by an insulating tunnel barrier. Electrons may tunnel across the barrier, preserving spin provided the spin flip scattering length is greater than the barrier width. Hence the resistance across the barrier will be much lower if there is a large density of states available for the tunnelling electrons on both sides of the barrier.

We now look at the densities of states on both sides of a tunnelling barrier for both spin components. We consider the parallel configuration; that is, the magnetizations parallel across the barrier. There will be a small amount of tunnelling across the barrier. For the parallel magnetization we have a large density of states tunnelling to a large density of states.

Early experiments in the manganites (oxide materials based on compounds such as LaCaMnO) showed that when the material was granular, rather than composed of large single crystals, there was a very large magnetoresistance at low temperatures that occurred for small magnetic fields. It was realised that this occurred because of tunnelling between the grains through the resistive material at the grain boundaries. One of the features of the grain boundary is that the exchange paths are broken. This means that very small fields are required to change the relative orientation of the two grains.

More recently structures have been made where an insulating layer is grown as a tunnel barrier between two ferromagnets. The insulator must be very thin, usually only a few lattice spacings. This is because the total probability of tunnelling is proportional to $\exp(-2d/\lambda)$, where d is the thickness of the tunnel barrier and λ is the decay length of the electron wavefunctions in the barrier. If the barrier is not perfect, then pinholes may appear which do not show the desired magnetoresistance.

In a TMR device the requirement is for one free layer and one pinned layer, as in a spin valve (Figure 4.21). One of the common ways to fabricate an insulating layer is to grow a layer of Al and then oxidize it to Al_2O_3. The quality of the device is lowered if there is an unreacted layer of Al, but on the other hand allowing too much oxidation so that transition metal oxides like CoO are formed is counterproductive because such oxides are antiferromagnetic and hence destroy the spin polarization. There is an asymmetry between the magnetic layers. One is grown on the substrate and is a high-quality epitaxial layer and the other is grown on top of the amorphous Al_2O_3. In most cases the layer that is magnetically soft is the one that is underneath the Al_2O_3 and was grown directly on the substrate.

Most of the tunnel junctions that are being developed for current applications are based on transition metals. Co is often used as it has a high spin polarization and it is combined with NiFe or FeCo as the soft layer. Figure 4.23 shows magnetoresistance data for a $Co/Al_2O_3/CoFe$ TMR device. The form of this data may be taken as generic for all tunnel junctions.

The advantages of using transition metals is the high transition temperatures, leading to robust devices at room temperature, and the ease of fabrication, similar to that for GMR spin valves.

Another possibility is hybrid devices in which an oxide is combined with a transition metal. These appear to be very promising as they combine the very large TMR expected for an oxide with the good performance at room temperature.

Although the ideas of TMR and GMR spin valves appear similar, there are many very important differences that result in replacing the metallic non-magnetic layer with an insulating layer. The first and most obvious experimental difference is that the overall resistance of the device is now much higher. Whereas it is very difficult to carry out GMR experiments in the current perpendicular to the plane geometry, as the resistances are too small to measure, it is absolutely impossible to perform TMR in anything other than the CPP geometry; no current at all will cross the barrier unless it is

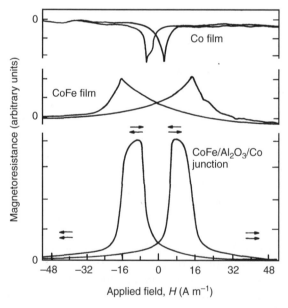

Figure 4.23 Anisotropic magnetoresistance in each individual CoFe and Co electrode and the junction magnetoresistance in $CoFe/Al_2O_3/Co$ spin tunnel junction versus applied field. After R. C. O'Handley, *Modern Magnetic Materials: Principles and Applications*, Wiley, New York, 2000, p. 607. Copyright 2000 with permission of John Wiley & Sons Inc.

driven by a voltage. In GMR the current is carried predominantly by the majority electrons that have a low density of states, hence a longer mean free path. In TMR the simple theory indicates that the effect depends only on the density of states. In fact there have been a number of experiments that have shown that this is not correct and that the way in which the mobile electron wavefunctions can match onto the empty conduction band states in the barrier is also a factor.

In GMR spin valves, one must choose a metal spacer layer that gives a very small exchange coupling between the magnetic layers. This is important if the device is to respond to low switching fields. This is not an issue in tunnel devices as the magnetic interactions do not easily cross an insulating layer. However, it is hard to ensure that one of the layers is really magnetically soft when there is a possibility of oxide formation and strain due to the amorphous layer.

The TMR junctions are particularly good candidates for magnetic random access memory (MRAM). This is because the two layers are not coupled magnetically, so the junction will stay almost indefinitely in one configuration. Another very attractive possibility is that the tunnel junctions will be used in three-terminal devices to make spin transistors, as well as other aspects of spin electronics, or spintronics. Spintronics is still in its infancy and it is impossible to foresee which directions will prove the most attractive.

4.4 PROBING NANOMAGNETIC MATERIALS

The power of the electron microscope has been amply discussed in Chapter 2. For nanomagnetic materials it is the use of scanning probe techniques, and in particular magnetic force microscopy, which is important.

A magnetic force microscope (MFM) is essentially the same as an atomic force microscope (AFM). However, the scanning probe is coated with a layer of magnetic material, which may be magnetically hard (e.g., CoCr) or magnetically soft (amorphous ferromagnetic alloy; e.g., FeBSiC); a tip is shown in Figure 4.24. It is usual to take a line scan in contact mode to give the topographic information, and then rescan the line at a fixed flying height of a few tens of nanometres to obtain the magnetic contrast. In contact mode van der Waals forces, which have a much shorter range than the magnetic forces, dominate and only topography is seen. At the flying height, the longer-range

Figure 4.24 Scanning electron microscope image of an MFM-coated tip. The scale bar is 10 μm

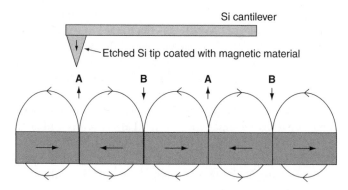

Figure 4.25   A schematic of the principle of MFM. The sample is magnetized in-plane, giving rise to stray fields above the plane vertically upwards (A) and vertically downwards (B). The tip is shown magnetized downwards, so it will see repulsion at A and attraction at B. Magnetic field lines are indicated by arrows

magnetic interactions between the tip dipole moment and the stray field from the sample are detected (Figure 4.25).

An oscillating cantilever (typically 70 kHz) would typically be used for high sensitivity. The tip is sensitive to the force gradient $\partial F_z/\partial z$, which modifies the effective cantilever spring constant k_{eff}: this produces a change in angular frequency

$$\omega_0 = \sqrt{\frac{k_{\text{eff}} - \partial F_z/\partial z}{m_e}} \qquad (4.36)$$

where m_e is the mass of the cantilever. If, as is usual, $\partial F_z/\partial z \ll k_{\text{eff}}$ then we can write:

$$\omega' = \omega_0 \left(1 - \frac{\partial F_z/\partial z}{2k_{\text{eff}}}\right) \quad \text{so} \quad \Delta\omega = \omega' - \omega_0 \approx -\frac{\omega_0}{2k_{\text{eff}}} \frac{\partial F_z}{\partial z}. \qquad (4.37)$$

For an attractive force there is a decrease in resonance frequency, and vice versa; this change is used to generate the contrast. As an example, Figure 4.26 shows data from a

Figure 4.26   AFM/MFM image pair taken on a hard disk. The 'black' data bits, visible in the MFM tracks, are where there is an attractive force between head and disk, and the 'white' data bits are where the force is repulsive. The images are of an area having side 18 μm, and the height scale on the AFM image is up to 100 nm (lightest regions). Image courtesy of Dr M. Al-Khafaji, University of Sheffield

simultaneous AFM and MFM experiment on a magnetic storage disk. The AFM image shows the scratches that come from the disk polishing process. These do not appear in the MFM image, but instead we can see the tracks and data bits (single bits, dibits and tribits).

4.5 NANOMAGNETISM IN TECHNOLOGY

Figure 4.27 shows a standard PC hard disk drive. Inside this particular drive there are two hard disks, coated on each side, and four read/write heads on the flyer. Currently (2004) you can buy drives offering capacities greater than 400 Gbyte for a desktop PC. As the data storage density increases, so the physical size of each bit (1 or 0) decreases. In order to read/write, the head must be capable of interacting with a smaller and smaller volume, and sensing a smaller and smaller absolute field value. e.g., at $10\,\text{Gbit/in}^2$ ($1.6\,\text{Gbit/cm}^2$) each bit is less than $1\,\mu\text{m}$ square. This also requires the head to get closer to the disk surface (currently the head travels a mere 12 nm above the disk surface). It is analogous to

Figure 4.27   The inside of a computer hard disk drive

Figure 4.28 The all-electric go-kart. Photo courtesy of Professor D. Howe, University of Sheffield

flying a Boeing 747 on full throttle (800 kph) 3 m above the ground and following the land contours! Spin-valve heads have now been introduced, and may be followed by TMR devices as discussed above. Methods of self-assembly may be used to increase the areal density of information stored on the disk surface. The challenge remains to keep to the demand anticipated in the road map shown in Figure 4.1.

It has only been possible to develop personal entertainment systems such as the Sony Walkman because of extensive research on nanomagnetic materials. The miniature, high-fidelity ear phones use NdFeB permanent magnets to give audio power at low battery drain and in a small volume. The motor to turn the cassette tape or the CD is also less energy intensive with the fitting of NdFeB magnets. This gives prolonged battery life. Developments in soft magnetic materials have allowed the production of more sensitive analogue heads for interrogating the tape, again requiring less battery drain.

The cordless screwdriver or power drill can deliver a torque beyond the capability of most home decorators. Again it is high-efficiency NdFeB-based motors that give the advantage in terms of power and battery life. There are also extensive programmes to use nanomagnetic materials in new forms of electric traction. Figure 4.28 shows an all-electric go-kart developed by a team in Sheffield. There is a motor in each wheel hub, with NdFeB magnets offering the power and efficiency gain.

4.6 THE CHALLENGES FACING NANOMAGNETISM

The grain size of magnetic particles cannot be reduced indefinitely without affecting the magnetic properties. Below a certain critical size the remanent magnetization is no longer fixed in direction by anisotropy, but can be flipped by thermal energy. If P is the probability of switching, then

$$P \propto \exp\left(\frac{-VK_\mathrm{u}}{k_\mathrm{B}T}\right), \tag{4.38}$$

where K_u is the local uniaxial anisotropy in the grain of volume V. The superparamagnetic radii for stability are typically 7 nm for 1 year's lifetime and 6 nm for 1s lifetime, for $K_\mathrm{u} = 10^5 \ Jm^{-1}$. This limits the simple option of increasing recording density by decreasing the size of particles currently used to coat disks. Materials of ever increasing anisotropy constant such as FePt and CoPt are under investigation, but eventually there will have to be a move towards lithography or self-assembly.

For extensive use of spintronics, there are processing compatibility issues between CMOS (complementary metal oxide semiconductor) electronics and metals, which will have to be resolved. The physics of magnetism and electron transport at interfaces must also be investigated further to resolve issues of spin flip and uncontrolled anisotropies at interfaces.

It is even fair to say that nanomagnetism faces the challenge of being in some senses already well developed. Alnico and NdFeB magnets have been around for a number of years and are already in applications at the low-cost, high-volume end of the market. Other nanotechnologies need to catch up.

Given the progress over the past 20 years, it would be a brave person who would go further in forecasting the impact of nanomagnetism on industry and commerce, but it will remain an integral part of many systems for years to come.

BIBLIOGRAPHY

General background

R. M. Bozorth, *Ferromagnetism*, IEEE Press, New York, 1993.

B. D. Cullity, *Introduction to Magnetic Materials*, Addison-Wesley, Reading MA, 1972.

D. Craik, *Magnetism: Principles and Applications*, John Wiley & Sons, Inc., New York, 1995.

R. C. O'Handley, *Modern Magnetic Materials: Principles and Applications* John Wiley & Sons, Inc., New York, 2000.

B. Heinrich and J. A. C. Bland (eds), *Ultrathin Magnetic Structures*, Vols I and II, Springer-Verlag, Berlin, 1994.

R. Skomski and J. M. D. Coey, *Permanent Magnetism*, IoP Publishing, Bristol, 1999.

S. Blundell, *Magnetism in Condensed Matter* Oxford University Press, Oxford, 2001.

Nanomagnetism

H. S. Nalwa, *Magnetic Nanostructures* American Scientific Publishers, Los Angeles, 2002.

A. Hernando, *Nanomagnetism*, NATO ASI Series E: Applied Sciences, Vol. 247, Kluwer, Dordrecht, 1992.

M. Vázquez and A. Hernando, *Nanostructured and Non-crystalline Materials*, World Scientific, Singapore, 1995.

G. Timp, *Nanotechnology*, Ch. 12, AIP Press (Springer-Verlag), New York, 1998.

A. S. Edelstein and R. C. Cammarata, *Nanomaterials: Synthesis, Properties and Applications*, IoP Publishing, Bristol, 1996.

G. C. Hadjipanayis and G. A. Prinz, *Science and Technology of Nanostructured Magnetic Materials*, NATO ASI Series B: Physics, Vol. 259, Kluwer, Dordrecht, 1990.

Review articles

J. Nogués and I. K. Schuller, Exchange bias. *Journal of Magnetism and Magnetic Materials*, **192**, 203–232 (1999).

R. Skomski, Nanomagnetics. *Journal of Physics: Condensed Matter*, **15**, R841–R896 (2003).

J. I. Martín, J. Nogués, K. Liu, J. L. Vicent and I. K. Schuller, Ordered magnetic nanostructures: fabrication and properties. *Journal of Magnetism and Magnetic Materials*, **256**, 449–501 (2003).

Q. A. Pankhurst, J. Connolly, S. K. Jones and J. Dobson, Application of magnetic particles in biomedicine. *Journal of Physics D: Applied Physics*, **36**, R167–R181 (2003).

P. Tartaj, M. del Puerto Morales, S. Veintemillas-Verdaguer, T. González-Carreño and C. J. Serna, The preparation of magnetic nanoparticles for application in biomedicine. *Journal of Physics D: Applied Physics*, **36**, R182–R197 (2003).

C. C. Berry and A. C. G. Curtis, Functionalisation of magnetic nanoparticles for applications in biomedicine. *Journal of Physics D: Applied Physics*, **36**, R198–R206 (2003).

5

Processing and properties of inorganic nanomaterials

5.1 INTRODUCTION

Materials science is concerned to a great extent with the interaction between phenomena operating on different levels of scale, from that of the atom through to the macroscopic scale of everyday engineering components. For the majority of the time, however, materials science focuses on the intermediate scale; that of the microstructure, its means of production and control and properties related to it. We usually take microstructure to describe structural features on the scale between about 0.3 nm and 1 m and to include, in the classical definition, composition, crystal structure, grain size and shape (morphology), spatial distribution of phases, etc. The properties exhibited by a material depend, to a great extent, on the scale of the microstructure or, to be more specific, the relationship between microstructural scale and its influence in constraining particular physical mechanisms. For example, the strength of material is very strongly dependent on the grain size at all length scales, the grain boundary acting to constrain the propagation of dislocations through the material. In magnetic materials the interaction of the grain boundary with the domain wall thickness is of central importance. In both of these examples there exists a characteristic length; i.e., the critical bowing radius for dislocation propagation in the first example and the magnetic exchange length in the second, defining the scale on which the phenomena operate. When the constraints imposed by microstructure and these characteristic length scales overlap, it is quite often found that well-known laws governing material properties at conventional length scales break down or require re-examination. This is the range in which the new class of nanomaterials reside. As discussed in Chapter 1, the term 'nanomaterial' is taken as a descriptor for any material with at least one dimension smaller than 100 nm. Another way of defining the nanostructured regime is to state that a nanostructured material is attained when the

Nanoscale Science and Technology   Edited by R. W. Kelsall, I. W. Hamley and M. Geoghegan
© 2005 John Wiley & Sons, Ltd

grains become equal to or smaller than those characteristic length scales related to their specific functionality. Control and manipulation of physical, mechanical and chemical properties as a function of the material microstructure on the nanometre scale often leads to the discovery of new materials, be they metallic, ceramic or polymeric in nature.

With the recent surge of interest in nanostructured materials, or to be more specific, the chemical, electronic and optical properties of nanomaterials, there is now a strong driving force to develop methods for their large-scale manufacture. Since the material properties of interest are very dependent on grain size, most applications will require the production of materials with well-controlled grain sizes and grain size distributions.

In this chapter we shall address the issues faced when processing inorganic nano-crystalline materials, the factors influencing their production and those governing the retention of nanostructure during further processing, and we shall briefly examine some of the interesting novel properties that are generated. The chapter could easily be retitled 'Processing *for* Properties in Inorganic Nanomaterials', as this is in fact our ultimate aim – the retention of the remarkable physical and chemical properties exhibited by materials on the nanoscale after processing to give useful engineering components.

5.1.1 Classification

The classification of bulk nanostructured materials is essentially dependent on the number of dimensions which lie in the nanometer range. Thus, depending on which dimensions fall within this range, they may be classified as: (a) zero-dimensional (0D); e.g., nanopores and nanoparticles; (b) one-dimensional (1D); e.g., layered or laminate structures; (c) two-dimensional (2D); e.g., filamentary structures where the length is substantially greater than the cross-sectional dimensions, and finally (d) three-dimensional (3D); e.g., structures typically composed of consolidated *equiaxed* (all three dimensions equally nanoscale) nanocrystallites. The classification system is useful in describing the type of structure we can manufacture. Figure 5.1 schematically illustrates all four types and, of these four types, 3D and 1D have so far received the greatest attention.

5.2 THE THERMODYNAMICS AND KINETICS OF PHASE TRANSFORMATIONS

5.2.1 Thermodynamics

Material microstructures are almost always unstable from a thermodynamic stand-point. That is, there is almost always some way in which the structure of a material can change to reduce the total energy. Because we strive to produce materials with structures that are optimised for specific applications, these changes will inevitably lead to deterioration in performance. This lack of thermodynamic stability is, however, often

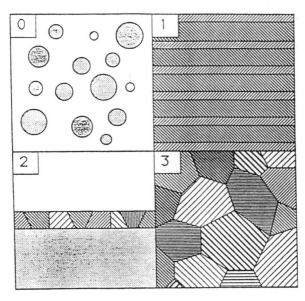

Figure 5.1 Schematic illustration of nanomaterials showing fine structural scale in different numbers of dimensions. These may be zero-dimensioned powders; thin deposited layers showing fine scale in one dimension; surface coatings with two-dimensional nanostructures; and three-dimensional nanostructures. Reprinted from *Nanostructured Materials*, Vol. 3, R. W. Siegel, Nanostructured Materials – Mind over Matter, pp. 1–18, copyright 1993, with permission from Elsevier

mitigated by very slow transformation kinetics which effectively impart an acceptable degree of stability.

Thermodynamic instability comes in two forms: genuine instability and metastability. Metastable structures must pass through an intermediate state of higher energy before a system can achieve the most stable state, thermodynamically speaking; i.e., that with the lowest energy. This necessitates that activation energy be supplied to the system before the reaction can take place, thereby lending a degree of 'stability' to the metastable state. In the case of genuine instability, however, no such activation barrier exists, with the result that the transformation is spontaneous.

In this chapter we are concerned with instability as related, for example, to the transformation from the liquid or vapour phase to the solid under favourable conditions, and structural metastability, which is particularly acute for nanostructured materials as a result of the excess energy associated with grain boundaries and interfaces. The distinction drawn between unstable and metastable transformations is due to Gibbs, and reactions are classified as being Gibbs type I or Gibbs type II transformations. For a Gibbs type I transformation, an energy barrier exists which is related to the nucleation of a stable nucleus of a new phase within the old before this new lower-energy structure may grow. In this process of nucleation and growth, the activation energy barrier results from the presence of an interface between the different phases. This type of transformation is highly localised and is characterised by dramatic changes in atomic arrangements on a small scale.

In contrast, Gibbs type II transformations involve the gradual change from an unstable phase to a more stable form via a series of intermediate steps, each possessing a steadily reducing free energy. Gibbs type II transformations are unlocalised; the whole of the structure transforms simultaneously. The well-known spinodal decomposition of a solid solution (Section 8.6.1) is perhaps the best-known example of this type of transformation.

Both Gibbs I and Gibbs II transformations are thermally activated; that is, they require a thermal activation to facilitate the movement of atoms to allow them to progress. This requirement means that at low temperatures the rate of change will be almost imperceptible; the transformation rates being significant only when atomic diffusion becomes significant (i.e., between $0.3T_m$ and $0.5T_m$, where T_m is the melting point).

The extent to which a particular reaction is feasible is given by the Gibbs free enthalpy function, G, which is defined as

$$G = H - TS, \tag{5.1}$$

where H is the enthalpy, T the absolute temperature and S the entropy of a system. There will be a free energy change associated with any change to the system. If a system changes from an initial state (i) to a final state (f) at constant temperature as a result of a chemical or physical process, there is a free energy change $\Delta G = G_f - G_i$ given by

$$\Delta G = \Delta H - T\Delta S, \tag{5.2}$$

where ΔH and ΔS are the corresponding enthalpy and entropy changes in the system. Reactions occur spontaneously at constant temperature and pressure when ΔG is *negative* and a system will tend to lower its energy by every available route until no further reduction is possible. At this stage $\Delta G = 0$ and we say that the system is in equilibrium. If ΔG is *positive* then the reaction is not possible.

Because it is the sign of ΔG that determines whether a reaction will or will not occur, the entropy change, ΔS, may be negative during a phase transformation. For example, in a transformation from the liquid or vapour phase to the solid, although there are fewer configurations available to the solid state, the overall decrease in enthalpy as a result of the transformation more than compensates for this, and the overall ΔG for the transformation is negative.

To examine its practical implication, let us consider a very familiar phase transition, that of liquid to solid. Upon freezing, a liquid changes from a condition in which there is no long-range order to one in which every atom is associated with a position in a crystal lattice. This type of reaction is defined as a first-order transition; this means that at the point of transition the change in entropy, ΔS, is discontinuous, leading to a discontinuous change in dG/dT. When a liquid turns into a solid at the freezing point T_f, a latent heat ΔH_f (also denoted L_V, the latent heat of fusion per unit volume) is released. Making the simplifying assumption that the enthalpy and entropy changes ΔH_f and ΔS_f do not change significantly as a function of temperature, this may be related to the change in entropy on freezing ΔS_f by

$$\Delta S_f = \frac{\Delta H_f}{T_f} = \frac{L_v}{T_f}. \tag{5.3}$$

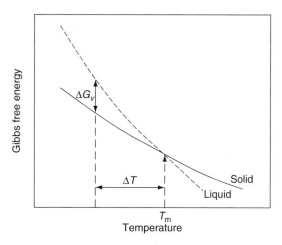

Figure 5.2 Schematic diagram of the variation of Gibbs free energy with temperature near the melting point T_m. At an undercooling ΔT the change in free energy per unit volume on transforming from the liquid to the solid state is ΔG_V

Figure 5.2 shows schematically how the free energy of a pure metal changes with temperature. At the equilibrium melting point, T_m, the free energies of the liquid and solid phases are equal. Below T_m, the solid has a lower free energy and is therefore the stable phase, whereas above T_m the reverse is true. The divergence of the free energy curves as the temperature falls below T_m, produces a steadily increasing driving force for crystallisation. The magnitude of the free energy change on crystallisation can be easily determined using this figure. In this case ΔG_v, the free energy driving the transformation at a small undercooling ΔT, is given by

$$\Delta G_v = \frac{L_v}{T_m} \Delta T \qquad (5.4)$$

A sufficiently large free energy change will produce the driving force for nucleation of a new phase. This is the subject of the next section.

5.2.2 Homogeneous nucleation

For a phase transformation to occur there must be a nucleation event. This process is of vital importance to us in the understanding of the formation of nanostructured materials as the rate at which it occurs will influence the scale of the grain structure that is ultimately developed. The simplest models for nucleation are concerned initially with thermodynamic questions related to the formation of a single stable nucleus. After determining that nucleation may occur, we may then specify a nucleation rate, J, which is the number of stable nuclei forming per unit volume of liquid per unit time. At this point we move from the realm of thermodynamics to that of reaction kinetics. Ultimately, as we shall see in Section 5.2.5, it is these kinetics that will determine the structural scale.

Let us start by asking what energy is required to nucleate a spherical particle of radius r in an undercooled liquid. There will be a negative contribution to the free energy which is proportional to the volume of the sphere; this is because the solid phase has a lower free energy than the liquid at all temperatures below T_m, as discussed previously, and this will lower the free energy by an amount ΔG_V per unit volume. However, formation of a nucleus requires that an interface is produced; there is an interfacial energy between the solid (S) and liquid (L) phases γ_{SL} and this leads to a positive contribution to the free energy which is proportional to the surface area of the nucleus. We can represent the net change ΔG in the free energy on nucleating a droplet of radius r as

$$\Delta G = \frac{4}{3}\pi r^3 \Delta G_V + 4\pi r^2 \gamma_{SL}. \tag{5.5}$$

This function is sketched in Figure 5.3. By minimising the derivative of this free energy change with respect to r, we find a maximum at a critical radius r^*, given by

$$r^* = \frac{2\gamma_{SL}}{\Delta G_V}. \tag{5.6}$$

If a spherical cluster of atoms forms by some thermodynamic fluctuation, but has a radius less than r^*, it will be unstable, in that growth of such a cluster will increase the overall free energy of the system. If however the sphere is of size equal to or greater than r^* these particles will be stable and will grow, leading to a lowering of the system energy.

The energy required to form this critical nucleus ΔG^* may now be easily calculated; it is given by

$$\Delta G^* = \frac{16\pi\gamma_{SL}^3}{3\Delta G_V^2}. \tag{5.7}$$

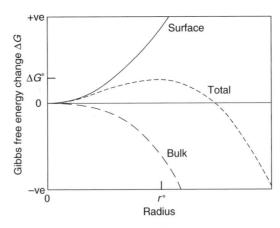

Figure 5.3 Schematic diagram of the variation in Gibbs free energy when a crystal of radius r is nucleated in a melt undercooled by ΔT below its melting point

Remembering from Section 5.2.1 that, for a small undercooling ΔT, ΔG_V may be represented by

$$\Delta G_V = \frac{L_V \Delta T}{T_m}, \tag{5.8}$$

where L_V is the latent heat of fusion per unit volume, we arrive at expressions for r^* and ΔG^* in terms of the undercooling:

$$r^* = \left(\frac{2\gamma_{SL} T_m}{L_V}\right) \frac{1}{\Delta T} \tag{5.9}$$

and

$$\Delta G^* = \left(\frac{16\pi\gamma_{SL}^3 T_m^2}{3L_V^2}\right) \frac{1}{\Delta T^2}. \tag{5.10}$$

Interestingly, these expressions show that the critical radius and the free energy barrier for nucleation decrease as the undercooling increases; nucleation becomes more likely under these circumstances.

Nucleation is an activated process; it can only occur if a thermal fluctuation occurs increasing the local free energy by an amount ΔG^*. The probability that such an event occurs is given by a Boltzmann factor, thus the rate of nucleation of droplets is proportional to

$$\exp(-\Delta G^*/k_B T), \tag{5.11}$$

where the concentration of critical nuclei per unit volume C^* is given by

$$C^* = C_0 \exp\left(\frac{-\Delta G^*}{k_B T}\right), \tag{5.12}$$

and C_0 is the number of atoms per unit volume.

The nucleation rate per unit volume, J, may then be obtained from the product of this concentration and the frequency of atom addition, f_0, to one such critical nucleus:

$$J = f_0 C_0 \exp\left(\frac{-\Delta G^*}{k_B T}\right). \tag{5.13}$$

In the case of a liquid–solid phase transformation where atoms are being added by a diffusive mechanism, f_0 may be approximated by

$$f_0 = \frac{4\pi r^{*2} D_L}{a^4}, \tag{5.14}$$

where D_L is the diffusivity in the liquid and a is the interatomic spacing.

Nucleation frequency, then, is a very strong function of temperature and also, via Equation (5.10) of ΔG^*, the undercooling; it changes from near zero to very high values over a very narrow temperature range. Using typical values, this equation predicts that significant rates will only be observed at significant undercoolings typically of the order of $0.2\,T_m$, yet we know from experience that crystals form at undercoolings of only a few degrees. This is because the theory outlined in this section describing homogeneous nucleation is relevant only to pure liquids and, as we shall see in the next section, the barrier for nucleation is greatly reduced by the presence of impurity particles, a substrate or indeed the walls of the vessel in which the material is placed. Such heterogeneous nucleation is, in practice, much more common than the homogeneous nucleation described here.

5.2.3 Heterogeneous nucleation

When a pre-existing surface or a solid particle is present in the undercooled liquid, this will reduce the activation energy required for the nucleation of a new crystal. This decrease can often be substantial, and in such cases the particle may act as a nucleation site.

Assuming for simplicity that the nucleus which forms on such a foreign body takes the form of a hemispherical cap as shown in Figure 5.4, the relationship between the contact angle, θ, and the interfacial tensions acting on the nucleus γ_{NS}, γ_{NL} and γ_{SL} between nucleant and solid, nucleant and liquid, and solid and liquid, respectively, may be derived by balancing forces at the interface:

$$\gamma_{SL}\cos\theta = \gamma_{NL} - \gamma_{NS}. \tag{5.15}$$

In the case of heterogeneous nucleation the critical energy required to form a heterogeneous nucleus, ΔG^*_{Het}, is reduced by a geometrical factor, $f(\theta)$, which defines the efficacy of the particle as a nucleation site and is defined as

$$f(\theta) = \frac{(1 - \cos\theta)^2(2 + \cos\theta)}{4}, \tag{5.16}$$

so that ΔG^*_{Het} becomes;

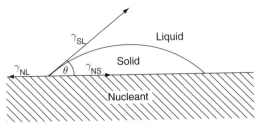

Figure 5.4 Heterogeneous nucleation of spherical cap on a solid surface which acts as a catalyst for crystallisation by lowering the activation energy for nucleation

$$\Delta G^*_{\text{Het}} = \left(\frac{16\pi\gamma^3_{\text{SL}} T^2_{\text{m}}}{3L^2_V}\right)\frac{1}{\Delta T^2} f(\theta). \tag{5.17}$$

Thus, for all contact angles less than 90°, this factor substantially decreases the degree of undercooling required to achieve a measurable nucleation frequency. The most effective nucleation sites are seed crystals of the desired phase, as the contact angle is effectively zero.

This is the principle employed in the refinement of as-cast microstructures in conventional alloys. In this case nucleant particles, termed grain refiners, are deliberately added to increase the nucleation frequency, hence reducing the grain size. This method does not normally allow us to refine structures to below the micron scale; however the concept is useful in understanding those factors influencing formation of nanostructures by devitrification of metallic glasses and we shall come across them again in a later section.

5.2.4 Growth

Beyond the critical nucleus size r^*, atomic addition to the nucleus causes a continuous decrease in the free energy of the system. This means that growth of these post-critical nuclei, or clusters, is spontaneous. The rate of growth is controlled both by interfacial attachment kinetics and by the rate of transport of atoms to the interface of the growing crystallite.

At a growing crystal interface, atoms are continuously being attached and detached via thermally activated processes; the rates of attachment being independent of departure rate. Because there is a reduction in free energy when an atom attaches to an interface, departure is a more difficult process. The probability of the successful attachment of an atom to a growing interface is, however, dependent on the interfacial roughness.

At an atomically smooth interface (Figure 5.5(a)) an adsorbed atom from the liquid that establishes less than half of the atomic bonds that it would have in the bulk will have a low probability of being attached to the growing crystallite and will rapidly detach and return to the liquid. Growth in this case is difficult and relies on mechanisms whereby ledges are produced and propagated, as these are more favourable sites for atomic attachment. It may be demonstrated that the growth rate is proportional to the number density of active sites.

At an atomically rough interface (Figure 5.5(b)) where there are very large numbers of favourable sites for atomic addition, an atom joining from the liquid will make more

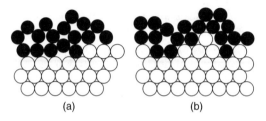

(a)	(b)

Figure 5.5 (a) Smooth and (b) rough solid–liquid interfaces: the black, liquid, atoms are restricted in terms of possible attachment sites in (a) whereas in (b) the number of possible attachment sites is high

than half the bulk atomic number of bonds and have a high probability of being incorporated into the growing crystal.

Although the example given here covers the basic concepts of crystallite growth, the actual growth mode and growth rate for a given material are very dependent on its method of synthesis and this is beyond the scope of this chapter. The interested reader should consult the bibliography at the end of this chapter for advanced study.

5.2.5 Overall transformation rate

The overall transformation rate df/dt, and hence the scale of the microstructure developed, will depend on both the nucleation frequency J and the growth rate R_G:

$$\frac{df}{dt} = f(J, R_G) \tag{5.18}$$

where f is the fraction of the whole that has been transformed. The coupling between these two processes depends on the microstructure and morphology of the structure being formed.

If we assume the transformation to be nucleated homogeneously, at small ΔT, transformation rates are controlled by the nucleation rate, since J is low and R_G is large. This condition results in large grains and coarse microstructures. When ΔT is high, transformation rates are controlled by the growth rate since J is large and R_G is low. This results in a fine-scale microstructure consisting of a large number of very small crystallites or grains. Intermediate to these extremes we would expect to see a situation where mixed control takes place.

If nucleation is heterogeneous, the situation is similar but these events will occur at lower undercoolings. If, however, heterogeneous nucleation occurs at very low undercoolings, all nucleation occurs prior to significant growth and the scale is determined to a great extent by the concentration of heterogeneous nucleants.

5.3 SYNTHESIS METHODS

Many of the techniques used in the synthesis of nanomaterials are basically those which have previously been employed in the manufacture of fine-grained and metastable materials. Modification and control of these processes has led to their being able to generate structures at the nanoscale and to facilitate this, it is vital that the underlying mechanisms responsible for the evolution of microstructure at a specific scale are well understood. For example, in the case of processes involving phase changes; e.g., from vapour or liquid to solid, control of the nucleation and growth kinetics is central to the control of the evolved microstructure. Processes involving the refinement of structure through application of mechanical work, on the other hand, may tend to a limiting minimum structural scale dependent on the maximum stresses that may be evolved during deformation and milling. This method can also be used to produce an intimate mix of particular compounds or elements, thereby allowing us to form nanoparticulate phases by chemical reaction via low-temperature heat treatments.

Synthesis of nanomaterials by all of these routes will be covered in what follows. Space dictates that only a very few of the multitude of synthesis techniques available for the preparation of nanomaterials can be covered in this chapter, yet those which are not covered explicitly are basically variations on the processes we do cover. We shall concentrate on techniques for producing bulk and particulate nanomaterials; techniques for the manufacture of thin films, wires and quantum dots have been covered in Chapters 1 and 3. The bibliography contains suggestions for further study.

5.3.1 Rapid solidification processing from the liquid state

Rapid solidification (RS) processes such as melt atomisation, melt spinning or planar flow casting and their derivatives all generate high cooling rates during solidification. The wide variety of rapid solidification methods available for the manufacture of non-equilibrium materials have been extensively reviewed on a number of occasions and several of these reviews are listed in the bibliography. In all of these RS processes, the liquid is manipulated such that at least one dimension is very small and in good thermal contact with a cooling substrate or fluid jet. The high cooling rates ($\sim 10^4 \, \mathrm{K \, s^{-1}}$ for atomisation and $\sim 10^6 \, \mathrm{K \, s^{-1}}$ melt spinning or planar flow casting) can lead to the formation of non-equilibrium microstructures, with metastable crystalline phases and metallic glasses being typical products.

The very large undercoolings obtained during RS mean that it is possible to generate nanostructured materials directly from the molten state; the nucleation rate being much enhanced and the growth rate suppressed. The latent heat evolved during nucleation and growth causes the temperature of rapidly solidified metals and alloys to increase from its maximum undercooled state. This phenomenon, termed recalescence, must be balanced by the rate of heat extraction if fine structures are to be generated by this technique. In practice this is difficult to achieve and the rise in temperature as a result of recalescence is often of sufficient magnitude to suppress further nucleation events, and hence it leads to the generation of relatively coarse microstructures. Whilst possible then, it is, generally speaking, quite difficult to control. Much greater control may be exercised if nanostructured alloys are derived from rapidly solidified precursors such as amorphous, or glassy, alloys by devitrification.

5.3.2 Devitrification

Controlled crystallisation, or devitrification, of amorphous alloys is a more convenient way of generating a nanostructured configuration than by direct quenching from the liquid state. Comparison of structures produced directly from the melt and those produced by devitrification show that there are no discernible differences between them. During conventional solidification the process is driven by an undercooling of only a few kelvins. Glassy or amorphous alloys, on the other hand, may be considered as deeply undercooled liquids when analysing the nucleation and growth kinetics. The transformation kinetics in this case are sufficiently slow, because of the much slower diffusion rates in the solid, that recalescence effects are negligible. Control of the crystallisation temperature remains critical, however. Glass-forming alloys are multicomponent systems; i.e., they are usually composed of more than two elements, meaning that additional factors such as solute redistribution must be taken into account when analysing nucleation and growth.

Certain factors have been identified as promoting devitrification on a fine, but not necessarily nanometre, scale: the crystallising phase should have a composition distinct from the amorphous precursor; the presence of a very fine dispersion of precipitated phases or quenched-in nuclei which may act as heterogeneous nucleation sites; the alloy should exhibit a tendency to undergo a phase separation prior to crystallisation. The relative rates of nucleation and growth are, as with all other processes, also important to the scale of the structure produced.

The excellent soft magnetic properties generated by devitrification of the FINEMET materials have been a key driving force in the study of devitrification as a route to nanostructured alloys. The composition is basically an FeSiB glass-forming alloy to which small amounts of Cu and Nb have been added, Cu to stimulate nucleation and Nb to suppress crystallite growth. The role ascribed to Cu was for a long time purely speculative and based on the knowledge that the Fe–Cu binary system has a miscibility gap in the liquid state.

It was generally supposed that Cu clusters, approximately 5 nm in diameter, would spontaneously appear in the fully amorphous matrix during the earliest stages of a devitrification heat treatment and that these would promote nucleation of the ferromagnetic α-Fe nanocrystallites. This could be achived either by local depletion of Cu from the glassy matrix or by acting as *heterogeneous nucleation sites* (Section 5.2.3). This process is shown schematically in Figure 5.6. The assumption that Cu clusters serve as heterogeneous nucleation sites for the primary crystals seems quite reasonable, but direct proof that this was indeed the mechanism was not easy to obtain. Studies conducted using atom probe field ion microscopy (APFIM) finally delivered such proof. The rejection of Nb and B during growth leads to an enrichment in the remaining amorphous phase, which becomes stabilised and contributes to approximately 20% of the total volume of material. The alloy is therefore not fully nanocrystalline, but is a nanocomposite consisting of grains of diameter ~10 nm separated by ~2 nm of a glassy

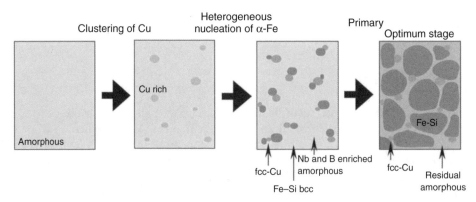

Figure 5.6 Schematic illustration of the microstructural evolution in melt-spun $Fe_{73.5}Si_{13.5}B_9Nb_3Cu$ amorphous alloy by primary crystallisation. Reprinted from *Progress in Materials Science*, Vol. 47, K. Hono, Nanoscale microstructural analysis of metallic materials by atom probe field ion microscopy, pp. 621–729, copyright 2002, with permission from Elsevier

matrix. The matrix further retards growth of the nanocrystals but also performs a crucial role in defining the magnetic properties (Section 5.8.2). Quenched-in nuclei, when present, fulfil a similar role but are not nearly as efficient at refining the grain structure and their presence does not guarantee the formation of a nanocrystalline structure. The miscibility gap responsible for the spontaneous formation of copper in the previous example may cause nanocomposite structures to form in a number of other alloy systems. Glasses based on palladium, copper, zirconium and aluminium have all shown a tendency to phase separation prior to crystallisation to produce two distinct amorphous phases; crystallisation then initiates at the interface between the two amorphous phases. The small fluctuation wavelength of the phase separation acts to minimise the grain sizes generated in this manner.

At present, there is no simple link between glass formation and the potential of an alloy system to yield a nanocomposite structure on devitrification. What is clear, though, is that this method of using metallic glasses as starting materials in the formation of nanostructured alloys has great potential, particularly when considering bulk metallic glass formers as precursors for the manufacture of nanostructured components with the ability of these alloys to be cast or warm forged to shape prior to devitrification being of particular interest.

5.3.3 Inert gas condensation

The inert gas evaporation–condensation (IGC) technique, in which nanoparticles are formed via the evaporation of a metallic source in an inert gas, has been widely used in the synthesis of ultrafine metal particles since the 1930s. A similar method has been used in the manufacture of carbon black, an ink pigment, since ancient times. The technique employed now for the formation of nanopowders, in reality, differs from that used to produce carbon and lampblack primarily in the choice of atmospheric composition and pressure and in the use of a chemically reactive source. Thus, although the technology is old, the application to the production of truly nanoscaled powders is relatively recent.

A schematic representation of a typical experimental apparatus for the production of nanopowders by IGC is shown in Figure 1.26. In its basic form, the method consists of evaporating a metallic source, using resistive heating (although radio frequency heating or use of an electron or laser beam as the heating source are all equally effective methods) inside a chamber which has been previously evacuated to about 10^{-7} torr and backfilled with inert gas to a low pressure. The metal vapour migrates from the hot source into the cooler inert gas by a combination of convective flow and diffusion and the evaporated atoms collide with the gas atoms within the chamber, thus losing kinetic energy. Ultimately, the particles are collected for subsequent consolidation (i.e., compaction and sintering, see Section 5.6), usually by deposition on a cold surface. Most applications of the inert gas condensation technique carry this approach to extremes by cooling the substrate with liquid nitrogen to enhance the deposition efficiency. Particles collected in this manner are highly concentrated on the deposition substrate. While the particles deposited on the substrate have complex aggregate morphology, the structure tends to be classified in terms of the size of the crystallites that make up these larger structures. The scraping and compaction processes take place within the clean environment to ensure powder surface cleanliness (i.e., to reduce oxide formation) and to minimise problems associated with trapped gas.

Although the IGC process generates a very narrow particle size distribution of 3D primary crystallites, typically a few nanometres in diameter, the exact size of the crystallites is very dependent on the type of carrier gas used, its pressure, the evaporation temperature and the distance between source and collecting position. Reducing the gas pressure and lowering the evaporation rate of the source and employing a light gas in the chamber produces a finer particle size. This effect is amplified by using He, which has a very high thermal conductivity. Higher evaporation temperature, inert gas pressure and inert gas molecular weight favours the formation of larger particles. The influence of the most important factors influencing the nucleation and growth of particles during IGC processing will be considered in the following sections.

5.3.3.1 Factors influencing nucleation and growth of fine particles

Nucleation kinetics

The production of fine particles requires that a high nucleation rate is achieved in the vapour whilst restricting the growth rate of these particles and preventing their coalescence. These requirements are conflicting.

In the case of a phase transformation involving homogeneous nucleation from a vapour phase, the volume free energy term is given by

$$\Delta G_V = \frac{RT}{V_m} \ln \Omega, \qquad (5.19)$$

where R is the universal gas constant ($8.314 \, \text{J K}^{-1}\text{mol}^{-1}$), V_m is the molar volume and Ω is the vapour supersaturation, defined as the ratio (P/P_e) where P is the vapour pressure of the element present in the gas and P_e is the vapour pressure in equilibrium with the solid phase at temperature T (K). Using this expression for ΔG_V in Equations (5.5) to (5.14) shows that the critical free energy for nucleation, hence the nucleation frequency of atom clusters from the vapour, are very strongly dependent on the degree of vapour supersaturation. As the metal vapour is convected away from the evaporation source, the dilution of the metallic vapour by inert gas will lead to a decrease in the partial pressure from the saturation condition at the source, seemingly reduction ΔG_V. However, the much more rapid decrease in the equilibrium vapour pressure as the temperature rapidly decreases will more than offset this, resulting in a high vapour supersaturation to drive the homogeneous nucleation process.

At high supersaturations the initial nucleation frequency, J, may be extremely high, leading locally to a swift depletion of metal atoms from the vapour. This has implications for further growth of the particles by condensation, as we shall see in the next section.

Particle growth

Growth by condensation from the vapour

In the case of condensation from the vapour, the growth rate is proportional to the net difference in condensation and evaporation of atoms from the critical nucleus.

The driving force for growth is given by the difference between the instantaneous vapour pressure and the equilibrium pressure local to the growing cluster. The growth rate is therefore

$$\frac{dr}{dt} = \xi V_c \frac{P - P_e}{\sqrt{2\pi M k_B T}}, \tag{5.20}$$

where ξ, the condensation coefficient, is related to the number and nature of attachment sites and varies between 0 and 1, V_c is the molar volume of the condensate and M is the molecular weight of the inert carrier gas.

Any decrease in vapour supersaturation, (P/P_e), will therefore also produce a decrease in the driving force, $(P - P_e)$, for growth. Under ideal conditions the particles will be rapidly transported away, or advected, from the nucleation zone above the source in order to restrict their growth. They will then be deposited as a loosely agglomerated nanopowder at a collector.

The main growth mechanism for nanopowders may not, however, be condensation from the vapour. Alternative mechanisms of particle growth during the synthesis of nanopowders by evaporation or condensation methods, namely coalescence and agglomeration, may be the dominant growth processes.

Growth by coalescence and agglomeration
When the concentration of nuclei above the source is very high, particles may collide. The frequency at which these collisions occur is a function of the particle density per unit volume of gas, the particle diameter and the diffusion coefficient for the powder particle in the gas which is, obviously, dependent on the temperature. The rate of coalescence, on the other hand, is proportional to the diameter of the colliding particles and their viscosities. These also exhibit strong temperature dependences.

When the particles are very small and the gases still relatively hot, they will coalesce quickly after the initial collision, resulting in the formation of larger particles with a smaller surface area. At lower T the collision rate is greater than the rate of coalescence, and if there is still sufficient thermal energy to drive diffusive processes, this results in the formation of fractal-like agglomerates consisting of small nanocrystallites (Section 5.6.2). Although, in this instance, the fine nanometric structure of the colliding particles may still be preserved, the resulting agglomerates are actually much larger and may be quite rigid. These particles will have a large surface area.

The strength of the bonds which form between the nanoparticles in agglomerates may vary widely depending on material properties, gas atmosphere, and the temperature–time history that the particle has experienced. They may be soft, van der Waals type bonds or solid connections in the form of small necks which will give rise to hard, sintered agglomerates. Soft agglomerates, being formed at temperatures too low for significant coalescence, may be separated into their component nanocrystallites with relative ease, thus facilitating subsequent consolidation. Consolidation of powders containing hard agglomerates, on the other hand, is much more problematic and their presence frequently results in consolidates with densities well below the theoretical maximum (Section 5.6).

5.3.3.2 Improving yield, powder quality and production rate

The IGC process is relatively inefficient, as a result of powder condensation on the chamber walls, and the production rate is low. Moreover, the process as it now stands is batch based. Modification of the process to simultaneously increase the number concentration of nanoparticles produced and decrease the average particle size would seem to be in order if the process is to be scaled up to large volume manufacture of nanopowder.

The most obvious way of improving yield would appear to be to increase the evaporation rate from the source. Unfortunately, this has the effect of increasing the number of clusters within the nucleation zone, hence increasing the probability of interparticle collision and therefore the formation of agglomerates or large coalesced particles. If, however, enhanced evaporation of the source could be coupled with a more efficient method than free convection for transporting the powders away from the nucleation zone, this could lead to a significant improvement in terms of process efficiency and flexibility, with respect to product size, as well as increased material production rates.

The aerosol-flow condenser is a technique which has grown from this line of reasoning. In this jets of a carrier gas flow over an evaporation source, carrying away metal vapour. As the mixture of vapour and carrier gas flows downstream and cools, particles form and grow by the processes of nucleation, condensation and coagulation in a continuous process. By control of evaporation rate and gas flow rates, the cooling rate of the aerosol (the mixture of gas and particulate) may be varied to limit the extent of coalescence and agglomeration whilst simultaneously allowing greater control of the particle size and of the particle size distribution (PSD) generated.

5.3.4 Electrodeposition

The formation of nanoparticles and bulk nanostructured materials (BNMs) by electrodeposition; i.e., the deposition of a conductive material from a plating solution by the application of electrical current, may be achieved using either two separate electrolytes or, more conveniently, from one electrolyte by exercising appropriate control of the deposition parameters. These processes can be applied to the synthesis of pure metals, alloys and composites alike, have few size and shape limitations, and require only modest initial capital investment. Furthermore, and rather importantly from the point of view of its future potential as a synthesis route, the material production rate is high.

While many of the processes associated with the crystallisation from an electrolyte are, as yet, not well understood, it is clear that the electrodeposition of nanostructured materials is possible if a large number of nuclei form on the surface of the electrode and their growth is severely restricted. These requirements may be met by employing a variety of techniques either singly or in combination, including the use of high current density for a short duration, the use of organic additives as growth inhibitors, control of bath temperature, and manipulation of bath pH. The role of each of these factors in the control of microstructure evolution during electrodeposition will now be briefly examined.

5.3.4.1 Pulsed electrodeposition

Pulsed electrodeposition (PED) differs from the more normal direct current (DC) electroplating techniques in that the peak current density employed is very high ($1-2\,A\;cm^{-2}$), much larger than that possible in DC electroplating, but that the duration of this high current density is very short. The deposition rate in the PED technique is governed by a number of controllable parameters, including the pulse duration, the time between pulses, and the pulse current. Cations are depleted from the electrolyte adjacent to the electrode during the application of the pulse and their levels are restored by the diffusive transport of material to this depleted region between pulses. The size of the nuclei is significantly dependent on both pulse duration and current density, whilst a balance must be struck between the replacement of cations close to the electrode by diffusion and the minimisation of grain coarsening when determining an appropriate interpulse time.

5.3.4.2 Growth inhibitors

Addition of suitable organic compounds to the electrolyte leads to the inhibition of grain growth. These adsorb on the electrodeposit, in a reversible way, and block active sites for growth, such as surface steps, hence leading to a reduction in the growth rate of crystallites. They also act to inhibit surface diffusion of adatoms, so that nucleation becomes the preferred method of electrolyte depletion. The crystallite size depends in this instance on the concentration of inhibitor in the electrolyte.

5.3.4.3 Bath temperature

At high temperature, adatom diffusion at the surface is much enhanced, thereby resulting in a much higher rate of grain growth. Inhibiting molecules also decrease in efficacy because the increased rate of desorption from growth sites as a result of the higher temperature will also increase the number of active attachment sites. On the other hand, reducing the bath temperature will contribute to the inhibition of growth. Changes in electrolyte temperature can also change the width of the grain size distribution as well as the average grain size.

5.3.4.4 Electrolyte bath pH

Control of the redox potential of the species by addition of a complex former which alters the redox of the desired species is particularly important if alloys with components of widely differing redox potentials are to be electrodeposited. If this equilibrium of the complexes depends additionally on the pH of the solution, a defined redox potential and therefore a defined alloy composition may be obtained by variation of the pH value.

5.3.4.5 Products

In addition to the obvious plating and surface coating of objects, nanostructured electrodeposits may be manufactured with complex geometries simply by altering the cathode shape. In a similar way, the process may also be modified to allow the production of different thicknesses of sheet or even powders.

The electrodeposition of nanostructures has an advantage over many other methods of synthesis in that it involves a logical extension of current industrial capability with respect to the commercial electroplating of materials. Nanomaterials can be manufactured easily using existing apparatus by a simple modification of bath chemistry and the plating methods used in current plating and electroforming operations. The industrial advantage of this method for the manufacture of fully dense nanomaterials via a one-step operation should not be underestimated.

5.3.5 Mechanical methods

5.3.5.1 Mechanical alloying or mechanical milling

Mechanical alloying (MA) or mechanical milling (MM) is a dry, high-energy ball milling technique. Strictly speaking, the term mechanical alloying is restricted to the formation of alloys or mixtures by mechanical means whereas mechanical milling is intended to describe the process of milling powders to reduce the particle size or for the refinement of structure. Since the original development of MA, as a way of incorporating oxide particles into nickel-based superalloys intended for application in gas turbines, it has been used in the preparation of a very wide range of materials from oxides to amorphous alloys and latterly, as MM, in the synthesis of nano-structured metals and alloys from atomised powders.

The technique is simple. Powders, typically 50 µm in diameter, are placed with hardened steel, ceramic or tungsten carbide (WC) balls in a sealed container and shaken or violently agitated. A high energy input is achieved by using a high frequency and small amplitude of vibration during milling; the collision time under these conditions is generally estimated to be of the order of 2 µs. There is a rise in the system temperature associated with such violent deformation but this is actually quite modest, typically 100–200 K.

Various ball mills are used to produce MA/MM powders, including tumbler mills, attritor mills, vibratory mills and planetary mills. They differ in capacity and milling efficiency but the process is effectively the same in each case. The simplicity of the apparatus allows it to be scaled up to the production of tonne quantities with relative ease and has the added advantage that both the starting materials and the product materials are in the form of relatively coarse (∼50 µm) powders.

Problems inherent to the technique, such as contamination by reaction with the atmosphere or by wear of the milling medium and also the need to consolidate the powder product after synthesis without coarsening the nanostructure, are often used as arguments against implementing it as a method of nanostructured material production. Such contamination can, however, be minimised by limiting the milling time or by using a protective milling atmosphere. Careful control of the processing conditions can also limit contamination by wear debris arising from the milling media to $\sim 1-2$ at% and oxygen and nitrogen levels for MA/MM materials can be less than 300 ppm. This, argue the proponents of the technique, is a higher purity of product than that achieved for materials synthesised by means such as IGC. In addition, the particle size of the product ($\sim 50\,\mu$m) means that there are fewer problems associated with porosity when consolidating than for true nanopowders.

In the case of nanostructured metals produced by mechanical milling, the structure is generated by the creation of a deformation substructure. During the initial stages of the milling process, a high dislocation density, ρ_d, is generated within each of the grains of the starting material as a result of repeated deformation by the milling medium. At a certain level of strain these dislocations start to annihilate and recombine, a process referred to as recovery, thereby reducing the overall dislocation density. The dislocations form loose tangles or networks which then evolve with increasing strain to produce low-angle grain boundaries (LAGBs) that separate individual subgrains. Each subgrain shows a small orientation difference or misorientation with respect to its neighbours. As the deformation proceeds, this misorientation increases by incorporation of further dislocations into the boundaries, causing them to gradually change in character, finally becoming high-angle grain boundaries (HAGBs). A high initial density of dislocations is therefore required to facilitate grain refinement by ball milling. It is through this process of continuous grain subdivision and dislocation accumulation in the boundaries that the observed three to four orders of magnitude reduction in grain size is achieved.

There is, however, a natural limit to the number of dislocations which may be present in a material. This limit is reached when the number of dislocations being produced by deformation and the number being annihilated by recovery are equal.

In metallic systems the grain size decreases with milling time, reaching a minimum grain size $\langle d \rangle_{\min}$, characteristic for each metal. The minimum grain size that may be achieved during room temperature milling is found to vary roughly as the inverse of the absolute melting point, T_m, and also to be a function of the crystal structure of the metal. This is illustrated by Figure 5.7, which shows the variation of $\langle d \rangle_{\min}$ with T_m for face-centred cubic (fcc) metals.

The present explanation for this behaviour is that, in metals with low melting points, the dislocation density is limited by recovery. Recovery is a thermally activated process, the activation energy for which is close to that for self-diffusion and it therefore scales with the alloy melting point (Section 5.5.1). In low melting point metals, recovery can be significant at room temperature and it is considered that it is this process which limits the final grain size rather than the deformation supplied by the mill. In contrast, for metals of high melting point such as the refractory metals, there will be almost no recovery at the milling temperature. The minimum grain size in this case is therefore limited by the stress required to generate and propagate new dislocations rather than the rate of recovery. The Hall–Petch equation (1.23) sheds some light on why a limiting value of grain size, generated by deformation, may occur. It is evident from Equation (1.23) that a decrease in $\langle d \rangle$ will give rise to an increase in yield strength and the most usual explanation of this relationship is based on a calculation of the stress ahead of a large array of dislocations piled up at a grain boundary. As shown in Figure 5.8(a),

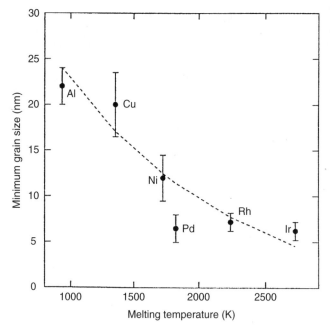

Figure 5.7 Minimum grain size obtained during mechanical milling for different fcc metals plotted against melting temperature. From J. Eckert, J. C. Holzer, C. E. Krill III and W. L. Johnson, *J. Mater. Res.* **7**, 1754 (1992)

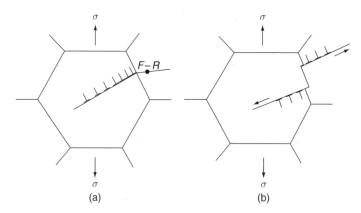

Figure 5.8 Two models proposed to explain the observed Hall–Petch behaviour in microcrystalline materials. F-R signifies a Frank-Reed type dislocation source is operating in this case. Details of both are given in the text: a) in this section and b) in section 5.7.1

the length of this pile-up can be assumed to be of the same magnitude as the grain size, d, and when the stress from the pile-up is large enough to induce plastic deformation in an adjoining grain, we say that the yield stress has been reached. If, during milling, a maximum deformation stress per collision, σ_{max}, can be generated, the Hall–Petch relation suggests that there exists a structural scale at which the material yield stress will exceed this. At this value of $\langle d \rangle$, deformation will no longer proceed by the generation and propagation of matrix dislocations; the minimum grain size will then have been attained. This competition between the deformation and recovery behaviour is often cited as the reason for the generation of a characteristic $\langle d \rangle_{min}$ for a given material.

Once an entirely nanocrystalline structure is formed, further deformation is accomplished by grain boundary sliding (GBS) and there is no further refinement of structural scale. GBS does, however, accommodate deformation energy; this is stored within the grain boundaries themselves and also in the form of strains within the nanograins which arise because of grain boundary stresses. Grain boundary energies may, as a result, be higher than expected, as much as 25% higher than the value for the grain boundary energy in a coarse-grained sample.

5.3.5.2 Mechanochemical processing

Mechanically activated exchange reactions have also been examined as possible methods for the synthesis of ultrafine and nanoscale powders. In this process, chemical precursors undergo reaction, either during milling or subsequent low-temperature heat treatment, to form a nanocrystalline composite consisting of ultrafine particles embedded within a salt matrix. The ultrafine powder is then recovered by removing the salt through a simple washing procedure. This process is termed mechanochemical synthesis and, to date, it has been successfully applied to the preparation of a very wide range of nanomaterials, including transition metals, magnetic intermetallics, sulfide semiconductors and oxide ceramics.

The technique is essentially equivalent to that for MA. In this process, however, the starting powders are typically in the form of reactant compounds. These are milled to form an intimate mixture of the reactants which either react during milling or during a subsequent low-temperature heat treatment, to yield a product and a by-product phase. Generally speaking, the product is finer if the reaction can be made to occur in a controlled manner during heat treatment.

The severe microstructural refinement imparted by the high-energy milling step, however, has the effect of decreasing the effective diffusion distance between the reactant phases. The net effect is that the chemical reactivity of the mixtures is increased, and if the energetics of the reaction is favourable, this will occur during the milling operation after a given milling time has elapsed.

To facilitate greater control of the reaction a diluent phase, usually the same compound as the reaction by-product, is often added to the milling mixture. This leads to a reduction in the overall frequency of reactant–reactant collisions during milling, thereby reducing the reaction rate and the rate of heat generation. The diluent also increases the diffusion distance between crystallites of the product phase, such that coarsening of the nanocrystallites during the heat treatment stage will be inhibited by its presence.

As with mechanical attrition, the process can be readily scaled up to the production of commercially viable quantities of nanopowder. Furthermore, the simultaneous formation of ultrafine particles with an intervening salt matrix suggests that agglomerate formation can more readily be avoided than is possible with other synthesis techniques since the salt matrix inherently separates the particles from each other during processing. The mechanochemical synthesis technique therefore allows significant control to be exercised over the characteristics of the final washed powder. The method has significant potential for the low-cost synthesis of a wide range of ceramic and metallic nanopowders, including the production of UV absorbers such as ZnO and TiO_2 for the cosmetics industry. The unique combination of the band gap structure of these semiconducting oxides, which results in their absorbing strongly in the UV part of the spectrum, and the very fine scale of the powders, so that the particles do not strongly scatter incident light, make them excellent alternatives to organic based sunscreens.

5.4 STRUCTURE

The total intercrystalline region – grain boundaries (where individual grains meet to form an interface), edge length (lines where three grains meet) and vertices or triple points (higher-order junctions) – is important in the microstructure of a material at any length scale. Grain boundaries act as rapid diffusion routes and barriers to dislocation motion; grain corners and edges may act as nucleation sites for recrystallisation or may have mechanical properties different from those of the bulk. As the average grain size, $\langle d \rangle$, falls below 100 nm, however, the relative proportion, and hence the overall influence of these geometrical components of the microstructure on the material behaviour, increases dramatically. If the mean diameter of grains in the assembly is $\langle d \rangle$, then the number of grains per unit volume, N_V, is proportional to $1/\langle d \rangle^3$ and the mean grain boundary area A_V, mean edge length L_V and mean number of corners C_V per unit volume are given by:

$$A_V = C_A/\langle d \rangle, \tag{5.21}$$

$$L_V = C_L/\langle d \rangle^2, \tag{5.22}$$

$$C_V = C_C/\langle d \rangle^3, \tag{5.23}$$

where the constants C_A, C_L and C_C are dependent on the assumed grain geometry. In the case of a grain boundary of thickness δ (typically 0.5 nm), the volume fraction of atoms in the grain boundaries, F_V, is given by

$$F_V = C_A\delta/\langle d \rangle. \tag{5.24}$$

What is immediately obvious, however, from this expression is that F_V becomes important only as δ and $\langle d \rangle$ become comparable in size, and gains greater significance the closer they become to each other. In the case of nanomaterials, for example, the fraction of atoms per unit volume considered to be existing within the boundary changes very rapidly, varying from 0.03 when $\langle d \rangle$ is 100 nm to greater than 0.5 when $\langle d \rangle$ is 5 nm. The distance between atoms in the boundary is slightly altered with respect to that in the bulk and the structure is somewhat looser. This presence of such a high fraction of 'loose' structure has been employed in the explanation for the deviation of many properties of nanomaterials from those observed on the conventional scale. In some cases, however, as we shall see later in this chapter, the properties were later found to result from poor compaction, the presence of defects or other causes related to processing, rather than as a result of structural relaxation effects.

In addition to the fact that grain boundaries are essentially non-equilibrium defects, such an excess of intercrystalline material makes nanomaterials thermodynamically unstable and provides a driving force for structural coarsening. It is of little surprise that the processing of nanomaterials is concerned, to a very great extent, with the suppression of grain growth.

5.4.1 Microstructure

5.4.1.1 Grain size and matrix strain

As stated, the grain size is the most important structural parameter in nanomaterials. The properties of nanomaterials depend on it and thus, an accurate method of determining grain size is of paramount importance. However, the uncertainty in grain size measurement for nanomaterials is much greater than for materials with conventional microstructural scale. As such, relationships for these materials derived directly from experimental data should be viewed with some caution.

There are a number of analytical techniques available to us for measuring grain size, microstrain and also the apparent nanoparticle size, including X-ray diffraction, transmission electron microscopy (TEM) techniques and measurement of specific surface area of nanopowders.

The width of the Bragg reflection in a standard X-ray diffraction pattern can provide information on the average grain size. The peak breadth increases as the grain size decreases, because of the reduction in the coherently diffracting domain size, which can be assumed to be equal to the average crystallite size. If the broadening arises entirely as a result of size effects, the full width at half maximum (FWHM) true peak breadth (corrected for instrumental broadening and measured in radians), β, is related to the average grain size $\langle d \rangle$ via the Scherrer equation:

$$\beta = \frac{0.9\lambda}{\langle d \rangle \cos \theta},$$

(5.25)

where λ is the wavelength of incident X-ray radiation, and θ is the Bragg angle for the peak being analysed. The error in $\langle d \rangle$ obtained by this technique is of the order of 10%, meaning that it is difficult to measure small changes in grain size with great certainty. On the positive side, the measurement is simple to perform and the volume sampled is large compared with other techniques such as TEM. Peak breadth may also be influenced by the degree of microstrain and the peak breadth for an exclusively strain-broadened profile is approximately related to the root mean square strain, $\langle e^2 \rangle^{1/2}$, thus

$$\beta = k \langle e^2 \rangle^{1/2} \tan \theta,$$

(5.26)

where k is a constant depending on the strain distribution in the specimen. Where both size and strain components are present these may be separated. These methods are quite involved and beyond the scope of this chapter, but several useful references are included in the bibliography for the interested reader.

TEM techniques are ideal for the direct measurement of grain size, grain size distributions and any structural non-uniformity such as a bimodal grain size distribution. This type of measurement is however, rather labour-intensive and a very large number of grain size measurements are needed before the results can be considered to be statistically significant. Nevertheless, it is a useful complementary technique.

5.4.1.2 Particle size measurement

The average particle size $\langle D \rangle$ of a nanopowder may be derived from a measurement of its specific surface area (area per unit mass), A_m, via the relation

$$A_m = \frac{6}{\rho_{th} \langle D \rangle}, \tag{5.27}$$

where ρ_{th} is the theoretical density of the material. This will give a value for the apparent particle size of a powder which may differ significantly from the nanocrystallite grain size derived by X-ray analysis.

Comparison of measured values of grain or particle size derived by different techniques is a good guide to the overall quality of a true nanopowder; close correlation between the results of different measurement techniques being indicative of a powder of good quality with respect to average particle size, size distribution and particularly the degree of particle agglomeration and coalescence.

5.4.2 Grain boundary structure

It has been proposed that the grain boundaries in nanomaterials are different in character from those in the same material in the coarse-grained condition. Those proposing these differences postulated that the boundaries between nanocrystallites might have unique characteristics resulting from the constraints imposed on the grain boundary atoms as a result of their synthesis by consolidation of nanopowders.

High-resolution TEM work has shown that there is in fact no difference in the type of grain boundary structures observed in nanomaterials when compared with the coarse-grained version of the same material produced by the same route. The low density of grain boundaries in consolidated nanopowders may therefore result from a high fraction of distributed nanoporosity. This issue remains unresolved.

5.4.3 Structural metastability

As well as the morphological metastability, nanomaterials may also exhibit additional topological metastabilities (e.g., as a result of a different crystal structure from that at equilibrium) or compositional metastabilities (e.g., as a result of the occurrence of extended solid solubilities) which impart unusual and sometimes useful characteristics to the material. There are numerous reviews of the techniques and effects of the processing of metastable materials, most of which are related to rapid solidification (RS) processing. This chapter is concerned with factors influencing the retention of nanostructure, so these other forms of metastability, although important, are beyond its scope. A number of reviews are included in the bibliography for further study.

5.5 MICROSTRUCTURAL STABILITY

As discussed in Section 5.2, most useful engineering materials are metastable to a greater or lesser extent. Thus all may revert to a lower energy state if sufficient thermal energy is imparted to allow the activation barrier for this transformation to be overcome. Nanostructured materials are highly metastable, largely as a result of the enhanced number density of grain boundaries. The properties that engineers and scientists wish to exploit are, however, strongly dependent on the microstructural scale. The manufacture of useful components from nanomaterials often necessitates a consolidation step, and such consolidation processes normally require that the powders are exposed to relatively high temperatures. At these temperatures, when diffusion becomes significant, the nanograins may exhibit rapid rates of growth leading to an unacceptable coarsening of the structure, thereby returning the material properties to those of their coarse-grained counterparts.

This section discusses several phenomena of importance to the processing of nanomaterials, namely, diffusion, grain growth and pinning mechanisms. In the sections that follow their relevance to consolidation processing will also be considered.

5.5.1 Diffusion

Chemically driven atomic motion in crystalline materials is described by Fick's laws of diffusion. Fick's first law states that the mass flux per unit area J is proportional to the concentration gradient present in the material:

$$J = -D\frac{dC}{dt}. \tag{5.28}$$

Local conservation of material is expressed by the equation of continuity:

$$\frac{dC}{dt} = -\frac{dJ}{dx}. \tag{5.29}$$

Combining this with Equation (5.28) leads to Fick's second law:

$$\frac{dC}{dt} = D\frac{\partial^2 C}{\partial x^2} \tag{5.30}$$

The constant of proportionality in these equations, D, is defined as the diffusion coefficient and is itself governed by an Arrhenius-type expression:

$$D(T) = D_0 \exp\left[\frac{-Q_D}{RT}\right] \qquad (5.31)$$

where D_0 is a pre-exponential factor and Q_D is the molar activation energy for diffusion. If this diffusion occurs by transport through the crystal lattice, we term this lattice or volume diffusion and the activation energy is denoted as Q_V.

Diffusion is, however, much more easily accomplished along surfaces and grain boundaries, or via dislocations, than through a lattice, because these features have a more open structure. This type of 'short circuit', grain boundary diffusion may also be well represented by Equation (5.31), where the pre-exponential term is now D_{B0} and the activation energy is for boundary diffusion, Q_B.

Where there is diffusion both through the lattice and via a high-diffusivity path, it is more appropriate to use an apparent diffusion coefficient, D_{eff}, which is given by the expression

$$D_{eff} = D_V + \frac{\delta}{\langle d \rangle} D_B, \qquad (5.32)$$

where $\langle d \rangle$ is the average grain size and δ is the grain boundary width (taken to be $\sim 0.5\,nm$) in the case of grain boundaries acting as the high-diffusivity path. When diffusion is enhanced by the presence of these dislocations, the expression for D_{eff} is

$$D_{eff} = D_V + a_c \rho_d D_p, \qquad (5.33)$$

where ρ_d is the dislocation density (with units m^{-2}), a_c is the area of the dislocation core associated with rapid diffusion (m^2), and D_p is the pipe diffusion coefficient.

From these expressions it is immediately obvious that grain boundary and dislocation pipe diffusion will dominate when the second term on the right-hand side of Equations (5.32) and (5.33) is greater at a given temperature than the lattice diffusion term. When the high-diffusivity path contribution to D_{eff} far exceeds that of lattice diffusion, the contribution of the lattice to the overall mass transport may be ignored.

It is found that, for a given crystal structure and bond type, the activation energy is approximately proportional to the melting temperature, and that the diffusion coefficient at the melting temperature and the pre-exponential factor, D_0, are approximately constant. Within each class and for each type of diffusion path, the diffusion coefficient exhibits the same relationship with respect to the absolute temperature T. In most close packed metals the activation energy for lattice diffusion and grain boundary diffusion may be approximated as $18RT_m$ and $10RT_m$, respectively with $D_0 = 5.4 \times 10^{-5}\,m^2\,s^{-1}$ and $\delta D_{B0} = 9 \times 10^{-15}\,m^3\,s^{-1}$; for bcc these activation energies are approximated by $18RT_m$ and $12RT_m$ with $D_0 = 1.6 \times 10^{-4}\,m^2\,s^{-1}$ and $\delta D_{B0} = 3.4 \times 10^{-13}\,m^3\,s^{-1}$.

Using these typical values we may deduce that the dominant mass transport mechanism for both crystal structures is indeed grain boundary diffusion for all $T \leq T_m$ when

$\langle d \rangle$ is less than 100 nm. In coarse-grained materials this tends to be the case only at lower temperatures, typically less than $0.5T_m$, where diffusion is, in any case, rather limited.

5.5.2 Grain growth

Grain growth occurs in polycrystalline materials to reduce the overall energy of the system by reducing the total grain boundary energy. Grain boundaries are a non-equilibrium defect; a single-phase material is most thermodynamically stable when all boundaries are removed. In nanomaterials, where the grain boundaries constitute a significant proportion of the total volume, this tendency for growth should be expected to be particularly strong and the main task during processing of nanomaterials is, as far as possible, to prevent this from happening.

If we assume that the mean curvature of boundaries in all grains in a polycrystalline aggregate to be inversely proportional to the average grain radius $\langle r \rangle$, then the mean driving force for growth is given by $\alpha \gamma_b / \langle r \rangle$, where γ_b is the grain boundary energy (usually assumed to be $\sim 1/3$ of the surface energy of the material γ_S) and α is a geometric constant (equal to 2 if the grains are considered to be spherical). Assuming that the grain boundary energy is isotropic (i.e., equal for all grain boundaries and in all crystallographic directions) and that the boundary velocity is proportional to this driving force, the average growth velocity of the grain assembly is given by

$$\frac{d\langle r \rangle}{dt} = \alpha \frac{\gamma_b}{\langle r \rangle}. \tag{5.34}$$

Integrating this expression gives the following relationship for grain growth as a function of time:

$$\langle r \rangle^2 - \langle r_0 \rangle^2 = Kt, \tag{5.35}$$

where K is given by

$$K = K_0 \exp\left(-\frac{Q_G}{RT}\right) \tag{5.36}$$

and Q_G is the activation energy for grain growth.

Thus, for the ideal case of an isotropic grain assembly consisting of equiaxed grains of radius $\langle r \rangle$, growth obeys a parabolic rate law. By determining the Arrhenius rate parameter K over a range of temperatures, the activation energy for grain growth, Q_G, may also be determined.

In practice, experimental data shows that grain growth can be described by a similar rate law expression, but that the observed behaviour is rarely parabolic. The general form of the expression for grain growth is therefore taken to be

$$\langle r \rangle^n - \langle r_0 \rangle^n = Kt, \tag{5.37}$$

where n is found to vary over a wide range and also to be a strong function of temperature.

At low normalised temperatures (T/T_m), n typically takes a high value, in common with conventional materials. n decreases towards 2, the theoretical value, as the annealing temperature increases towards the melting point. It is interesting to note that the values for the exponent show the same overall trend; i.e., tending to be high at low T/T_m and decreasing as T/T_m increases, irrespective of whether the material is metallic or ceramic, conventional or nanostructured. There are of course notable exceptions, such as n-Al_2O_3, where the exponent has been observed to be 4 at all T/T_m. Nevertheless, the general trend holds true.

This variation in the value of n occurs essentially because materials are non-ideal and, as the driving force for growth is in any case very small, any deviation from this idealised behaviour will exert a significant influence on the observed grain growth kinetics. In reality, grain assemblies may be non-equiaxed or there may be a wide grain size distribution present. Such departures from ideality may be responsible for the high values of n during the early stages of grain growth. It is also highly likely that there will be some degree of anisotropy and non-uniformity in the grain boundary energy, as nanoparticles may also exhibit faceting behaviour.

The activation energy for grain growth is, generally speaking, more difficult to determine reliably, particularly given the large deviation of n from ideality. However, to date, experimental studies have tended to show that the activation energies in a particular nanomaterial are very similar in magnitude to those for grain boundary diffusion. Here again, there are some notable exceptions, in particular nanocrystalline Fe, for which a low value is observed for low T/T_m (125 kJ mol^{-1}), whereas a much higher value is obtained at high T/T_m (248 kJ mol^{-1}). The higher value here is more representative of the activation energy for lattice diffusion and is similar to the value obtained for coarse-grained Fe, whereas the low-temperature value is actually smaller than expected for grain boundary diffusion in Fe.

That we have a metastable microstructure which is prone to coarsening is therefore not in doubt, and the issue that must be addressed is that of the retention of this nanostructure during subsequent processing (e.g., sintering and compaction) to form bulk materials.

5.5.3 Zener pinning

To prevent growth of nanomaterials during processing, the motion of the grain boundaries must be impeded. This can readily be achieved via the pinning effect of fine pores or second-phase particles. We will call both of these impediments obstacles. Moving boundaries will become attached to these obstacles and they, in turn, will exert a retarding force on the boundary, proportional to the length of the boundary attached to them. The maximum force which may be exerted by a single particle on such a moving boundary is

$$F_{max} = \pi r_p \gamma_b, \tag{5.38}$$

where γ_b is the grain boundary energy and r_p is the obstacle radius. If the number of obstacles per unit volume is f_r, then the number of obstacles per unit area intersecting a randomly drawn plane is given by $3f_r/2\pi r_p^2$, and the retarding force per unit area of that plane is then approximated by

$$\frac{3f_r}{2\pi r_p^2} \pi r_p \gamma_b = \frac{3f_r \gamma_b}{2r_p}. \tag{5.39}$$

This force opposes the driving force for grain growth (5.34) so that, for growth to occur, the driving force must be greater than the retarding force; i.e.:

$$\frac{\alpha \gamma_b}{\langle r \rangle} > \frac{3f_r \gamma_b}{2r_p}. \tag{5.40}$$

In otherwords, there exists a critical ratio f_r/r_p above which no grain growth may occur. Thus, if such a stable dispersion of obstacles were to be introduced into a microstructure (a) by compaction to produce a fine distribution of pores, (b) by precipitation of a second phase or (c) by contamination from surface oxides, for example, then the scale of that microstructure could be stabilised. This will only succeed, however, if the particles present on the boundaries do not coarsen; coarsening would obviously lead to a change in the ratio f_r/r_p as a result of an increase in r_p, hence reducing the efficacy of pinning. Pinning may also, under certain conditions, lead to what is termed abnormal growth. This occurs when rapid coarsening of precipitates (i.e., the reduction in their number density and increase in the average pinning particle size) on some grain boundaries allows them to break free and move. In this situation a bimodal grain size distribution is rapidly established in which very large grains are surrounded by, and are growing into, fine-grained regions. Therefore care must be taken to select a suitable pinning phase when employing this effect.

5.5.4 Solute drag

A second commonly used retarding effect is that of solute drag. The degree to which a solute segregates to a grain boundary is inversely related to the solid solubility of that element in the matrix. The solute drag effect is, however, quite a complex phenomenon; at low boundary velocities the drag force is inversely proportional to the solute concentration at the boundary whilst, when the solute concentration is low or there is an increased driving force for growth, the solute can exert no net influence on the boundary mobility. The apparent activation energy for grain growth will also decrease as the temperature is increased since the effect of the solute is less pronounced at higher temperatures.

As the boundaries of nanomaterials exhibit higher solute atom solubility than those of coarse-grained materials, solute drag effects may be expected to be even more pronounced. A further possible influence of this enrichment is that it may lead to a significant decrease in the driving force for grain growth by reducing γ_b and thus, according to Equation (5.15), reducing the driving force for grain growth. A good example of this type of growth inhibition is that exhibited by nanocrystalline Cu–Bi. The very low solid solubility of Bi in Cu promotes its segregation to grain boundaries and the grain growth rate of Bi-doped nanocrystalline Cu is therefore much reduced in comparison to the non-doped nanocrystalline Cu. This effect has been attributed to the action of Bi in reducing γ_b. Similar strategies are also effective in the retention of ceramic nanostructures, for example the low-level doping of TiO_2 by yttrium has been found to be useful in retaining sub-100 nm grain sizes during sintering.

There are, however, some some negative aspects to adopting doping and solute drag strategies for retaining fine grain sizes. Their presence may for example be highly detrimental to properties (e.g., mechanical, magnetic or catalytic) and, as with all processing, this must be considered when designing the material composition and selecting the most suitable processing route.

5.6 POWDER CONSOLIDATION

Techniques employed to date in the consolidation of nanopowders to form bulk nanostructured materials (BNMs) have been borrowed, to a large extent, from the powder metallurgical and ceramic processing industries. However, because of the very small size of the powder particles, special precautions must be taken to reduce the interparticle friction, and hence heating, and also their activity, to minimise the danger of explosion or fire.

Close control of the consolidation stage is essential to the scientific study of nanomaterial properties and to their use in engineering applications. A number of apparently remarkable changes in mechanical and magnetic properties of materials attributed to the nanoscale structure during the early stages of nanomaterial development were later found to be a processing artefact. A good example is the Young's modulus of elasticity, E, which was initially thought to be significantly lower for nanomaterials than for those same materials at conventional grain sizes. It is now almost universally accepted that this observation was entirely attributable to the presence of extrinsic defects such as pores or cracks which occurred during the consolidation stage of processing. This example highlights the greatest problem associated with the study of the properties of nano-materials – the production of high-quality, fully dense BNMs. The consolidation stage need not be viewed in such negative terms, however, as at this stage we may chose to combine different nanopowders to form composites, either directly by the combination of desired phases in desired quantities or by mixing precursor nanopowders and promoting reactions by subjecting these mixtures to heat treatment under controlled conditions.

The various stages of the consolidation process will be considered briefly here. In addition, novel sintering techniques that have been applied to nanomaterial consolidation and comment on their potential for exploitation will be briefly considered. As with other areas of the topic, this is presently a very active research area and our knowledge and understanding of the phenomena are growing and improving continuously. A number of excellent reviews of the field, suitable for advanced study, are listed in the bibliography.

5.6.1 Compaction of nanopowders

Problems which arise during the compaction and sintering of nanopowders may adversely influence the overall quality of the product. Generally speaking, most defects can be traced to one of several causes, including the presence of hard agglomerated particles, high plastic yield and fracture resistance, resistance to particle motion under pressure and contamination of particle surfaces. Irrespective of the consolidation method employed, the influence of each of these factors is essentially the same; they render the processes of compaction and densification more difficult.

Ideally, it would be desirable to have as starting materials for consolidation, clean, unagglomerated nanostructured powders. The powders themselves may have a micron-scale average particle size, $\langle D \rangle$, or they may be true nanopowders, depending on their synthesis route. They would be compacted at low or moderate temperature to produce a so-called green body with a density in excess of 90% of the theoretical maximum. Any residual porosity would be evenly distributed throughout the material and the pores would be fine in scale and have a narrow size distribution.

Unfortunately, most powders show some degree of deviation from this ideal. Most sintering defects can be related to the microstructures of the compacts in this 'green' state. Such defects are far more problematic when the powders are in the nanosize range than in the case for conventional powders. In particular, the elimination of pores with sizes greater than or equal to the size of the nanopowder particles is particularly troublesome. Such pores may result from the presence of particle agglomerates.

Pressing of metallic powders to produce a green body would, on the conventional (micron) scale, also involve a certain amount of plastic yielding of the particles or their fracture. The pressure required to produce a high degree of compaction is normally several times the yield stress of the material being compacted. Due to the dependence of yield stress on $\langle d \rangle^{-1/2}$ (i.e., the Hall–Petch relationship (1.23)), the cold compaction of metal nano-powders is likely to require stresses in the gigapascal (10^9 Pa) range, for which non-conventional compaction routes would be necessary.

Compaction on the conventional scale would also involve a certain amount of sliding and rearrangement, both of which become increasingly difficult as particle size decreases. On the nanoscale, the frictional force between particles becomes a very significant obstacle to their relative motion and rearrangement. These factors contribute to the formation of green bodies with a lower density than would be considered acceptable for conventional micron-scale powders.

Rearrangement of the particles may be facilitated by the use of lubricants or via wet compaction, either individually or in combination with ultrasonic agitation or centrifuging to promote good powder packing prior to pressing and/or sintering. Warm compaction methods which aim primarily to remove surface contamination and impurities also have the effect of reducing the material yield stress. Their application to the compaction of nano-materials has also led to improvement of green body density. However, the distribution of the residual porosity still strongly depends on the state of agglomeration of the powders.

5.6.2 Sintering

Nanoscale powders are characterised by an extremely high surface area; for example, 1 kg of a copper powder in the form of spheres of radius 5 nm has a total surface area calculated using Equation (5.27) of around $75\,000\,\text{m}^2$. If the radius were $50\,\mu\text{m}$, the surface area would be $15\,\text{m}^2$. Associated with this large surface area is a very substantial amount of energy, and it is this energy that drives sintering.

Sintering is a process which occurs when powders are packed together and heated to a high temperature, typically $\sim 2T_\text{m}/3$, when diffusion becomes significant. Necks begin to form at powder contact points which grow, thereby reducing the overall surface area and densifying the powder. Atoms diffuse from the grain boundaries established at intersections between differently oriented particles and deposit on the interior surface of the pore, thereby filling up the remaining space. It is worth noting that the 'green' component will shrink during normal sintering as the porosity progressively disappears and the density increases.

The driving force for sintering is given by γ_S/r, where γ_S is the surface energy (generally considered in this type of analysis to be isotropic, but not necessarily true for nanopowders) and r is the radius of curvature of the particle (taken as the actual particle radius in unagglomerated spherical powders). The main mechanism facilitating mass transport is usually considered to be grain boundary diffusion. The assumption of isotropic surface energy is not necessarily true for nanopowders; as each particle is composed only of a limited number of atoms, they may exhibit faceting behaviour as a result. Differences in local atomic arrangement at the surface of nanoparticles, compared with the same material at a conventional particle size, may also contribute to differences in surface energy. Stabilisation of particle surfaces, such as by oxidation, will also tend to reduce the driving force for sintering. Our present understanding does not allow us to incorporate these additional influences into an analysis of sintering in a simple way.

From the argument presented so far, it is possible to infer that the rate of sintering in fine powders should be very much enhanced as a result of the reduced diffusion distances involved as well as the increased surface area. We can also infer that it will vary with temperature in a similar way to that of the diffusion coefficient. Many equations have been suggested for the rate of densification by sintering, but their general form is well represented by

$$\frac{d\rho}{dt} \propto \frac{1}{\langle D \rangle^n} \exp\left[\frac{-Q}{RT}\right], \tag{5.41}$$

where n is a constant, ρ is the density, Q is the activation energy for sintering and $\langle D \rangle$ is the mean powder particle diameter. The exponent n is typically about 3 and Q is usually considered equal to the activation energy for grain boundary diffusion.

An increased surface area such as that possessed by nanopowders should result in an enhanced densification rate at a given sintering temperature. Indeed it would suggest that full densification of a particular material should be achievable at much lower sintering temperatures. This is very attractive from the viewpoint of restricting grain growth during sintering but, as with many of our other expectations related to the nanoscale, it is unusual to observe rapid low-temperature densification in practice. This may be related to anisotropy of surface energy, as any anisotropy effect will be much more pronounced at low sintering temperatures; the surface energy becomes much more isotropic as T is increased.

5.6.3 Role of impurities

The ideal way of reducing impurities in nanopowders would seem to be to conduct all handling and compaction processes within a controlled environment. The IGC apparatus described in Section 5.3.3, for example, was designed to allow all operations, including powder formation and consolidation, to be conducted within the same chamber. In a manufacturing environment, however, powders may be consolidated at a location remote from the powder production facility, be that as a result of cost or logistical factors, so that contamination in the form of adsorbed gases and oxide films is highly likely. Alternatively the addition of surfactants to reduce agglomeration during wet processing will also inevitably give rise to a surface layer of impurities. Because sintering is controlled by surface processes, however, the densification kinetics are highly dependent on particle surface purity and quality. In the case of the sintering of nanopowders, with their enormous surface area, this issue becomes even more critical.

The extent of particle contamination is generally speaking a strong function of the powder synthesis route. For example, powders produced by the MA route are generally considered to be more contaminated than those made by IGC, largely as a result of pick-up from the milling medium. This is not an insurmountable problem. Impurities in metallic powders produced by MA are distributed throughout the bulk of the large nanostructured powder particles rather than being concentrated solely on the surface. In the case of ceramics such as oxides and nitrides which have been milled to the nanoscale, the presence of certain contaminants may be useful in preventing grain growth if they are present as fine dispersions (Section 5.5.3).

Given their enormous specific surface area, the concentration of adsorbed gas is, perhaps unsurprisingly, much higher than for micron-scale powders. Oxygen contamination, in particular, may cause problems for sintering of metallic powders as this oxygen tends to be present in the form of a surface oxide film. The stability of the oxide is of great importance in the processing of metals as such a barrier will inhibit sintering. The presence of an oxide film may, however, be desirable from a handling viewpoint as it can reduce the risk of powder explosion on contact with air. Oxides may then be considered as desirable or undesirable, depending on the stage of processing. The ideal scenario is one where an oxide is present during powder handling and subsequently removed either before or during the consolidation step. This can be achieved by careful selection of process conditions. For example, employing a reducing atmosphere such as hydrogen during sintering, which is also commonly used in the sintering of conventional sized metal powders, leads to a reduction of the oxide and, hence, a reduction in the sintering temperature required for metallic nanopowders. Sintering metallic nanopowders in a vacuum may also reduce the required sintering temperature if the oxide breaks down under these conditions, although the effect is less pronounced than for hydrogen. Ceramic nanopowders are found to densify at lower temperatures, or to achieve greater densities at high temperatures, when sintered in a vacuum rather than in air. So the choice of sintering atmosphere would seem to be of central importance to the overall kinetics of densification.

Further improvements in densification may be possible by using novel sintering processes, such as field-activated sintering, which seems to offer a good alternative route for the processing of nanopowders in cases where very high purity products are required.

5.6.4 Porosity

As densification is effectively the removal of residual porosity, the size and spatial distribution of pores will exert a significant influence on the overall densification kinetics. If the porosity is unevenly distributed, for example as a result of the presence of hard agglomerates in the powder, it may be impossible to achieve full, or near-full, density. The pores in this case become defects and they may significantly influence both the mechanical and physical properties of the material in which they reside. As a result care must be taken during the consolidation stage of processing to ensure that the porosity is as fine and as well distributed as possible.

Equation (5.41) can be reformulated to account for the presence of a well-dispersed low volume fraction of pores, thus

$$\frac{1}{\rho(t)(1 - \rho(t))}\frac{d\rho}{dt} \propto \frac{1}{r_p(t)}\frac{1}{\langle D\rangle^n}\exp\left[\frac{-Q}{RT}\right], \tag{5.42}$$

where $\rho(t)$ and $r_p(t)$ are the instantaneous density and the instantaneous pore radius, respectively.

Equation (5.42) demonstrates that the highest sintering rates will be achieved for materials containing the finest pores. These same small pores are, however, also those responsible for the inhibition of grain boundary motion, and hence grain growth (Equations 5.38 to 5.40). As long as the rate at which the pores disappear is balanced by the rate of reduction in pore size; i.e., that the ratio f_r/r_p remains constant, no grain growth will occur. On achieving full or near full density, however, with the restraining influence of the pores no longer operative, the rate of grain growth can be quite spectacular. This is demonstrated quite clearly in Figures 5.9 and 5.10.

5.6.5 Non-conventional processing

A number of non-conventional methods have been applied to the densification of nanopowders, including microwave sintering; shock or dynamic consolidation; and field-assisted sintering. All have been applied with the objective of enhancing densification rate, thereby allowing processing temperature and/or time to be reduced, with the ultimate aim of retaining the initial nanostructure. Of the methods listed above, the one that is currently generating most interest for the manufacture of BNMs is field-assisted sintering (FAS), sometimes referred to as spark/plasma sintering (SPS).

FAS methods involve the combination of the application of a pressure, P, and heating by the application of an electric current. In the initial stages of sintering, the multiple discharges of electrical current are passed through the powder compact. This is believed to promote the removal of surface adsorbates via generation of plasma and, hence, to enhance the process of bonding during the sintering process thus leading to a higher rate of densification. A typical FAS arrangement is shown in Figure 5.10; this press is usually contained in a chamber that allows the sintering operation to take place under vacuum or a protective gas atmosphere. The major advantage of this consolidation method is the very rapid rates of densification which may be achieved at relatively low temperatures. The cycle time from powder to component is relatively quick and the process is very attractive from the view point of commercialisation.

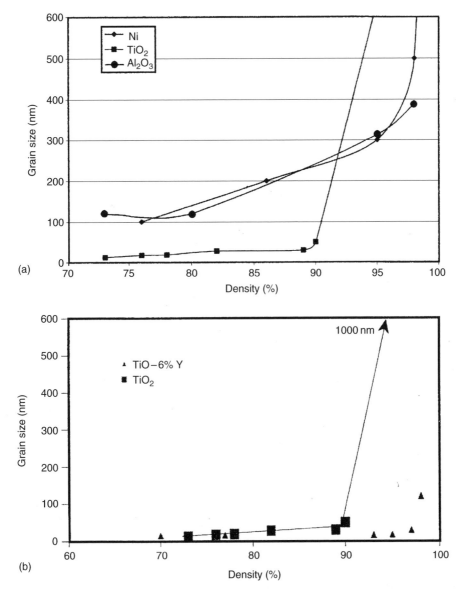

Figure 5.9 The effect of compact density on grain growth for (a) nanocrystalline Ni, TiO$_2$ and Al$_2$O$_3$ and (b) for TiO doped with 6% yttrium and TiO$_2$ showing a possible solute drag effect on grain growth. After J. Groza, Consolidation methods, in *Nanostructured Materials: Processing Properties and Applications*, ed. C. C. Koch, 2003. Reproduced by permission of IOP Publishing Limited

In terms of the types of materials that may be processed in this way to date, a variety of nanostructured pure metals, intermetallics, oxide, carbides, nitrides and composite systems have been manufactured. In all cases the nanostructure was retained after FAS was complete.

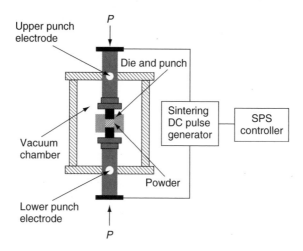

Figure 5.10   Schematic diagram of a spark/plasma sintering unit

5.7 MECHANICAL PROPERTIES

Extrapolation of mechanical properties from the conventional to the nanoscale led to an expectation amongst scientists and engineers that a significant improvement in both the strength and ductility of materials could be achieved by continued refinement of microstructural scale. In reality, however, whilst an increase in hardness and strength have been realised for nanostructured materials, the ductility is disappointingly low, showing an overall decrease when compared with materials of conventional grain size.

It would appear that when the length scale of the microstructure and the character-istic lengths relevant to deformation become comparable interesting new properties are indeed obtained. These may not, however, be quite as useful as with other cases. An extreme example is the change in mechanical behaviour of Cu from a ductile metal with moderate to low strength at conventional microstructural scales, to an extremely strong metal, with a yield strength approaching 1 GPa, but having a low ($\sim$2%) ductility, almost bordering on brittle, at grain sizes of less than 20 nm.

In this section some of the reasons for such high strengths and low ductilities are considered. As yet, there is no definitive answer to this problem and this may remain the case until defect-free BNMs become easier to manufacture reproducibly.

5.7.1 Hardness and strength

As discussed in Section 5.3.5 in connection with mechanical milling, the hardness and strength for materials of conventional grain size ($d > 1\,\mu$m) may be expressed by the empirical Hall–Petch equation (1.23). This $\langle d \rangle^{-1/2}$ dependence of the yield strength is also valid well into the nanosize range. However, it is obvious that the Hall–Petch relationship cannot be extrapolated to arbitrarily small grain sizes, and some form of lower limit to this behaviour must exist.

The pile-up model for the Hall–Petch expression requires that a large number of dislocations be present to form a dislocation pile-up. The most fundamental 'characteristic length' associated with a dislocation is the Burgers vector, **b**, which indicates the extent to which the crystal lattice is distorted by the presence of a single dislocation. There is also a further characteristic length associated with dislocation propagation; that of the equilibrium loop diameter $d(\tau)$. This is the minimum diameter that a curved dislocation line may assume under a shear stress τ, and it is given by

$$d(\tau) = \frac{Gb}{\tau},\qquad(5.43)$$

where b is the magnitude of the Burgers vector. If the shear stress is equal to that required for propagation, $\langle d \rangle$ must be at least equal to this critical length scale if deformation is to proceed by dislocation-mediated plasticity. The number of dislocations that can be contained within a grain is a function of grain size. According to the pile-up model, however, many such dislocations are required, and as the dislocation number within a grain decreases, an increasing deviation in yield strength from the plot of σ_y versus $\langle d \rangle^{-1/2}$ is to be expected. Based on this argument it has been suggested that the smallest possible 'pile-up' would consist of two dislocations and, hence, the minimum grain size for which this type of deformation mechanism would be valid would be that which could support the presence of just two dislocations. Below this, strength would be expected to remain constant or even to decrease, perhaps as a result of the dominant influence of other microstructural features. It has been proposed that as the fraction of the material associated with grain corners increases, this would produce a softening effect. Equally, it has been suggested that the increase in number of fast diffusion paths available to nanomaterials should facilitate the accommodation of deformation via grain boundary creep.

There are alternative explanations of the Hall–Petch relationship that do not depend on a dislocation pile-up mechanism. One of these alternatives suggests that ledges on grain boundaries are the source of dislocation activity (Figure 5.8(b)). The flow stress of a material is, in this case, proportional to the square root of the dislocation density, ρ_d, via the classical Taylor relationship for work hardening; i.e.,

$$\sigma_y = \alpha Gb\rho_d^{1/2}\qquad(5.44)$$

where α is a constant and G is the shear modulus. The dislocation density is proportional to the number of grain boundary ledges and this, in turn, is a function of the grain size. This reasoning also leads to the Hall–Petch σ_y versus $\langle d \rangle^{-1/2}$ relationship. Again this strengthening effect must be limited to a finite level as the number of boundary ledges scales with $\langle d \rangle$.

A deviation from normal behaviour would be expected for Cu at grain sizes below 20 nm in both conceptual frameworks. Experimentally, hardness data reveals a number of different material responses at grain sizes below ~20 nm. These include a normal Hall–Petch behaviour, no dependence of hardness on grain size ('zero slope') and in some cases a negative slope (so-called negative Hall–Petch behaviour).

It is emphasised that most negative Hall–Petch data obtained for nanocrystalline materials has been generated using nanomaterials annealed to increase their grain size prior to testing. This has led to the suggestion that the effect could be more readily ascribed to changes in microstructure such as densification, stress relief, phase transformation or changes in grain boundary structure rather than as a result of some underlying 'new' physical process.

5.7.2 Ductility and toughness

Studies to date indicate that there are three factors limiting the ductility of nanostructured materials: the presence of structural artefacts arising from processing, such as porosity or cracks; crack nucleation or propagation instability; plastic instability in tension.

Upon deformation, many nanomaterials are found to fracture within or only slightly beyond the elastic range. As fully dense BNMs are difficult to process, an obvious reason for this low fracture strength would be the presence of defects such as porosity or cracks or poor interparticle bonding. These may act as crack initiation sites resulting in easy crack nucleation and growth, promoting brittle behaviour in tension and causing premature failure before yielding has a chance to occur. The ductility of nanomaterials may therefore be severely compromised by their presence.

Factors other than the presence of defects may also act to limit the fracture strength and ductility of nanomaterials; for example, it is well known that grain size has a strong effect on ductility and toughness in materials of conventional grain size. In the case of nanomaterials, as the yield strength is dramatically elevated as a result of the very fine grain size, it is quite possible that it will become equal to or even exceed the fracture strength of the material. This is further exacerbated by the possibility of the fracture stress being reduced for nanoscale materials, which may arise as a consequence of the combination of reduced dislocation activity and increased grain boundary density. The high energy and the large number of these regions allow cracks to propagate along the boundaries, or intergranularly, with relative ease.

In general, there is also a tendency for nanomaterials to deform non-uniformly in tension by the production of shear bands. These are highly localised regions of material where all deformation is accommodated. As deformation proceeds, these regions become severely strained and may pull apart, leading to failure in the initial stage of straining in tension and very low apparent ductility. Once again, this may be related to the inability of nanometals to store dislocations and to uniformly work-harden, thereby sustaining the applied load without the formation of plastic instabilities.

One recently applied method of improving the ductility of nanomaterials is to manufacture materials with a bimodal grain size distribution. If large grains or dendrites are introduced into a nanomaterial matrix, either by an in situ process or by the mixing of powders of different different particle sizes prior to consolidation, these larger grains will be significantly softer than the nanostructured component during the initial stages of deformation. If the grain size of this 'soft' component is sufficiently large to allow the storage of dislocations and hence to experience work hardening, the net effect will be that, whilst the initially hard nanostructure becomes softer by the formation of shear bands, these large grains will become harder. Yield stress will therefore depend on the nanocomponent and ductility will depend on the work-hardening behaviour of these larger crystals.

Such use of widely or bimodally distributed grain sizes in structural applications is one way in which the excellent potential of high-strength nanomaterials may finally be realised, although research in this area is very much in its infancy at the time of writing (2004).

5.7.3 Creep and superplasticity

Creep is time-dependent, plastic deformation at constant stress. It occurs at all temperatures but usually becomes important only as the temperature approaches $\sim 0.5T_{\mathrm{m}}$. For conventional grain sizes it has also been found to be strongly dependent on grain size. Superplasticity, on the other hand, is the ability of some polycrystalline materials to exhibit very large tensile deformations without necking or fracture. Elongations of 100% to over 1000% are typically considered as defining features of this phenomenon. In the conventional microcrystalline regime it is found that, as the grain size is decreased, the temperature at which superplasticity is observed is decreased and the strain rate at which it may occur is increased.

The generic equation for creep and superplasticity in the steady-state regime (i.e., when a stable microstructure is attained and the strain rate $\dot{\varepsilon}$ becomes constant with time) is given by an exponential law:

$$\dot{\varepsilon} = A \frac{DGb}{k_{\mathrm{B}}T} \left(\frac{b}{\langle d \rangle} \right)^{s} \left(\frac{\sigma}{G} \right)^{n},\tag{5.45}$$

where A is a constant depending on the creep mechanism, σ is the applied stress, s is the grain size exponent and n is the strain exponent. The values of the exponents s and n depend on the creep mechanism in operation, which in turn depends on the temperature and stress conditions and the initial microstructure of the material. The equation is phenomenological but is found to work well for both creep and high strain rate creep, or superplasticity. For example, for high-temperature ($\sim 0.9\,T_{\mathrm{m}}$), low-stress Nabarro–Herring creep, in which deformation takes place by vacancy diffusion through the lattice, $n = 1$ and $s = 2$: for Coble creep, in which deformation is accommodated by vacancy diffusion at grain boundaries and which may occur at lower temperatures since the activation energy for grain boundary diffusion is smaller, $n = 1$ and $s = 3$. At intermediate to high stresses and temperatures between about $0.4T_{\mathrm{m}}$ and $0.7T_{\mathrm{m}}$, creep is governed by the movement of dislocations. Dislocation movement comprises glide along slip planes followed by climb (a diffusion-assisted process) over any obstacles encountered. Values of s and n in this intermediate regime reflect the dominance of one of these mechanisms of dislocation motion over another. When glide dominates then $n = 3$ and $s = 0$; there is no grain size dependence. When climb dominates then $n = 5$.

Superplasticity is a special case of high strain rate creep that occurs for some fine-grained materials in which grain growth is severely inhibited by the presence of a dispersion of pinning particles. As a result of the large grain boundary fraction (F_V), grain boundary sliding (GBS) becomes the predominant deformation mechanism. In this case, $n = 2$ and $s = 2$ if GBS is accommodated by dislocation movement controlled by lattice diffusion, or $n = 2$ and $s = 3$ if GBS is accommodated by dislocation movement controlled by grain boundary diffusion.

From this description of the phenomena, we might reasonably expect, at low grain sizes and low T/T_m, that Coble creep would be the dominant mechanism in nanomaterials as mass transport is dominated by grain boundary diffusion. If this were the case, enhanced creep rates, up to nine orders of magnitude greater than for the microcrystalline regime, would be expected in the case of nanomaterials. Similarly, we should also expect superplasticity to occur in nanomaterials at lower temperatures, perhaps even at room temperature, and higher strain rates than for the equivalent materials on the micron scale. In reality, echoing the observations for strength and ductility, the creep and superplastic behaviour observed for BNMs have not yet matched these predictions; the creep rate observed for a given nanomaterial is comparable to or lower than that of its coarse-grained counterpart whilst convincing evidence for superplasticity has yet to be reported for temperatures lower than $0.5T_m$. Why this should be is not yet clear, but what emerges from the scientific literature is that the investigation of these phenomena in BNMs has, in the past, been hindered by the presence of processing defects and impurities. As the processing of nanomaterials to form BNMs with controlled composition and microstructure becomes easier, their true behaviour with respect to creep and superplasticity should become clearer.

5.8 FERROMAGNETIC PROPERTIES

The properties of ferromagnetic nanomaterials are discussed fully in Chapter 4. In this section we will concentrate on the influence of material processing and defects on the properties developed. We will also look at nanocomposite magnets with remarkable soft and hard magnetic properties, examining the main structural features responsible.

5.8.1 Fundamental magnetic properties

Saturation magnetisation M_s was one of the properties of nanomaterials that were touted as showing a significant change in magnitude compared with conventional materials. Rather like the Young's modulus, early reports of a reduction of $\sim 40\%$ in saturation magnetisation M_s for consolidated nanocrystalline Fe compared with coarse-grained Fe were found, after considerable experimental and modelling effort, to result from the presence of processing defects. The initial reasoning put forward to explain the decrease, based on an enhanced degree of structural disorder in the system, resulting from a high fraction of particles within the boundary, was not well supported theoretically, and comparison of the behaviour of fully dense electrodeposited nanocrystalline nickel and fully dense coarse-grained nickel showed that there was little or no change ($<5\%$) in M_s down to grain sizes of $\sim 10\,nm$. The presence of oxides and other impurities on the surface of nanocrystalline nickel, however, can lead to significant changes in M_s.

In the case of very small nanoparticles, for which the fraction of surface atomic sites is large, some change in the Curie temperature, T_C, compared with the values for larger grain sizes on both the nano and conventional scales may be expected. The high fraction of atoms associated with either the particle surface or the grain boundaries should in fact alter T_C. In practice, however, only small deviations from the value of T_C for the bulk have been reported up to now.

5.8.2 Nanocomposite soft magnetic materials

The structure of soft magnetic nanocomposites is of utmost importance in determining their overall properties. Their crystallisation behaviour is discussed in Section 5.3.3, where it was noted that these alloys are in fact two-phase nanocomposites. In Finemet (FeSiBNbCu) and Nanoperm (FeZrBCu) the Curie temperatures of both the nanocrystalline and amorphous intergranular phase are important in describing the magnetic response. The intergranular phase, which is typically richer in non-magnetic species, generally has a lower T_C. The soft magnetic properties of the nanocomposites are dependent on the strength of the exchange coupling between the ferromagnetic grains through this amorphous layer which, in turn, is a strong function of T. As the temperature approaches the Curie point for the amorphous layer, coercivity increases dramatically, reflecting a decrease in the exchange length, L_{ex}, as a result of the magnetic decoupling of the grains. Controlling the composition and stability of this intergranular layer is therefore of paramount importance in generating and maintaining good soft magnetic characteristics.

On the other hand, the nanocrystalline structure of FeTaC soft magnetic materials derived by devitrification of a precursor metallic glass differ significantly from that for typical FINEMET or Nanoperm compositions. In this type of alloy, a refractory carbide phase (in this case tantalum carbide) tends to form at grain boundary triple points. These carbides tend to pin grain boundaries, which prevents structural coarsening. This fully nanocrystalline structure leads to a simplification of the magnetic behaviour with respect to T. Interestingly, because of the structural scale of the carbides and the nanocrystallites, this type of structure would also seem to be a promising candidate for hard surface coatings.

5.8.3 Hard magnetic materials

Exchange spring magnets are nanocomposite structures comprising hard and soft magnetic phases that interact via magnetic exchange coupling. These two-phase mixtures can exhibit a remanence ratio M_r/M_s (where M_r is the remanent magnetisation) greater than the theoretical upper limit of 0.5 predicted by the Stoner–Wohlfarth theory, hence they are referred to as remanence-enhanced materials. Exchange coupling between the hard and soft magnetic grains that form the microstructure forces the magnetisation vector of the soft phase to be rotated parallel to easy directions of adjacent hard-phase crystallites. Because the maximum energy product of this type of hard magnet is so large, they show great promise for advanced permanent magnet applications. Significant manufacturing problems are posed by the requirement that both the hard and soft phases are nanostructured and the requirement that the structure needs a sufficient degree of coherence across the interphase boundaries to allow the grains to be exchange coupled.

Originally observed for Fe-rich hard magnetic compositions (e.g., $Fe_{78}Nd_4B_{18}$), where a low volume fraction of the hard $Nd_2Fe_{14}B$ phase was surrounded by a magnetically soft Fe_3B phase, this has since been observed in systems where the magnetically soft phase α-Fe is present and also in entirely different alloy systems such as FePt(hard)–Fe_3Pt (soft).

Preparation has generally been via more conventional non-equilibrium processes such as melt spinning, either to form the desired structure directly or by devitrification of an amorphous precursor, or mechanical milling. More recently, however, FePt–Fe_3Pt

exchange spring magnets have been produced by sintering self-assembled mixtures of nanopowder precursors (FePt and Fe_3O_4) in a reducing atmosphere. The advantage of this method is that any relative combination of the two phases may be produced without the need for a suitable glass-forming composition being available. Thus, the magnetic properties can be engineered with great precision.

5.9 CATALYTIC PROPERTIES

The very high, very controllable, reactive surface area of nanopowders renders them ideal for use as catalysts; in fact, during the 1970s this was one of their first real applications. The properties of nanocatalysts may be explained, generally speaking, by simply extending well-known relationships for the sensitivity of catalytic activity and the selectivity to particle size into the nano range. There are exceptions, such as the much enhanced efficiency of ultrafine (below 6 nm) nanoparticles of Pt in the catalytic reduction of NO_2, and the enhanced catalytic activity of rutile (TiO_2) nanoparticles associated with the oxygen-deficient compositions which may exist as a result of processing. Hydrogen-absorbing alloys also show improved sorption properties over their coarse-grained counterparts. Enhanced surface diffusion also plays a role, particularly for gas sensor sensitivity and in hydrogen storage applications.

5.10 PRESENT AND POTENTIAL APPLICATIONS FOR NANOMATERIALS

Nanomaterials are relatively new developments and have received significant attention only during the past one or two decades. The number of existing applications for nanomaterials is therefore quite small. Those industries currently using nanomaterials range from the more obvious (e.g., the magnetics industry), to the surprising (e.g., the cosmetics industry). In fact, the cosmetics industry was one of the first to see the potential of nanomaterials and an interesting example is described below. There is potential for wider use in the production of coatings with tailored moduli and some possible applications are briefly discussed. If and when the mechanical properties of BNMs become more reproducible and problems associated with ductility and toughness are addressed nanomaterials should find niche markets in these applications also, but the present examples will concentrate on the most probable fields of application.

5.10.1 Ultraviolet absorbers

The cosmetics industry has used nanomaterials for some time. A particularly interesting example is the use of semiconducting oxide ceramics as UV absorbers or sunscreens. The conflicting issues of cosmetic appeal and UV absorption efficiency make conventionally sized oxide powders unsuitable because they impart excessive whitening to the skin upon application, although they offer effective UV protection for much longer than the competing organic-based screens.

The degrees of diffuse and specular reflection and scattering that are largely responsible for the whitening effect may be significantly reduced by using a nano-powder with a suitable particle diameter $\langle D \rangle$. On length scales small in comparison with the wavelength of incident light, scattering shows a sixth-power dependence on particle size, indicating that a relatively small reduction in average particle size and close control of the PSD will lead to a significant decrease in intensity of scattered light and therefore whiteness.

The refractive index and band gap of the ceramic component are also of great importance for good performance. Low values of the refractive index are necessary, as is absorption of light over the widest possible range of the UV part of the spectra. The oxide semiconductors ZnO and TiO_2, which have both low refractive indices and suitable band gap energies corresponding to wavelengths of $\sim$380 and $\sim$365 nm, respectively, are prime candidates for this role.

Unless $\langle D \rangle$ becomes sufficiently small that it begins to exhibit quantum dot behaviour; i.e., if the particles become smaller than the exciton Bohr diameter ($\sim$5 nm for both ZnO and TiO_2), where the band gap will become smaller and the UV absorption efficiency will be compromised; there is no size effect on absorption efficiency. Optimisation of $\langle D \rangle$ and PSD are driven by the need to compromise between protection via UV absorption and reflection mechanisms.

5.10.2 Magnetic applications

There are a number of exciting developments in magnetic materials that are dependent on materials possessing nanoscale microstructures. In particular, ultrasoft nanocomposite magnets and exchange-enhanced permanent magnets have been a key driving force the behind developments in the whole of the inorganic nanomaterials field.

The extremely soft magnetic properties and low cost of the new nanocomposite and nanocrystalline Fe-based magnetic alloys are being exploited in power electronics applications and for earth leakage devices, with further potential applications in magnetic shielding and sensors. The material producers and the method of production are so similar to those used in the manufacture of amorphous ferromagnetic alloys that these materials have already found some niche applications.

The market for exchange-enhanced nanocrystalline magnets is potentially very large. Bonded rare earth magnets are increasingly used in motors and actuators in many consumer products, and increasing their maximum energy product will enable the manufacture of smaller components with the same functionality. Flexibility of manufacture and the ability to tailor properties to specific applications are further enticements for manufacturers to utilize them.

5.10.3 Coatings

There is presently considerable interest in the development of nanostructured and nanolayered coatings. The methods used in their production (e.g., plasma-assisted vacuum techniques) make it is possible to synthesise materials with extreme mechanical properties that would be difficult by other means. The formation of coatings with ultra-high hardness and an associated high elastic modulus has been the target of much R&D activity in this area. This is largely because conventional fracture mechanics theory suggests this as a development vector and also that these properties are desirable for improving tribological properties. There is an alternative approach that has not yet been widely investigated – nanocomposite coatings with high hardness and *low* elastic moduli. These coatings may exhibit improved toughness and may therefore be better suited for optimising the wear resistance of 'real' industrial substrate materials, such as steels and light alloys, by providing a better modulus match. These nanocomposite coatings, although not necessarily exhibiting extreme hardness, may also provide superior wear resistance when deposited on the types of substrate material employed by industry.

Good examples of this type of coating are in the TiAlBN system for which the Al content appears to govern the elastic modulus; chromium nitride and tungsten carbide formed by 'doping' of elemental metallic systems also seem to offer promise. Thermal barrier coatings also appear to be particular candidates for nanostructured coatings.

BIBLIOGRAPHY

General overview

M. F. Ashby and D. R. H. Jones, *Engineering Materials*, Vols 1 and 2, Pergamon Press, New York, 1980, 1986.

J. C. Anderson, K. D. Leaver, R. D. Rawlings and J. M. Alexander, *Materials Science*, Van Nostrand Reinhold, London, 1987

Thermodynamics and kinetics

J. W. Martin, R. D. Doherty and B. Cantor, *Stability of Microstructure in Metallic Systems*, Cambridge University Press, Cambridge, 1997.

D. A. Porter and K. E. Easterling, *Phase Transformations in Metals and Alloys*, 2nd edn, Van Nostrand Reinhold, London, 1992.

J. W. Christian, *The Theory of Transformations in Metals and Alloys*, 1st edn, Pergamon Press, 1965.

The first two are also good sources of information on the stability of microstructures.

Synthesis

I. Brodie and J. J. Murray, *The Physics of Micro/Nano Fabrication*, Plenum Press, New York, 1992.

C. Surayanarayana (ed.), *Non-Equilibrium Processing of Materials*, Pergamon, Oxford, 1999.

C. Koch (ed.), *Nanostructured Materials: Processing, Properties and Potential Applications*, William Andrew Publishing, New York, 2002.

Stability

D. S. Wilkinson, *Mass Transport in Solids and Fluids*, Cambridge University Press, Cambridge, 2000.

F. J. Humphreys, *Recrystallization and Related Annealing Phenomena*, Elsevier, Oxford, 1996.

See also the first two references under 'Thermodynamics and Kinetics'.

Consolidation

J. Groza, in C. Koch (ed.), *Nanostructured Materials: Processing, Properties and Potential Applications*, William Andrew Publishing, New York, 2002.

Mechanical properties

D. G. Morris, *Mechanical Behaviour of Nanostructured Materials*, Trans Tech, Switzerland, 1998.

Magnetic properties

M. E. Mchenry, M. A. Willard and D. E. Laughlin, *Progress in Materials Science*, Vol. 44, 1999, pp. 291–433.

6

Electronic and electro-optic molecular materials and devices

In Chapter 3 the electronic behaviour of inorganic semiconductors with reduced dimensionality was described. Here we consider the related, but more recent, field of organic electronics and optoelectronics, as well as the recent discovery carbon nanotubes. We shall also introduce organic devices in the present chapter, specifically field effect transistors, light-emitting diodes and photovoltaics. Although in this chapter we are primarily interested in the basic science behind organic electronics and optoelectronics, the knowledge gained will be used elsewhere in this volume, notably Section 8.8.1, for its application in nanotechnology.

6.1 CONCEPTS AND MATERIALS

6.1.1 The solid state: crystals and glasses

Organic matter often crystallizes only in part, or not at all. In particular, polymers often solidify in the form of a *glass*, but there are a number of low molecular weight organic materials that also form glasses, and there are inorganic glass formers as well; e.g., silicon oxide used for windows. Roughly speaking, a glass displays the mechanical properties of a solid (similar to crystalline solids), but the structure is disordered as in a liquid. In thermodynamic terms, the glassy state is a non-equilibrium state. However, particularly for polymeric glasses, the relaxation time to reach equilibrium can diverge to infinity at the Vogel–Fulcher temperature, well above absolute zero; thus the non-equilibrium state is no longer transient. Thermodynamic theories have been tailored to accommodate the permanent non-equilibrium nature of the glassy state, and to describe the *glass transition* above which the material regains its fluidity and reapproaches thermodynamic equilibrium. These are powerful theories, but currently no comprehensive

Nanoscale Science and Technology   Edited by R. W. Kelsall, I. W. Hamley and M. Geoghegan
© 2005 John Wiley & Sons, Ltd

molecular theory of the glass transition based on first principles is available. The situation is similar to that in superconductivity at a time when the powerful, yet phenomenological Landau–Ginzburg (thermodynamic) theory was available, but not the microscopic Bardeen–Cooper–Shrieffer theory. The most important fact about the glassy state in the context of organic semiconductors is that the glass is structurally disordered like a liquid. Consequently, unlike crystalline materials there is no translation symmetry and Bloch's theorem (Section 1.2.6) does not apply.

For our context, the key difference between organic and inorganic solid matter is that excitations in inorganic matter are delocalized and best described by a wave vector $\mathbf{k}$, whereas in organic matter, excitations are localized and k is not a good quantum number. To understand *organic semiconductors*, and maybe *synthetic metals*, we have to understand how something like a band gap can arise within a single molecule. The key to this understanding lies in the chemistry of carbon.

6.1.2 Chemistry of carbon

The most common carbon isotope is ^{12}C (the nucleus has six neutrons and six protons), but there is a natural abundance of 1.2% of the ^{13}C isotope with seven neutrons. This nucleus has a nuclear magnetic moment, which is used in NMR. In atomic carbon the six electrons occupy the orbitals in Table 6.1.

1s refers to the electronic state in which the principle quantum number (QN) is $n = 1$, the orbital QN is $l = 0$, and the magnetic QN is 0; there are two electrons occupying 1s due to their two spins. In a similar way, the 2s and 2p orbitals are filled with two electrons each. The electronic configuration is written as $1s^2 2s^2 2p^2$.

6.1.2.1 Hybrid orbitals

Carbon, like many chemical elements, forms covalent bonds. The driving force for chemical reaction is the desire to share electrons between different atoms to complete electronic shells. Thus, usually:

$$\text{atomic orbitals} \rightarrow \text{molecular orbitals.}$$

So C should form two bonds to add two electrons to complete the vacancies: one to p_x and one to p_y. Carbon should be divalent – it should form two single bonds. But in reality it forms four bonds. In C, and some other atoms, chemical bonding proceeds via intermediate steps: *promotion* and *hybridization*:

$$\text{atomic orbitals} \rightarrow \text{hybrid orbitals} \rightarrow \text{molecular orbitals.}$$

Table 6.1 Orbitals of carbon

Orbital	1s	2s	$2p_x$	$2p_y$	$2p_z$
Number of electrons	2	2	1	1	0

For hybridization, C promotes one 2s electron into the empty p_z orbital, so we arrive at $1s^2 2s^1 2p^3$. Then C combines (hybridizes) the remaining 2s electron and either:

three 2p orbitals $\rightarrow sp^3$ hybrids, or
two 2p orbitals $\rightarrow sp^2$ hybrids, or
one 2p orbital $\rightarrow sp$ hybrid.

sp^3 hybrid orbitals have four fingers pointing into the corners of a tetrahedron. The tetrahedral bond angle is 109.5°. In this form C can form four bonds by sharing electrons with hydrogen 1s shells (e.g., CH_4, methane) or with other sp^3 carbons (e.g., $H_3C–CH_3$, ethane). The C–C bond in ethane is called a σ bond. σ bonds are very strong; diamond entirely consists of carbon held together by σ bonds.

sp^2 hybrid orbitals have three fingers in a plane, at angles of 120° to each other, plus one remaining p orbital perpendicular to the plane. This configuration is found in graphite, in which the sp^2 bonding results in sheets of hexagonally ordered carbon atoms (known as graphene sheets); the sheets are stacked one on top of the other, with the sheets only weakly bonded together. In this form C needs another sp^2 hybrid C to form a molecule; for example ethene ($H_2C=CH_2$) in which two of the three fingers of each C bond to H as before, and the third overlaps with another C sp^2 orbital to form a bond (σ bond). The remaining p orbitals of either C overlap, as well, to form another carbon–carbon bond; the so-called π bond. A $1\sigma + 1\pi$ bond constitutes a carbon double bond. This is a weaker bond, and the respective orbital is more *delocalized*; i.e., it occupies a relatively large space rather far away from its original carbon.

sp hybrid orbitals have two fingers along one axis (say the x axis) at 180° to each other, plus two remaining p orbitals (along the y and z axes). In this form C can bond with two hydrogens and another sp hybrid. It forms one σ bond between the sp orbitals, plus the remaining two p orbitals of each molecule overlap to form two π bonds (a carbon triple bond): this is ethyne ($HC\equiv CH$), also called acetylene.

6.1.2.2 The benzene ring

A regular hexagon is a planar structure with six sides and six corners, each with internal angle of 120°. sp^2 hybrid orbitals have an angle of 120° with respect to each other. Hence by σ bonding six sp^2 carbons, we can form a regular hexagon. Each carbon will form two σ bonds, one with each of its neighbours. There remains one sp^2 orbital per carbon to be capped; e.g., by a hydrogen. The remaining p orbitals will again overlap to form π bonds. The resulting structure may look like Figure 6.1.

Figure 6.1 The chemical structure of benzene can be seen to be a mixture of two possible 'borderline' structures

It is not quite clear where the π bonds should be. In reality, an intermediate state is adopted, where the π electrons are completely delocalized, so that it is impossible to assign double bonds (Figure 6.1). The π electrons form a cloud that spans the entire molecule. The side of a ring is 1.39 Å in length, intermediate between the C–C and C=C bond lengths.

6.1.2.3 Conjugated molecules

The benzene ring is the prototype of a *conjugated* molecule – a molecule with alternating single/double or single/triple carbon bonds. In conjugated molecules, π electrons delocalize throughout the entire molecule and are relatively loosely bound. We have been thinking of conjugated molecules being built step by step, by binding hybridized carbons together in a *linear combination of atomic orbitals* (LCAO). In the *molecular orbitals* (MO) description (introduced in Section 1.2.2), we instead imagine a given, rigid set of points at which atomic nuclei are fixed, and fill that skeleton with electrons to arrive at the molecule. The LCAO and MO approaches correspond to two schools of quantum chemistry and computer simulation. Both approaches should lead to the same molecules, but in the MO picture, the correspondence to semiconductors is easier to see. We need N electrons in the molecule to balance N positive charges. The first electrons will cluster closely to the atomic nuclei, resulting in almost undisturbed atomic orbitals; this is equivalent to saying that the carbon 1s electrons, for example, do not participate in chemical bonds, but the last few electrons will enter what we have called delocalized π orbitals. Although we can trace π orbitals to the hybridized atomic orbitals of carbon, we have seen that in a conjugated molecule, they may delocalize far from their original carbon, hence for the π cloud, the MO picture is more appealing.

The last pair of electrons (one for each spin state) to be filled into the molecule occupy a molecular orbital that is called the *highest occupied molecular orbital* (HOMO). Note that a half-filled HOMO would imply an unpaired spin; i.e., the molecule would be a *radical*. Since the completion of electronic shells is the driving force behind chemical reactions, radicals usually appear at intermediate stages of chemical reactions, but not as end product. The next molecular orbital beyond the HOMO is called the *lowest unoccupied molecular orbital* (LUMO). The HOMO and LUMO are collectively known as *frontier orbitals*.

Due to the delocalized and weakly bound character of π electron clouds, conjugated molecules can be ionized relatively easily, and the electron vacancy or surplus electron can travel along the molecule with relative ease. This situation is reminiscent of the VB/CB description of inorganic semiconducting crystals. Due to this analogy, the energy difference between LUMO and HOMO in a conjugated molecule is called the *band gap*. However, we will soon address the limits of the analogy between inorganic and organic semiconductors.

6.1.2.4 Buckminsterfullerene

Until 1990 it was universally thought that only two allotropes of carbon existed, diamond and graphite. The first evidence for the existence of an alternative form was the observation of enhanced absorption of stellar radiation, at a specific frequency corresponding to the predicted optical response of a 60-atom carbon cluster. The cluster has the same topography as a common soccer ball, with 20 hexagonal faces and 5 pentagonal faces. Molecular dynamics simulations have shown that this is the most stable possible configuration for a closed carbon cluster. The C_{60} molecule was originally christened buckminsterfullerene, after the architect Buckminster Fuller, who had designed a geodesic dome in the shape of a soccer ball. Subsequently, C_{60} molecules have been nicknamed buckyballs, and the whole family of carbon clusters (70-, 76-, 80- and 84-atom clusters have also been discovered) are referred to as fullerenes. In the solid state, C_{60} molecules assemble into a face-centred cubic lattice with a lattice constant of around 1 nm. Small atoms, such as potassium, can be introduced to interstitial sites in the lattice by heat treatment, and this changes the crystal from an insulating state to an electrically conducting state. Atoms can also be introduced inside the C_{60} cages.

6.1.3 Examples of organic semiconductors

The appendix to this chapter comprises a list of important conjugated molecules and polymers in the field of organic semiconductors. The acronyms by which they are referred to in this book are also provided along with a brief discussion of their properties. Readers are encouraged to familiarize themselves with the variety of conjugated materials that have been explored.

Historically, the organic semiconductor discipline has distinguished between polymeric and low molecular weight organic semiconductors. This distinction is nowadays blurred due to the advent of a number of hybrid materials, which combine properties and attributes of low molecular weight and polymeric materials. A few examples of these are included in the appendix.

The sheer length of the appendix – which is still incomplete – underscores one of the key assets of organic semiconductors: the practically unlimited diversity of synthetic organic chemistry allows the tailoring of materials with a large portfolio of properties. Besides this variety, it should be remembered that materials which are nominally the same often show very different performance in devices. Device performance can be very sensitive to low levels of impurities and/or chemical defects, and different chemical routes that lead to the same material often introduce different levels and types of defects and/or impurities. Even if the chemistry is ideal, the same material can still display very different properties when processed in different ways. The solvent and casting method used for solution processing, or deposition rate, type and temperature of substrate for evaporated films, thermal treatment cycles, and the presence or absence of even trace amounts of oxygen can have a decisive impact on the resulting device.

6.1.4 Excitations in organic semiconductors

In any semiconductor application, the material will not be in its ground state. To transport charge and/or emit light, the semiconductor needs to sustain excitations, and in the case of charge transport, these excitations also need to be mobile. In this section the fundamental properties of these excitations are discussed, whilst in Section 6.1.5, the generation and migration of excitations is described.

6.1.4.1 Polarons and excitons

When an electron is taken away from the HOMO or added to the LUMO of a molecule, the resulting molecule is termed a *radical ion*; namely a radical cation for positive charge, and radical anion for negative charge. The word 'radical' refers to the net spin the molecule will have due to the unpaired remaining (or added) electron. After removal or addition of the electron, molecular orbitals and the positions of nuclei will respond by a relaxation to a new position of minimum energy. Radical ions are often called *polarons*, analogous to the term used for inorganic semiconductors. Again, however, the inorganic polaron is delocalized with an associated wavevector **k** describing its coherent movement; the radical ion is not delocalized.

Due to the localized character of the polaron, its charge strongly couples to molecular geometry. Bond distances and angles will be distorted compared to the neutral molecule. This distortion will always reduce the energetic cost of forming a polaron. Therefore removing an electron costs somewhat less energy than the HOMO suggests, and an electron joining the molecule gains somewhat more energy than the LUMO suggests, because HOMO/LUMO levels are calculated for undistorted molecular geometries. Instead, the energies required for polaron formation are called the ionization potential I_p, and electron affinity E_a, respectively (we define these two parameters in Figure 6.7). As a practically important example, a neutral and a positively charged polythiophene segment are sketched in Figure 6.2. Note how the missing electron, or *hole*, leads to a redistribution of the π-bonds and hence to different nuclear distances and positions.

When an electron is removed from the HOMO but is placed into the LUMO instead of being removed entirely from the molecule, we arrive at an electrically neutral excitation, the so-called *exciton*. A typical way of lifting an electron from the HOMO into the LUMO is via the absorption of a photon. The π electrons redistribute into the excited π^* orbitals, which are also known as antibonding orbitals, as they destabilize the molecule. Nevertheless, the strong σ bonds are crucial in keeping the molecule intact. Again, the excitation leads to a related relaxation of the surrounding crystal lattice or molecule. Figure 6.3 shows the geometric relaxation and redistribution of electron density in an excited phenylene–vinylene segment.

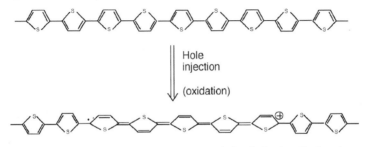

Hole
injection

(oxidation)

Figure 6.2 A polythiophene segment and the derived radical cation

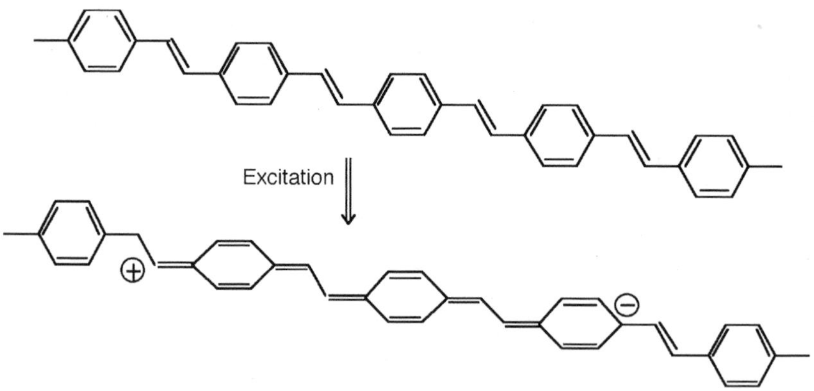

Figure 6.3   The transition from an aromatic, 'bonding' π phenylene–vinylene system to a quinoidal, 'antibonding' π^* system on optical or electrical excitation

The size of the exciton is about three repeat units, or 10 nm, and the exciton has clearly intramolecular, one-dimensional character. This makes organic excitons *Frenkel* excitons. Typical exciton binding energies E_b are in the range 0.2–0.5 eV. Note how considerable ambiguity arises in the term 'band gap', which can mean either the energy difference between LUMO and HOMO, or $I_p - E_a$ or $(I_p - E_a) - E_b$.

Figure 6.4 is an alternative, more schematic, representation of the electronic ground state, radical ions (here called polarons), and excitons in organic semiconductors. It shows two different types of exciton: *singlet* and *triplet* excitons. There are three ways in which hole and electron spin can combine so that the resulting overall spin part of the wavefunction is symmetric under particle exchange, and has total spin $S = 1$, namely $|\uparrow\uparrow\rangle, |\downarrow\downarrow\rangle$, and $(1/\sqrt{2})(|\uparrow\downarrow\rangle + |\downarrow\uparrow\rangle)$. Excitons with that property are called triplet excitons. The combination of spins $(1/\sqrt{2})(|\uparrow\downarrow\rangle - |\downarrow\uparrow\rangle)$ results in a spin part of the wavefunction that is antisymmetric under particle exchange, and total spin $S = 0$. This combination is called a singlet exciton.

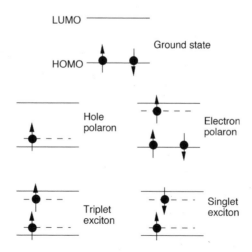

Figure 6.4   Energy level diagrams for excitations in organic semiconductors

6.1.4.2 Light emission from organic molecules

In simplified terms, the absorption of a photon that has generated an exciton on an organic molecule can in some cases be reversed (ignoring some intermediate steps discussed below). The excited electron drops back from the LUMO into the HOMO, emitting a photon in the process. This phenomenon is known as *fluorescence*. Since excitons can also be generated electrically by combination of polarons, this paves the way to organic electroluminescence (EL). First, we shall discuss fluorescence in some detail, using a framework based on ground- and excited-state molecular orbitals. Then we shall return to discuss a striking difference between fluorescence and EL, which is best understood using the more schematic picture of Figure 6.4.

Fluorescence in the MO picture

Figure 6.5 presents a schematic view of fluorescence in the MO picture. The two curves in the potential energy–distance diagram describe the ground state (S_0) and the first excited state (S_1) of a chemical bond; the equilibrium distance is the bond length, which is longer for the excited state. The horizontal lines represent the vibrational states of the bonds; i.e., nuclear distance oscillates around the equilibrium bond length within the limits of the intersection of the horizontal line with the potential curve.

The difference between vibrational levels is the vibrational-electronic or *vibronic spacing*, of order 0.1 eV ($\sim$1100 K). Consequently, at room temperature, almost all bonds are in the lowest vibronic level of the ground state. From there they may be lifted into the first excited state by absorption of a photon. This process is governed by the *Franck–Condon principle*, which states that electronic transitions are much faster than nuclear rearrangement. Hence transitions occur vertically in the diagram, from ground state equilibrium position to the most probable interatomic spacing of a

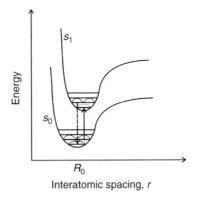

Interatomic spacing, *r*

Figure 6.5 Energy diagram for molecules in ground (S_0) and excited (S_1) electronic states. An absorption transition and an emission transition between these levels are shown for the first two vibronic states. These transitions occur to minimize any change in interatomic spacing, and this can only rapidly occur when the 'probability' of a given spacing is large in both the initial and final levels. Example probability functions are shown in the figure. Note that the interatomic spacing r is equal to the equilibrium spacing R_0 only in the ground vibronic state of the ground electronic state

vibrational mode of the excited state that is close to the ground state equilibrium distance (Figure 6.5). The relative intensity of these is controlled by the overlap integral between the electronic ground state and the vibronic ground state (0 vibronic state) and excited state and vibronic state $0, 1, 2, \ldots$ (Franck–Condon factors). In the excited state, however, there will be a rapid radiationless relaxation (typically of order 10^{-12} s) of the excited state into its lowest vibronic level (internal conversion). From there the photon may be re-emitted after typically between 1 and 10 ns with a transition to the lowest vibronic level of the ground state (0–0 transition), the first vibrational state (0–1 transition), second vibrational state (0–2 transition), and so on. This gives rise to the occurrence of several peaks in the absorption and fluorescence spectra, the so-called vibronic structure or *vibronic satellites*. Vibronic spacing is similar in the ground and excited states, so they often appear like mirror images (Figure 6.6). The onset of absorption or the intersection of the absorption and photoluminescence (PL) spectra in a plot such as Figure 6.6 is often called the *optical band gap*.

The *photoluminescence quantum yield* η_{PL} is defined as the ratio of the number of photons in to the number of photons out. Fluorescent decays of an excited state back to the ground state compete with a number of non-radiative decay processes (fluorescence quenching). The fluorescence quantum yield can be expressed as the result of a competition between fluorescent processes with rate Γ and non-fluorescent processes with rate k_n:

$$\eta_{PL} = \frac{\Gamma}{\Gamma + k_n}. \tag{6.1}$$

In a good dye this should be as high as possible. Perylene has $\eta_{PL} = 0.94$ in cyclohexane solution, and many modern conjugated polymers exceed a 50% quantum yield even in the solid state. However, some conjugated polymers display rather low fluorescence quantum yields. PATs, for example, have rather moderate quantum yields and are used widely in non-emissive transistors, but not in light emitting devices (see the Appendix for the chemical formula of PAT). This is because they display a molecular dipole moment that enhances non-radiative decay channels.

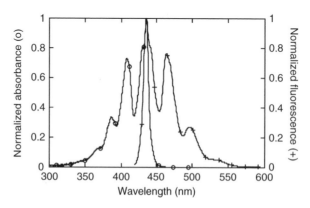

Figure 6.6 Absorption and emission spectra of perylene. Data from http://omlc.ogi.edu/spectra/ PhotochemCAD/html

Fluorescence, parity, electroluminescence and phosphorescence

The observation that poly(arylene vinylene)s or poly(arylene)s such as PPV, PPE, PF, PPP and MeLPPP do display bright fluorescence, whereas PATs have somewhat lower quantum yields, can be understood through the different strength of non-radiative decay channels. However, polyenes such as PA and PDA display $\eta_{PL} \approx 0$, without strong molecular dipoles. A different mechanism must be at work that forbids fluorescence in the polyenes.

A necessary condition for a molecule to be fluorescent is *parity alternation* between the ground state and the lowest-lying excited state. The simplest quantum mechanical model of bound electrons is a single electron bound to a harmonic potential (a harmonic oscillator). The harmonic oscillator has discrete energy eigenstates that can be labelled by a quantum number n. For all odd-numbered eigenstates of the harmonic oscillator, including the ground state $n = 1$, eigenstates obey $\Psi_n(-x) = \Psi_n(x)$; i.e., the state is said to have *even* (*g*) parity. For even-numbered eigenstates, including the lowest-lying excited state $n = 2$, eigenstates obey $\Psi_n(-x) = -\Psi_n(x)$; i.e., the state is said to have *odd* (*u*) parity. This property is called parity alternation.

A molecule is rather different from a harmonic oscillator. Nevertheless, molecular orbitals do display either *g* or *u* parity as well. However, parity alternation may be broken, mainly because of strong electron–electron interactions. Note the harmonic oscillator Hamiltonian contains no such interactions.

Fluorescent transitions can take place only between states of different parity. This is called a *selection rule*, and is related to the fact that a photon carries unit angular momentum $\hbar$ needs to be supplied from the emitting molecule. Only in molecules that retain parity alternation between ground and first excited state can we observe fluorescence. Experience (as well as extensive quantum chemical computations) shows that the strong electron–electron interactions in the polyenes break parity alternation and are therefore not fluorescent. In polyarylenes and polyarylenevinylenes, however, parity alternation is preserved.

In light-emitting devices, light is generated not by the absorption of a photon, but by the combination of an electron and a hole polaron. Figure 6.4 clearly shows that an electron in the LUMO and an electron vacancy (hole) in the LUMO can combine to form an exciton. The fluorescence emitted from an electrically generated exciton is called *electroluminescence* (EL). However, there is a fundamental difference between the formation of excitons by absorption of light and by combination of polarons. The ground state of a molecule carries net spin $S = 0$, and is thus a singlet state. The angular momentum $\hbar$ of a photon interacts with the orbital angular momentum of the molecular wavefunction (this leads to the parity alternation selection rule). However, to a first approximation, photon angular momentum does not interact with spin angular momentum, and thus cannot flip electron spins. Therefore absorption of a photon can only generate singlet excitons. Conversely, only singlet excitons have a *dipole-allowed* transition to the molecular ground state; fluorescence links an excited singlet state to the singlet ground state.

On electrical generation, however, polarons can combine to form triplet as well as singlet excitons. In fact, there are three combinations leading to a triplet, and only one leading to a singlet. If polaron–polaron capture were independent of mutual spin orientation (as one naively expects), only one in four electrically generated excitons would be able to yield electroluminescence: $\eta_{EL} = \eta_{PL}/4$. This limits the efficiency of EL devices, and has been subject to extensive applied and fundamental research (Section 6.2.3).

In some cases, light can be emitted from triplet excitons, although that transition is dipole-forbidden. This is facilitated by so-called *spin–orbit coupling*, which results from the fact that both spin and orbital angular momentum imply a magnetic moment, which interact via a product term $\mathbf{L} \cdot \mathbf{S}$ in the molecular Hamiltonian. In effect, the angular momentum $\hbar$ required for emission of a photon can be provided from the triplet spin, instead of orbital angular momentum. Remember that $S = 1$ for a triplet, thus the transition from a triplet to a singlet releases unit angular momentum $\hbar$. To facilitate spin–orbit coupling, atoms are required with electrons in states with high orbital angular momentum $\mathbf{L}$, as this strengthens the $\mathbf{L} \cdot \mathbf{S}$ term. Generally, these are relatively heavy atoms (heavier than carbon) that have higher atomic shells occupied, such as sulphur and phosphorus. The resulting dipole-forbidden emission is called *phosphorescence*. The rate of a dipole-forbidden transition is much lower than for a dipole-allowed transition. Typical phosphorescence lifetimes can be up to $\sim$1 ms.

6.1.4.3 Controlling the band gap

The wavelength of light emitted from an excited molecule, and therefore the colour of emission, is determined by the band gap. In particular, for light emission applications it is paramount to control the band gap. One of the key assets of organic semiconductors is that the band gap can be fine-tuned by synthetic chemistry. In fact, E_a and I_p can often be manipulated rather independently of each other. This subject cannot be discussed systematically here, and the reader is referred to the review by Kraft and colleagues listed in the bibliography. Instead, the key concepts are illustrated by some examples.

The band gap of PATs and PDAs can be tuned rather widely by the attachment of *side chains*. We will see in Section 6.1.6.2 that side chains are essential to promote polymer solubility. Side chains lead to twisting of adjacent ring systems out of the coplanar conformation. The twist angle depends on the length, type and bulkiness of the side chains. The larger the twist, the poorer the overlap between π systems, thus the larger the band gap becomes; a *blueshift* occurs. PATs and PDAs are the outstanding examples for *steric* band gap tuning. Another approach to blueshifting a band gap is to reduce π coherence by deliberately introducing meta- instead of para-linked phenyl rings into the main chain of phenylene vinylenes.

When alkoxy rather than alkyl chains are attached to a ring system, the respective polymer ends up with a reduced band gap due to the influence the oxygen has on the benzene π orbitals. This explains the reduced band gap in MEH-PPV as compared to PPV. A much stronger influence on the band gap is exercised by attaching strongly electron-withdrawing cyano groups to the vinylene bonds in phenylene vinylene. The resulting CN-PPV has a higher electron affinity than PPV or MEH-PPV. The band gap is reduced to give a red-emitting polymer, and the character of CN-PPV is changed to electron transporting, as compared to hole-transporting PPV. These are examples of *electronic* band gap tuning.

The side chains of polyfluorenes are not directly attached to the π system, and thus can exercise only very little influence on the band gap. Band gap tuning is achieved by *copolymerization* instead. Alternating copolymers of fluorene with two thiophene units (F8T2), or of fluorene with a bisthiadiazole unit (F8BT), both have a similar band gap in the green, compared to the blue band gap of the polyfluorene homopolymer. However, in F8T2 the reduction in band gap results from a reduced ionization potential, whereas in F8BT it results from a higher electron affinity.

6.1.5 Charge carrier injection and transport

The injection and transport of charge carriers is an issue of practical importance for semiconductor devices: Transistors require the injection of one type of carrier from an electrode, and rather fast mobility of that carrier. Light-emitting devices require the injection of carriers of both types from different electrodes. Photovoltaic devices need to separate excitons and transport the resulting carriers to opposite electrodes. It is thus paramount that we discuss the factors controlling carrier injection and transport.

6.1.5.1 Charge carrier injection

The characteristics of carrier injection from a metal electrode into a semiconductor are controlled by the work function Φ of the metal relative to the electron affinity E_a of the semiconductor for electron injection, and relative to the ionization potential I_p of the semiconductor for hole injection. An energy level diagram such as Figure 6.7 is often used

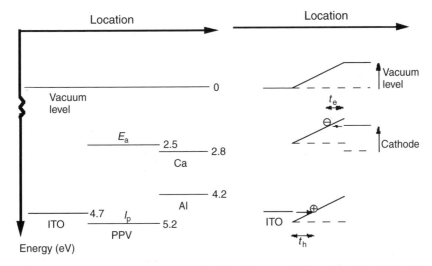

Figure 6.7 Energy levels for a PPV layer sandwiched between unlike electrodes. Indium tin oxide (ITO) is a transparent metallic material that is commonly used as anode for OLEDs; the anodes are typically metals such as aluminium or calcium. Left: no bias voltage applied (numbers indicate associated energies relative to the vacuum level in eV). Right: a voltage is applied in forward bias. t_e and t_h are the respective distances that electrons and holes have to tunnel to be injected into the semiconductor. Note that under this bias, the vacuum level of the cathode moves relative to the anode

to illustrate carrier injection into an *organic light-emitting device* (OLED). Note that due to the molecular nature of organic semiconductors, there are no surface dangling bonds that can distort bulk energy levels. Also, organic semiconductors are generally not deliberately doped. Hence there is no band bending that could distort the level diagram.

The left-hand part of Figure 6.7, shows an *injection barrier* of 0.5 eV for holes from ITO into PPV, and 0.3 or 1.7 eV for electron injection from Ca or Al into PPV, respectively. For electron injection from ITO there would be a large barrier of 2.2 eV. Thus, the use of electrodes made from unlike metals defines a *forward* and *reverse bias* for the OLED. To minimize barriers, a high work function anode and a low work function cathode are required. However, for barriers of 0.5 eV, as shown here, the current density in a device under forward bias will be controlled by the injection of carriers across the barrier, rather than by the transport of carriers across a device.

The right-hand part of Figure 6.7 shows the same device (assuming a Ca cathode) under a forward bias. Carriers can now overcome injection barriers by tunnelling or by thermionic injection. The tunnelling current density j ($j = I/A$ with current I, area A) against the applied voltage characteristic $j(V_{bias})$ is described by *Fowler–Nordheim (FN) tunnelling* and satisfies the following equation:

$$j_{FN} = \frac{C}{\Delta V}\left(\frac{V_{bias}}{d}\right)^2 \exp\left(-B\frac{d\,\Delta V^{3/2}}{V_{bias}}\right), \tag{6.2}$$

where B and C are constants, and ΔV is the injection barrier. Note that j_{FN} depends on applied voltage V_{bias} and film thickness d only in the combination V_{bias}/d such that j_{FN} scales with the applied field E. Thermionic injection is described by the *Richardson–Schottky* equation:

$$j_{RS} = AT^2 \exp\left(\frac{-(\Delta V - V_m(E))}{k_B T}\right), \tag{6.3}$$

where A is a constant and $V_m(E)$ describes the field-dependent lowering of the injection barrier by the attraction of the injected carrier to its mirror charge; $V_m(E) = \sqrt{eE/4\pi\varepsilon_r\varepsilon_0}$.

Generally, RS injection will dominate at low fields, and FN tunnelling at large fields. Experimental $j(V)$ results on MEH-PPV diodes can qualitatively be described by Equation (6.2), however, the absolute current density is several orders of magnitude lower than described by Equation (6.2). This is due to a large backflow of carriers from the semiconductor into the metal.

In the case $\Delta V \to 0$; i.e., when a work function is matched to the respective semiconductor level, the FN equation predicts a divergence of current density. Then the current density will be controlled by carrier transport across the film, rather than by injection, and the metal–semiconductor contact is termed *ohmic*. Practically, a contact can be considered ohmic whenever carrier transport is a more restrictive limit on current density than carrier injection. Roughly, barriers of 0.3 eV at the very most can be considered ohmic.

Ohmic contacts are generally desirable, and considerable effort has been devoted to increase the work function of the transparent ITO anode by a variety of physicochemical treatment cycles. For example, recently it has become common to coat ITO with a thin film of the high work function synthetic metal PEDOT/PSS ($\Phi = 5.2\,\mathrm{eV}$). The cathodes are commonly low work function materials such as Ca. These materials require protection from the ambient atmosphere, otherwise they would degrade rapidly. This can be provided by encapsulation or by capping with a more stable metal such as aluminium.

6.1.5.2 Charge carrier transport: mobility

After injection, charge carriers in a semiconductor will generally move in the direction of the applied field (at least on average). Two types of motion are common: coherent (band-like) motion, described by a wavevector **k**, and incoherent (hopping-type) motion. In disordered materials with localized excitations, such as organic semiconductors, incoherent motion will prevail. This can be visualized as the superposition of a thermally driven random diffusion and a directed motion in the direction of the electric field. The first of these motions will not be considered here, as it is undirected and does not result in a net current when averaged over time or the ensemble. As a starting point, we shall consider the directed motion of charge carriers under an applied field in a fashion similar to the motion of a solid sphere in a viscous medium under the influence of a constant force (e.g., gravity). The sphere (or the carrier) will rapidly accelerate up to a steady-state velocity v, at which the friction exerted by the viscous medium balances the constant force. This velocity is given by

$$v = \mu E, \tag{6.4}$$

where E is the electrical field, $E = V_{\mathrm{bias}}/d$, and μ is the charge carrier mobility, typically quoted in units $\mathrm{cm^2 V^{-1} s^{-1}}$. Besides I_p and E_a, the carrier mobility is one of the most important characteristics of an organic semiconductor, and can span a wide range of orders of magnitude, from around 10^{-7} to $10^{-1}\,\mathrm{cm^2 V^{-1} s^{-1}}$, within the regime of hopping-type transport. Considerably higher mobilities of up to $10^2\,\mathrm{cm^2 V^{-1} s^{-1}}$ or more can be observed for coherent motion in organic single crystals. However, applications based on organic single crystals are uncommon and will not be discussed here.

We also note that the analogy to the sphere in a viscous fluid cannot be pushed too far: it is deceptive to assume $v \propto E$ from the apparently simple Equation (6.4) as μ is a function of field E. Bässler in Marburg has carried out extensive computer simulations of hopping-type transport. In his model, he assumes that HOMO levels are not all equal in energy, but display a Gaussian distribution around the average HOMO. This energetic distribution is termed *diagonal disorder* and is characterized by a variance σ^2, or the dimensionless quantity $\hat{\sigma} = \sigma/k_\mathrm{B} T$. The hopping rate is also affected by positional or *off-diagonal disorder* that is quantified by another variance, Σ^2. Bässler proposed the following Σ-dependent equation for μ:

$$\mu(E, T) = \mu_0 \exp\left(-\left(\frac{2\hat{\sigma}}{3}\right)\right) \begin{cases} \exp\left(\left(\hat{\sigma}^2 - \Sigma^2\right)\sqrt{E}\right) & \text{for } \Sigma < 1.5 \\ \exp\left(\left(\hat{\sigma}^2 - 2.25\right)\sqrt{E}\right) & \text{for } \Sigma \geq 1.5 \end{cases} \tag{6.5}$$

6.1.5.3 Charge carrier transport: space charges

The current density j that results from the drift of carriers in an applied field should be given as

$$j = qn\mu E = \sigma E \tag{6.6}$$

where $q = \pm e$ is the charge per carrier, n is the carrier density, and $\sigma = qn\mu$ is the conductivity. Equation (6.6) is equivalent to Ohm's law with resistance $R = d/\sigma A$.

However, Equation (6.6) assumes that the electric field E acting on each carrier at any position within the film is equal to the externally applied field. That will be true for low current densities, when the total charge of all carriers within the film at any one time, Q_{tot}, can be neglected with respect to the capacitance charge, Q_{cap}, of the film. Q_{cap} is the charge the electrode/film/electrode structure can hold at the respective applied voltage as a capacitor, and $Q_{tot} \ll Q_{cap} = CV$, with $C = \varepsilon_0\varepsilon_r A/d$; thus $Q_{tot}/A \ll \varepsilon_0\varepsilon_r E$. When Q_{tot} becomes comparable to Q_{cap}, the externally applied field becomes screened inside the film by *space charge*, and the local field acting on a carrier will be less than the applied field and will depend on position. To estimate the *space charge limited current* (SCLC) in that situation, we assume that we have charge/area, Q_{cap}/A, crossing the film per unit *transit time* τ. τ is the time a carrier would require to cross the film in the space charge free limit, $\tau = d/\mu E$. This leads to an estimate of space charge limited current density $j_{SCLC} \approx Q_{cap}/A\tau = \varepsilon_0\varepsilon_r\mu V^2/d^3$.

The precise treatment of the current density in a semiconductor in the SCLC regime solves the relevant electromagnetic equations in a self-consistent manner, balancing the two competing effects of field screening by space charges and dispersal of space charges by field. This calculation leads to the following results:

$$E(x) = -\frac{3}{2}V\sqrt{\frac{x}{d^3}}, \tag{6.7a}$$

$$n(x) = \frac{3\varepsilon_r\varepsilon_0 V}{4\sqrt{xd^3}}, \tag{6.7b}$$

$$j_{SCLC} = \frac{9}{8}q\varepsilon_r\varepsilon_0\mu\frac{V^2}{d^3}, \tag{6.7c}$$

where x is a spatial coordinate, electrodes are at $x = 0$ and $x = d$, and $E(x)$ and $n(x)$ are the position-dependent field and carrier density, respectively. Equation (6.7c) is known as *Child's law*. Note that $j \propto E(x)n(x)$ does *not* depend on x. The derivation of Equation (6.7) assumes mobility to be independent of E, and it cannot therefore be exactly correct for a Bässler-type $\mu(E)$ (Equation 6.5). Space charges screen the applied field in such a way that $E(0) = 0$; there is no field at the injecting electrode. Strictly, therefore, no current should be drawn across the electrode–semiconductor junction. Carrier diffusion, which we had ignored at the very beginning, comes to the rescue. However, we rely heavily on the assumption of an ohmic contact at the injecting electrode. The calculation also relies on the assumption that only one type of carrier is injected from one electrode, and none from the other.

The above discussion was also based on the assumption that the only carriers present in the semiconductor are injected carriers. Intrinsic carriers do not contribute to space charge, and thus transport based on intrinsic carriers will follow Ohm's law, which is the case in metals. At very low voltages and current densities this will also be true for real semiconductors, which always have a certain level of intrinsic carriers. We shall thus see a transition from $j \propto V$ to $j \propto V^2$ as the voltage is ramped up.

To discriminate experimentally between barrier type and ohmic injection, it is common to compare $j(V_{bias})$ characteristics of devices having different thickness d. FN tunnelling depends only on the applied field $E = V_{bias}/d$, thus in a plot of j against E all characteristics will coincide regardless of d. In SCLC, j will follow Child's law, $j \propto V_{bias}^2/d^3$; thus in a plot of jd as a function of E^2, all characteristics will coincide. The experimental procedure has to be carried out with symmetric electrodes; e.g., gold/ semiconductor/gold for holes and Ca/semiconductor/Ca for electrons. This will ensure single-carrier currents.

SCLC, as described by Equation (6.7c) has been observed for PPV and dialkoxy-substituted PPV. The carrier mobility can be revealed from the slope of the jd versus E^2 plot, and in the case of PPV and dialkoxy PPV it was $\sim 10^{-6}$ cm^{2}V^{-1}s^{-1} or less.

6.1.5.4 Charge carrier transport: traps

The most serious deviation from ideal SCLC behaviour is caused by the presence of *traps*, a fact of life in all real semiconductors. A trap is defined as a site with an ionization potential lower than the ionization potential of the bulk material (hole trap), or with electron affinity higher than that of the bulk material (electron trap). The fact that most real organic semiconductors transport only one type of carrier (holes or electrons), with only few *ambipolar* materials known that transport both types, is thought to be the result of carrier-specific deep traps. However, even for the carrier type that is transported, there will be traps.

In a trap, a carrier is immobilized and will therefore contribute to space charge, but not to the current. j_{SCLC} has to be reduced by a factor $\Theta = n_{trap}/n_{tot}$ with trapped and total charge carrier densities n_{trap}, n_{tot}. Alternatively, we may leave Equation (6.7c) unchanged and introduce an effective mobility μ_{eff}:

$$\mu_{eff} = \Theta\mu. \tag{6.8}$$

The introduction of Θ, however, is not a simple concept. Firstly, the boundary between trapped and mobile carriers is somewhat blurred for shallow traps, which lie energetically only a few $k_B T$ below the HOMO. Carriers in such traps can be thermally activated back into the conduction band. Secondly, carrier release from deeper trap sites can be field activated, thus making Θ and μ_{eff} complicated functions of the applied voltage. Equation (6.8) also opens the door to some confusion between μ and μ_{eff}.

To discuss the effect of traps on current densities, it is instructive to extend the concept of the Fermi level E_F from metals to semiconductors. The Fermi level of a semiconductor (without applied bias) is located halfway between the HOMO and LUMO. This is because the chemical potential for electrons as $T \rightarrow 0$ approaches the midgap. In a semiconductor containing mobile and trapped carriers, the Fermi level (similarly to the HOMO and LUMO) becomes a function of location x and applied bias V, rising with higher V. Generally, traps will influence the j/V characteristics in a very different manner if the local Fermi level is above or below the trap depth. This may result in a very rapid rise of j with V within a small range of voltages, with power laws $j \propto V^m$ where m is often larger than 10. Nevertheless, in the presence of traps, current density j will always be lower than in the trap-free SCLC case.

6.1.6 Polymers versus small molecules

All of our previous discussions on conjugated molecules considered a given, defined molecule. Now we shall discuss aspects that are specific to polymeric organic semiconductors. A polymer is often viewed as infinitely long, but this is often far from true in conjugated polymers. This is due to the, often limited, solubility of conjugated polymers, which results from their stiff backbone. When the polymer reaches its solubility limit during the polymerization reaction, it precipitates and cannot grow any longer. This often happens at relatively low molecular weights. In the organic semiconductor field, there is a tendency to call a material a polymer if it has a distribution of molecular lengths, whereas molecules with a defined number of repeat units (such as 6T) are called oligomers. It does not necessarily mean that the polymer is very long.

6.1.6.1 The band gap in a conjugated polymer

One of the most important properties of a semiconductor is its band gap. We therefore need an understanding of what controls the band gap of a conjugated polymer. Let us consider an oligomer series of molecules with a defined length of n_r repeat units, and monitor the band gap as the number of repeat units grows. Obviously, larger n_r means better π electron delocalization and smaller band gap. Will the band gap close altogether for a polymer, making the material metallic instead of semiconducting?

Experimentally, this is not observed. Instead, one finds the following scaling of band gap with n_r:

$$E_g(n_r) = E_\infty + \frac{E_1 - E_\infty}{n_r} \quad \text{for} \quad n_r < n_{ECS}, \tag{6.9a}$$

and

$$E_g(n_r) = E_\infty + \frac{E_1 - E_\infty}{n_{ECS}} = \text{constant} \quad \text{for} \quad n_r > n_{ECS}. \tag{6.9b}$$

For short oligomers, the band gap reduces proportional to $1/n_r$ but there is a critical n_r (n_{ECS}) beyond which E_g remains constant even when n_r continues to grow. In terms of photophysics, n_{ECS} is the boundary between low molecular weight material and polymer. As an example, we look at a fluorene oligomer series. n refers to the number of fluorene units; the absorption maximum (instead of E_g) is reported. The $1/n_r$ relationship holds approximately for all data reported; i.e., up to $n_r = 10$ (Figure 6.8). However,

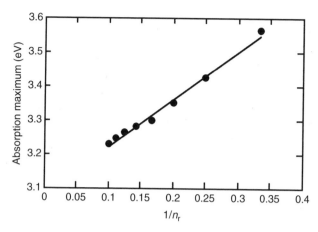

Figure 6.8 Optical gap plotted as a function of $1/n_r$ for oligofluorenes. The data used in this figure is taken from G. Klaerner and R. D. Miller, *Macromolecules* **31**, 2007 (1998)

extrapolating this relationship, $n_r \to \infty$ or $1/n_r \to 0$ leads to an absorption maximum $E_\infty = 3.08$ eV. However, a polyfluorene ($n_r \gg 10$) displays a larger absorption maximum of 3.20 eV. This is consistent with $n_{ECS} \approx 12$.

This behaviour can be understood with the concept of the *effectively conjugated segment* (ECS) as initially introduced for PPV.[1] In this picture, a conjugated polymer consists of a string of effectively conjugated segments. At the end of each effectively conjugated segment, one photophysical unit ends and another begins. This is reminiscent of the concept of persistence length L_p in polymer physics, which describes the length over which a polymer chain roughly maintains a given direction; L_p is a measure of chain stiffness. In fact, for polyfluorene, L_p is found to be approximately 11 repeat units, which compares well to $n_{ECS} \approx 12$ repeat units. Hence we see that the presence of conformational degrees of freedom in a polymer chain may limit the coherence of the π electron cloud and thus limit effective conjugation length. This enables band gap tuning via chain backbone architecture.

However, conformational disorder is not the only limiting factor of effective conjugation length. The backbone of MeLPPP has no conformational degrees of freedom, and all rings are forced to be coplanar. This provides excellent geometric conditions for π overlap. Nevertheless, the MeLPPP effectively conjugated segment length is finite, and is in fact surprisingly short. There is an intrinsic limit to the coherence length of π-conjugated molecules. As the example of MeLPPP teaches, this limit may be approached sooner for better π overlap.

When the ECS is limited by conformational disorder, there will be a certain distribution of ECS lengths within a sample. Consequently, absorption spectra will be the superposition of the spectra of ECSs of slightly different lengths, hence different E_g; as a result, the vibronic structure of the absorption spectrum is lost. This is known as *inhomogeneous broadening*. However, in MeLPPP, with its perfectly ordered backbone,

[1] H. H. Hörhold, M. Helbig, D. Raabe, J. Opfermann, U. Scherf, R. Stockmann and D. Weiβ, *Z. Chem.* **27**, 126 (1987).

inhomogeneous broadening is absent; the MeLPPP absorption spectrum displays clearly resolved vibronic structure that mirrors the emission spectrum.

In emission, however, vibronic structure is usually maintained better, even for polymers with a significant degree of conformational disorder. This phenomenon is understood as the result of *exciton diffusion*. An exciton can be transferred from a shorter ECS (larger band gap) to a longer ECS (smaller band gap), but not the other way round. In PPV an exciton can sample a sphere of radius ~5 nm radius before emitting fluorescence. Consequently, for polymers the term 'band gap' becomes even more ambiguous, as the average conjugated segments control absorption, but the lowest band gap conjugated segments control fluorescence.

6.1.6.2 Polymer solubility

Much of the appeal of molecular electronics comes from the fact that organic materials can be processed from solution. However, solubility of stiff, elongated molecules rapidly decreases as they get longer, and at the same time, the melting point rises. This is due to the reduced entropy of solution and entropy of melting, a problem that was first encountered probably by Vorländer in the 1920s in his quest to synthesize larger and larger liquid crystalline molecules. Typically, molecules with five or six repeat units such as pentacene or 6T are beyond solubility.

There are two common approaches to this problem. Historically the first, but now not so much in use, is the *precursor approach*. Instead of synthesising a finished, conjugated polymer, one prepares a flexible and soluble precursor polymer, which is processed into a thin film in the envisaged device situation. Then, via a polymer analogous reaction (typically driven by heat), the precursor is converted in situ into a conjugated, and usually insoluble, polymer. A precursor route to PPV based on thermal tetrahydrothiophene elimination is shown in Figure 6.9.

Note that the precursor is a water-soluble polyelectrolyte with a flexible, non-conjugated backbone.

Another approach to soluble, stiff polymers is known from the synthesis of liquid crystalline polymers and can be directly transferred to conjugated polymers. When a stiff backbone is decorated with long, flexible side chains, these side chains gain considerable entropy on dissolution, hence the polymer becomes soluble. This approach has led to the *hairy rod* family of liquid crystalline polymers as well as to a large number of soluble conjugated polymers. Many side chain decorated, soluble conjugated polymers are included in the appendix.

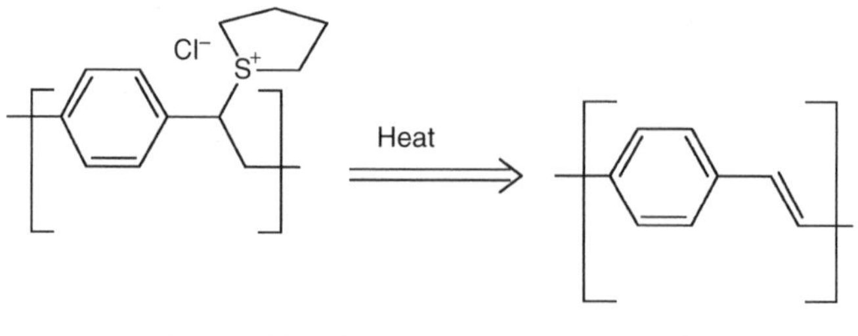

Figure 6.9   The precursor route to PPV

The resulting soluble conjugated polymers also melt at moderate temperatures, since melting as well as dissolution is entropy-driven. In fact, most conjugated polymers share the basic architecture of stiff backbone with flexible side chains with the hairy rod liquid crystals. It is therefore not surprising that several of them exhibit liquid crystalline phases, most prominently alkyl-substituted PFs and its copolymers such as F8T2 and F8BT.

6.1.7 Organic metals?

Considering polyacetylene (PA) in a similar fashion as we did the benzene ring, we find it can have two borderline structures, as shown in Figure 6.10. Again, it is unclear where the double bonds (π bonds) should be and, similarly to the benzene ring, they may be completely delocalized along the whole chain. This would lead to metal-like behaviour, as the π cloud can move continuously (i.e., in response to moderate fields): inject electrons at one end and they drop out at the other.

However, this does not happen. In the 1950s, Rudolf Peierls proved a general theorem, stating that there can be no one-dimensional metal. Conjugated polymers were not known in Peierls' day, but the correspondence between the then hypothetical 'one-dimensional metal' and materials like PA is obvious.

Peierls showed that a one-dimensional metal would be unstable against a metal–semiconductor transition, the *Peierls transition*. Atoms will move closer to each other pairwise, leaving a wider distance to the next pair, rather than remaining spaced equally (Figure 6.11). Also, the π electron density will redistribute to be higher between close

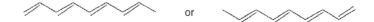

Figure 6.10   The double bond assignment in PA is ambiguous

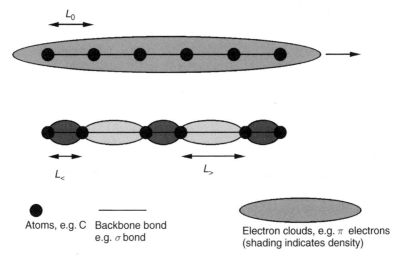

Figure 6.11   The Peierls transition shown schematically

atoms, and lower between distant atoms. There will be a limit to the pairing due to the elastic distortion of the one-dimensional chain. This 'elastic force' results from the σ bonds. The balance between these forces is given by a bond alternation parameter δ:

$$\delta = \frac{L_> - L_<}{L_> + L_<} \tag{6.10}$$

The description with alternating single/double bonds having different bond lengths is therefore adequate. Note that the size of the unit cell has doubled due to the Peierls transition. The number of valence electrons in the double unit cell is now even, hence the material can be a semiconductor. In fact, the doubling of the unit cell opens up a band gap between the HOMO and the LUMO, which refers to an antibonding state, where the π electron density is high between the long bonds and low between the short bonds.

It should be noted that a Peierls transition would only lead to band gaps of order 0.2 eV in a conjugated polymer, whereas gaps of order 3 eV are often found. The Peierls transition is an oversimplification. The true band gap is due to the simultaneous presence of a *Mott–Hubbard* transition. Nevertheless, McDiarmid, Sarikawa and Heeger were awarded the 2000 Nobel prize in chemistry for the discovery of synthetic metals. As so often in the applied sciences, a theoretical limit can be sidestepped rather then overcome. Their discovery will be discussed in Section 6.2.1.

6.2 APPLICATIONS AND DEVICES

The expected massive added value of organic semiconductor products once they are manufactured in bulk quantities and sold competitively on the markets has fuelled the recent rapid development in organic semiconductor materials. It is therefore impossible to sensibly discuss materials without devices and applications. We will see how the demands on new or improved materials originate in the attributes and requirements that the respective application demands. Therefore we will here introduce the key applications and devices that are being addressed currently. This discussion excludes the one application that is already dominated by organic semiconductors, namely, photoconductors for xerography and laser printers.[2]

6.2.1 Synthetic metals

Typical examples of so-called synthetic metals are polyacetylene (PA), polydiacetylene (PDA), polyaniline (PANi), and poly(3,4-ethylene dioxythiophene) (PEDOT). In their pristine state, these materials are organic semiconductors and the term 'synthetic metal' is not entirely appropriate. Synthetic metals become quasi-metallic only as a result of *doping*. Doping with heteroatoms (electron acceptors near the VB, or electron donors

---

[2] This topic is covered in much detail in, for example, P. M. Borsenberger and D. S. Weiss, *Organic Photoreceptors for Imaging Systems*, Marcel Dekker, New York, 1993.

near the CB) is a well-known concept in inorganic semiconductors, and can make a semiconductor quasi-metallic at room temperature, although strictly, the defining characteristics of a metal are not met.

Doping of organic semiconductors is somewhat different as it involves a chemical reaction between semiconducting polymer and dopant, wherein the dopant changes the number of π electrons in the backbone. Both redox and acid/base reactions between semiconductor and dopant have been used successfully; dopant concentrations of between 1% and 50% are typically used. When referring to a synthetic metal, it is not sufficient just to quote the (intrinsically semiconducting) polymer used, but also the type and concentration of dopant, together with the relevant processing conditions. Room temperature conductivities in the range 10^{-3} to $10^{7}\,\mathrm{S\,m^{-1}}$ have been reported; the highest of these conductivities are comparable to those of metals: For example, gold has a conductivity of approximately $5 \times 10^{7}\,\mathrm{S\,m^{-1}}$ at room temperature.

Historically, the first discovery of metallic conductivity in doped PA was an accident. An iodine-containing catalyst was used for PA synthesis. By accident, 1000 times the intended catalyst concentration was added. It turns out that iodine is a useful (redox) dopant for PA, and it was this fact which led to the award of the 2000 Nobel prize in chemistry.

6.2.1.1 Solution-based synthetic metals

The first synthetic metals existed in the form of intractable films only. The breakthrough from laboratory curiosity towards industrial application for synthetic metals came with the advent of modern, aqueous solution based preparations that can be applied to most surfaces, or can be processed by an inkjet printer. Doped PANi and PEDOT are now commercially available in that form. The fact that synthetic metal preparations are water based, but do not dissolve in common organic solvents, makes them very attractive for use as metallic electrodes or interconnects in conjunction with organic semiconductors, which are typically processed from organic solvents but do not dissolve in water.

As the outstanding example, we here discuss the manufacture and applications of PEDOT preparations. For high conductivity, PEDOT is acid doped (rather than redox doped) with poly(styrene sulfonic acid) (PSS) to form the highly conductive PEDOT/PSS complex. The key idea to arrive at a water-based preparation is to synthesize PEDOT from EDOT monomer in an aqueous medium that already contains the PSS. The reaction is shown schematically in Figure 6.12.

Note how the chemical bonding pattern of EDOT/PEDOT changes under acid doping: A C=C π bond opens up and the C bonds to a proton donated by the acid. As a result, there is a net positive charge on the PEDOT chain that will strongly attract the negative charge left on the acid. Since this happens at many points along the chain, PEDOT and PSS become closely intertwined and float as a fine dispersion or colloid in water, rather than precipitate. Also, an unpaired π electron remains on the main chain that is highly mobile along the chain. The removal of one π electron from the HOMO is akin to the removal of one electron from an inorganic semiconductor VB by an electron acceptor just above the VB level, but results from a chemical reaction, not the presence of a heteroatom.

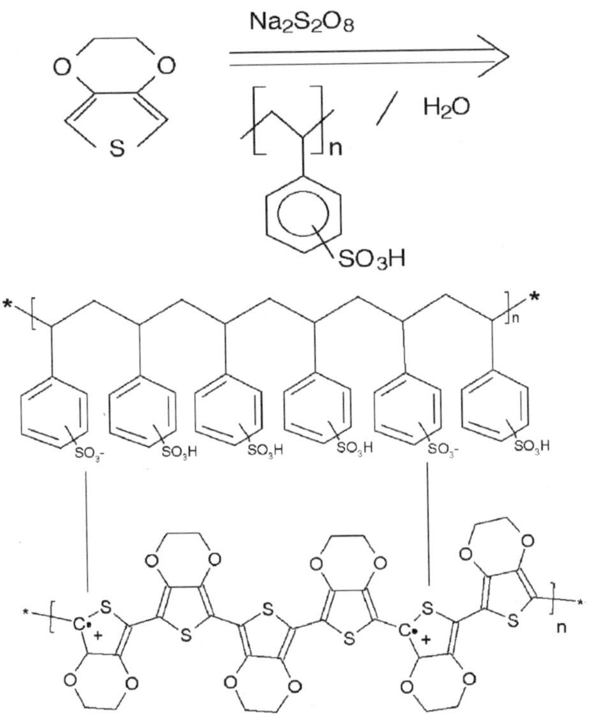

Figure 6.12 The synthesis of PEDOT from EDOT in aqueous medium in the presence of poly(styrene sulfonic acid). $Na_2S_2O_8$ acts as oxidizing agent for EDOT polymerization

Films cast from commercial PEDOT/PSS solution (Bayer trade name Baytron P) display resistivities typically between 10^{-2} to $10^{-3}\,\Omega\,m$; i.e., conductivities of between 100 and $1000\,S\,m^{-1}$. For surface coating applications, it is common to quote the *sheet resistance*, expressed in $\Omega/\square$ (ohms per square). Sheet resistance is defined as resistance between two electrodes of length L separated by distance L; this is independent of L. A typical sheet resistance for cast PEDOT/PSS is $1\,M\Omega/\square$ when $20\,g/m^2$ material is used. Recently, PEDOT/PSS coated polyester sheets have become available with much lower sheet resistance of order $1\,k\Omega/\square$ (Agfa Orgacon).

6.2.1.2 Applications of synthetic metals

PEDOT/PSS was originally developed as an antistatic coating for photographic films. Large-scale film processing, or 'development', leads to static charging, with potentials up to several thousand volts. Discharge sparks can expose the film. Similarly, static charges have to be avoided in the packaging of electronic components. A PEDOT/PSS coating with surface resistance of $\sim 1\,M\Omega/\square$ is sufficient to disperse charges before high voltages build up. Today more than $10^8\,m^2$ per year of photographic film is coated with PEDOT/PSS. Another interesting application of PEDOT/PSS is as counterelectrode in anodically prepared electrolytic capacitors.

In the context of organic semiconductor devices, the most interesting applications of PEDOT/PSS are as coatings for the commonly used ITO anodes, and as printable electrodes and connectors in organic transistor circuits. PEDOT/PSS has a very high work function ($\Phi \approx 5.2\,\text{eV}$), considerably higher than that of ITO ($\Phi \approx 4.7\,\text{eV}$). Thus, ohmic hole injection can be achieved for many common organic semiconductors from a PEDOT/PSS coated anode, when it would be non-ohmic from uncoated ITO. It is tempting to replace ITO on glass anodes altogether by flexible, transparent Agfa Orgacon sheets. These can be patterned by a photolithographic etching procedure, and are expected to have a great impact in the development of novel small-scale organic electroluminescent (EL) devices. However, the sheet resistance is still too large for large-scale displays.

In the field of organic transistors, the impact of PEDOT/PSS results from the option to process from solution, in particular via inkjet printing. PEDOT/PSS can serve as gate metal, as inkjet printed PEDOT/PSS source and drain electrodes, and as printed vias (interconnects).

6.2.2 Organic field effect transistors

The aim of organic FET (OFET) research is not to outperform current inorganic semiconductor processors in terms of integration density or processing speed. Instead, the idea is to make low-performance integrated circuits based on organics at extremely low cost, cheap enough to be discarded after single use (disposable electronics). One target application is an electronic RF price tag on a food wrapper that a supermarket checkout can read remotely. The target of extremely low cost imposes strict limits on the processing technologies that can be used practically. For example, any vapour deposition step in the device manufacture has to be avoided; vacuum is too expensive.

6.2.2.1 The principle of FET operation

Figure 6.13 is a sketch of a thin film FET with electric connections. A voltage between source (S) and drain (D), called the *drain* or *source–drain voltage* V_D, attempts to drive a current through the semiconducting transistor channel. However, a *drain current* I_D will only flow if there are mobile charge carriers in the channel.

A *gate voltage* V_G will extract carriers from the source into the semiconducting channel. However, since the *gate dielectric* is an insulator, these carriers cannot reach the gate metal. Instead, they will form an *accumulation layer* at the channel–insulator interface. The channel semiconductor thus gets doped by applying V_G. Unlike chemical doping, this is quickly reversible, simply by switching off V_G. V_G switches channel conductivity, in some (desirable) cases by several orders of magnitude. Accordingly, I_D at a given V_D will change with V_G; the FET is an *electronic switch*. Note that an OFET is switched on by applying V_D and V_G of equal polarity *opposite* to the sign of the mobile carriers. Thus, OFETs can easily be used to determine the type of carriers a particular material sustains. If positive V_G, V_D switch the transistor on, then the carriers are negative (electrons); if negative V_G, V_D are required, then the carriers are holes. To minimize injection barriers from source to the semiconductor, high work function

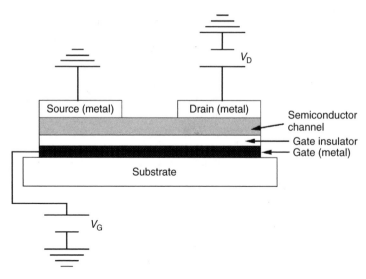

Figure 6.13 A FET with bottom-gate architecture. Transistors may also be built the other way round, with top-gate architecture

metals such as gold or PEDOT/PSS are used for hole channel semiconductors, and low work function metals such as Ca for electron channel semiconductors.

6.2.2.2 Quantitative description of FET operation

To characterize the gate and source – drain voltage-dependent drain current in a FET beyond the qualitative on/off description, usually two types of measurements are carried out: the *output* and *transfer* characteristics. The output characteristic of a FET is the family of curves I_D against V_D at different V_G, but V_G fixed during every single V_D scan. The transfer characteristic is the family of I_D against V_G curves at different V_D, but V_D fixed during every single V_G scan.

The following formulae apply (the derivation is rather technical):

$$I_D = \frac{Z}{L} C_i \mu (V_G - V_T) V_D \text{ for small } V_D \ (|V_T| < |V_D| \ll V_G) \qquad (6.11a)$$

$$I_D = I_{sat} = \frac{Z}{2L} C_i \mu (V_G - V_T)^2 \text{ for large } V_D \ (V_D \geq V_G). \qquad (6.11b)$$

where Z/L is the channel width/length, C_i is the capacitance per unit area of the insulator, V_T is a threshold voltage, and μ is the charge carrier mobility. For a dielectric, $C_i = \varepsilon_r \varepsilon_0 / d$, where ε_r is the dielectric constant and d is the film thickness. ε_r describes how much a dielectric shields the applied field. The large V_D regime is known as the *saturation regime*, I_D does not rise with higher V_D any more but levels off at a (V_G-dependent) saturation current I_{sat}. Saturation will be reached at $V_D \approx V_G$, because at $V_D = V_G$ the field between gate and drain is zero, and the accumulation layer pinches off. The theory of the threshold voltage V_T is rather intricate, and has specific differences between organic and inorganic FETs.

From experimental transistor characteristics, the charge carrier mobility can be determined with the help of Equation (6.11). A robust method is to plot $\sqrt{I_{sat}}$ against V_G (for $|V_D| > |V_G|$), which should produce a straight line with a slope directly related to μ.

For a low-noise measurement, we wish to have a relatively high drain current, which at given mobility, dielectric and operating voltages is determined by the geometry factor, written as Z/L. The maximum geometry factor leads to higher currents that can be measured more comfortably. Often, source and drain are therefore evaporated through an interdigitating comb shadow mask, with many channels in parallel. Practical transistors, however, have to be small and should work with only one channel. Also, L controls the time τ that a carrier requires to cross the channel:

$$\frac{1}{\tau} = f \approx \frac{\mu V_D}{L^2}. \tag{6.12}$$

L enters the equation as L^2; one factor of L comes from the linear dependence of the transit time of carriers through the channel on L, the other factor of L arises from the dependence of the electric field on $1/L$. τ is one of the limiting factors of switching speed. A genuine engineering challenge is therefore to achieve, at low cost, the smallest possible L without inadvertently creating electrical short circuits.

6.2.2.3 Requirements on OFET materials

Good OFETs should display high drain current at low drain and gate voltages, without relying on the optimized geometry factor; they should also have high on/ off ratios, which means drain current at $V_G = 0$ should be extremely low. We will discuss the materials requirements for the different components of an OFET to achieve these goals.

Requirements on OFET semiconductors

Equations (6.11) and (6.12) underscore the importance of high carrier mobilities for good drain current at moderate gate and drain voltages. Hence much effort in synthetic chemistry and physical chemistry is geared towards high-mobility materials. In this section we shall discuss several aspects of charge carrier mobility in organic semiconductors. We assume throughout that we are in the incoherent (hopping) transport regime.

In an OFET, carriers are confined to a very thin interface layer between gate insulator and semiconductor, typically only ~5 nm thick. The resulting interface mobility may differ from the bulk mobility as measured, for example, by the time-of-flight (ToF) method. The approach closest to practical application is thus to build an OFET, measure saturated transfer characteristics, and determine μ from the $\sqrt{I_{sat}}$ versus V_G plot.

In a given material, carrier mobility can depend very strongly on the material morphology, which in turn may change as a result of different preparation conditions. Besides the synthesis of suitable materials, the control of morphology is the key to high mobility. Hence there is no such thing as *the* mobility of a material with a given chemical formula. In fact, there are charts for a few materials showing the evolution of carrier mobility over the years as a result of improved preparation technique.[3] Therefore a wide range of data may be quoted for the same material. This is meant as a caution before considering a few examples in Table 6.2.

Different applications require minimum mobilities of 0.01 to 0.1 $cm^2V^{-1}s^{-1}$. The high mobilities found for some single crystals (e.g., single-crystal pentacene in the coherent transport regime), or for certain thin films carefully deposited onto specific substrates under carefully controlled conditions, are often not useful for practical devices. The challenge is to develop materials that display reasonably high mobilities after a realistic deposition method, such as inkjet printing. As a rule of thumb, thiophene or thiophene-containing materials are the most promising, both for low molecular weight materials (6T and soluble derivatives thereof) or polymers (polyalkylthiophenes

Table 6.2 Mobilities for various molecules at room temperature. The molecules are either hole transporters (h) or electron transporters (e)

Material	Morphology	μ ($cm^2V^{-1}s^{-1}$)	h/e
Pentacene	Large single crystal, vapour deposited, no grain boundaries, along c axis (crystallographic axis perpendicular to the molecular plane)	100	h
Pentacene	Spin coated	0.001	h
Pentacene	Evaporated onto substrate, substrate at room temperature	0.002 to 0.03 (different studies)	h
Pentacene	Evaporated onto substrate, substrate at 120 °C	0.62	h
F8T2	Aligned liquid crystal, parallel to alignment	0.01 to 0.02	h
F8T2	Aligned liquid crystal, orthogonal to alignment	0.001 to 0.002	h
PPV	Several morphologies and methods	10^{-7} to 10^{-4}	h
PAT	Regioregular, edge-on substrate	0.1	h
PAT	Regioregular, face-on substrate	0.001	h
6T	Single crystal	0.1	h
6T	Polycrystalline	0.02	h
DH6T	6T with (α, ω)-dihexyl endchains	0.05	h
C_{60}	In ultrahigh vacuum	0.08	e
C_{60}	Under air	$< 10^{-5}$	e

[3] See, for example, Figure 14.13 in G Hadziioannou and P F van Hutten (eds.), *Semiconducting Polymers*, Wiley-VCH, Weinheim, 2000.

or F8T2). Generally, a high degree of molecular order such as regioregularity in polyalkylthiophenes, or liquid crystalline alignment in F8T2, does result in enhanced mobilities. However, crystallization is usually undesirable. Organic single crystals may display very high mobility coherent transport but, for realistic preparation techniques, materials end up in a polycrystalline morphology, or for polymers, partially crystalline morphology. A crystalline grain boundary, or the interface between a crystalline and an amorphous region, typically represents a deep trap that must be avoided, as it reduces effective mobility μ_{eff} (6.8). Since the channel current is localized in a very thin interface layer close to the gate insulator, trap states at the gate insulator surface may be just as bad as in the semiconductor. The preparation (i.e., synthesis and processing) of materials with high degrees of intra- and inter-molecular order, but without crystallinity, is one of the major challenges in the design of OFET materials.

Another requirement is a certain degree of environmental stability. As an example from Table 6.2, C_{60} may display very high electron mobilities when oxygen is strictly excluded, but electron mobility is poor in the presence of oxygen. Low-cost devices cannot be encapsulated in a way that excludes oxygen completely, thus C_{60} plays no role in current OFET technology.

Generally, it is fair to say that p-type (hole-transporting) OFET materials are now developed to a much higher state of the art than n-type (electron-transporting) materials. C_{60} will not plug that gap, owing to its oxygen sensitivity. A single OFET can operate with a p-type channel, and it would thus be tempting to abandon research into n-type materials altogether. However, the popular CMOS (complementary metal oxide semiconductor) architectures of inorganic semiconductor circuits require n-type and p-type transistors on the same chip. The desire to mimic them with OFETs means that the quest continues for a high-quality n-type OFET material.

High carrier mobility ensures that, for the operation of an OFET as a switch, relatively moderate values of V_G/V_D lead to a decent I_D, and the switch can be switched on easily. However, this is not the only requirement on an OFET. A switch is useless if it cannot be switched off as well as on. An off-current may result if the OFET semiconductor is permanently (chemically) doped, thus giving channel conductivity even at $V_G = 0$. Practically, the on/off ratio can be read directly from the transfer characteristic with V_D applied at the level at which the respective device is meant to operate, and V_G applied or switched off again at whatever level is available. The required on/off ratios depend on the application; e.g., an on/off ratio of 10^6 is required for active matrix addressed liquid crystal displays. For high on/off ratios, OFET semiconductors have to show high mobility μ at very low (intrinsic) conductivity σ. However, since $\sigma = nq\mu$, where n is the (intrinsic) charge carrier density, a high mobility (desirable) implies a high conductivity (undesirable) unless we keep n extremely low, emphasizing the need for ultrapure materials to avoid accidental doping. Here vapour-deposited films have the edge over solution processing, because vapour deposition is a purification step (an evaporation–condensation cycle is similar to recrystallization). Ultrapure preparation is a great challenge to synthetic chemistry. As an example, the conventional preparation of 6T couples two 3T by oxidative dimerization with ferric chloride (Fe(III)Cl$_3$). However, traces of ferric chloride make it into the final product, and chloride is known to be an oxidative dopant; for example, Cl$^-$ can turn PA into a synthetic metal. This intrinsic doping leads to unwanted conductivity, and OFET grade 6T is now made by coupling lithiated 3T with ferric acetylacetonate instead. Thiophenes also suffer from oxygen exposure, but in a different way from C_{60}; thiophenes form a weakly bound charge transfer complex with molecular oxygen that acts as dopant, leading to unwanted conductivity. This limits the practical usefulness of the otherwise very promising thiophenes. A certain level of protection is afforded in the top-gate architecture.

Requirements on the other OFET materials

The source metal should show good carrier injection into the semiconductor. This favours high work function metals such as gold (Au), or indeed PEDOT/PSS, which is a boon for all-plastic electronics. Usually it is convenient to use the same material for the drain metal as well. Then the source and drain become interchangeable and are defined by the wiring. The electrical requirements for the gate metal are not very strict. The sheet resistivity need not be extremely low, because ideally no current flows to the gate. This again favours a synthetic metal.

The gate insulator is critical and may be the least well developed aspect of OFET technology. We desire transistors that can operate with rather low gate voltages: in a simple application, no high-voltage source will be available. To achieve low gate voltage switching, we require high capacitance per unit area, C_i, for the gate insulator. Since $C_i \propto \varepsilon_r/d$, a good gate insulator should provide pinhole-free insulation at very low thickness, have a high dielectric constant ε_r, have high electric breakdown strength, and be insoluble in the solvent of the channel material, because channel and insulator have to be prepared on top of each other. The solubility requirements are usually fulfilled by insulating polymers such as poly(vinyl phenol) (PVP) or poly(methyl methacrylate) (PMMA) that dissolve in alcohols, but not in less polar organic solvents. Unfortunately, the dielectric constant of organic materials is usually rather low ($\varepsilon_r \leq 4$), and relatively thick layers are required ($d \geq 300\,nm$) to obtain pinhole-free films. The inorganic gate insulator SiO_2 that is conventionally used for inorganic chips also has a rather poor dielectric constant ($\varepsilon_r = 3.9$) but provides insulation in much thinner films. Other inorganic insulators have a much higher ε_r ($\varepsilon_r = 9$ for Al_2O_3, $\varepsilon_r = 22$ for ZrO_2, $\varepsilon_r = 23$ for Ta_2O_5).

Of course, OFETs can be (and have been) built with SiO_2 and other inorganic gate insulators (e.g., Ta_2O_5), and these devices display lower gate voltage switching than comparable devices with organic insulators. However, these gate insulators are more expensive, so they are not an option for disposable electronics. Hence, there is room for improvement.

6.2.2.4 Integrated circuits based on OFETs

The discussion so far has focused on the properties of single OFETs, however, logic circuits always require a large number of transistors wired up to each other. The desire for high packing density is one driving force to achieve OFET architectures as small as possible. The other is a direct consequence of Equations (6.11) and (6.12), which show that a short channel length L is required for fast OFETs with high drain current at moderate V_G and V_D.

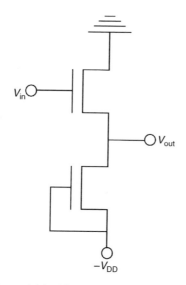

Figure 6.14 The principle of an inverter

The basic building block of a logic circuit based on FETs is the *inverter*. It can be built from two FETs, as shown in Figure 6.14. $-V_{DD}$ is a supply voltage. When V_{in} is high, V_{out} will be low, and vice versa; the inverter performs the logical operation NOT. Note that the drain of the second FET is connected to its gate. This introduces a key feature of circuits as compared to individual transistors, which poses a considerable challenge to device engineering: to make useful circuits, there needs to be interconnects from the source/drain level to the gate level, going through the semiconducting channel/ and through the gate insulator. These interconnects are called *vias*. The preparation of via interconnects represents another great engineering challenge for the manufacture of practical organic circuits. Patterning on the source/drain and the gate level are also necessary. Currently, two approaches to plastic circuit engineering are competing: photolithography and inkjet printing techniques.

The group at Philips Research Laboratories in Eindhoven have used camphor sulphonic acid doped PANi as the source/drain and gate metal. To pattern the source/drain level, a photoinitiator molecule was added; on exposure to UV light, it cross-linked the PANi and led to an increase in PANi resistivity of 10 orders of magnitude – practically turning PANi into an insulator. Hence the negative of a mask could be patterned into the PANi. The insolubility of PANi in organic solvents was crucial for deposition of the semiconductor poly(thienylene vinylene) (PTV) and the gate insulator PVP. Vias were punched mechanically with pins, providing sufficient mixing between the gate and source/drain PANi to give electrical contact. A 15-bit all-plastic programmable code generator was demonstrated that still worked when the device was sharply bent.

The photolithography approach to organic electronics has its limitations. Apart from probably being too costly, when plastic substrates are used, dimensional stability is not given over large areas; there may be some warp between the gate and source/drain level, which leads to a loss of registration. An alternative approach that does not suffer from this problem uses inkjet printing to put an inverter in the top-gate architecture. The source/ drain and gate were made from PEDOT/PSS, with F8T2 as channel semiconductor and

PVP as gate insulator. Since the inkjet printer can be mounted with an optical system, local registration between source/drain and gate is possible and, in principle, circuits of unlimited size can be manufactured. Importantly, an effective method for making via holes was also devised. When a droplet of plain solvent is printed onto an existing semiconductor, it will dissolve the semiconductor. When the solvent then evaporates, semiconductor is deposited mainly at the edges of the droplet. This phenomenon is known as the 'coffee stain effect'. Effectively, the solvent droplet has etched a hole into the semiconducting film which can then be filled with a synthetic metal droplet to make a via interconnect.

6.2.3 Organic light-emitting devices

The huge optical display market is currently dominated by cathode ray tubes (CRTs) and liquid crystal displays (LCDs). Both technologies are rather dated and have severe drawbacks; for example, CRTs are bulky and consume large amounts of energy, and LCDs are passive devices that require backlighting and often suffer from poor viewing angles. Replacing them with organic light-emitting devices (OLEDs) is tempting commercially as well as technologically, and the desire to do so has provided much of the momentum for organic semiconductor research throughout the 1990s.

6.2.3.1 Overview of basic processes

As discussed in Section 6.1.4, during fluorescence excitons are created by the absorption of light before fluorescing, whereas during EL excitons are created by electron and hole polaron capture. Polarons first have to be injected from the electrodes, and migrate towards each other. They then form an exciton that can sometimes decay under the emission of light. Figure 6.15 shows the basic architecture of an organic light-emitting device (OLED). The variety of electrical and photophysical processes involved are summarized in Figure 6.16.

Assuming that carrier traps can be avoided, charge carrier mobility is considerably more important for OFETs than it is for OLEDs. This is simply because of the much shorter distances carriers need to travel across a device to the other electrode. In the planar OFET geometry, typical channel lengths are several micrometres, while in the vertical OLED architecture, layer thickness is of order $\sim$100 nm. Apart from special devices, such as organic lasers or OLEDs that emit short pulses, we can therefore use conjugated materials with much lower carrier mobilities in OLEDs than in OFETs. The PPV family of OLED materials is a case in point. However, to operate OLEDs as efficiently as possible, all processes shown in Figure 6.16 need to be optimized as discussed below.

6.2.3.2 Bipolar carrier injection

The basic physics of carrier injection from electrodes into organic semiconductors were discussed in Section 6.1.5.1. The key feature that sets apart an OLED from an OFET is that an OFET can operate with one type of carrier, whereas an OLED always requires

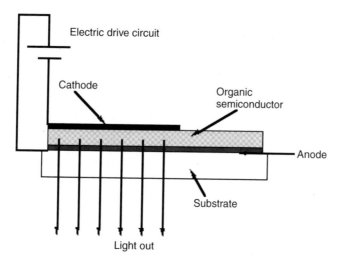

Figure 6.15 OLED architecture

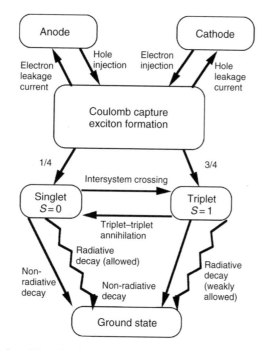

Figure 6.16 A chart describing the formation and decay of excitons in organic electroluminescent devices. Adapted from D. D. C. Bradley, *Current Opinion in Solid State & Materials Science* **1**, 789 (1996)

both electrons and holes to be injected, otherwise no excitons can form. Ideally, we wish to have ohmic injection of holes at the anode, and electrons at the cathode, simultaneously. As a consequence, OLEDs will use unlike metals for the anode and cathode, namely a high work function anode and a low work function cathode, whereas OFETs normally use the same metal for source and drain electrodes. This defines a *forward bias* and a *reverse bias* for OLED operation. However, it is still difficult to achieve ohmic (i.e., barrier-free) injection at both electrodes for a given organic semiconductor.

The breakthrough towards efficient organic EL devices came from a device engineering approach at the Eastman Kodak Research Laboratories, rather than from materials development. Here a bilayer device was manufactured that consisted of a low ionisation potential, hole-transporting diamine layer and a high electron affinity, electron-transporting Alq_3 layer (see the Appendix), which is also an efficient green emitter. This idea has since been widely adapted and modified. Figure 6.17 shows the level diagram for a (fictitious) double-layer device consisting of a hole-transporting layer (HTL) and an electron-transporting layer (ETL). Both layers are assumed to have a band gap $(I_p - E_a) = 2.5$ eV, however the HTL has lower I_p than the ETL, and the ETL has higher E_a than the HTL. It is immediately obvious that a single-layer device using either HTL or ETL alone would necessarily have one large (1 eV) injection barrier. In the double-layer architecture both barriers are moderate (0.5 eV).

In addition to the injection barrier, both holes and electrons will encounter an internal barrier at the HTL–ETL interface, but this barrier is not detrimental to device performance. Instead, it can help to improve the balance between the electron and hole currents. Assuming a slightly smaller injection barrier for holes than for electrons, or a higher hole mobility than electron mobility, a carrier imbalance with a larger hole current than electron current would be expected even in a bilayer device. However, since holes will encounter an internal barrier, they will not simply cross the device and leave at the cathode as a 'blind' leakage current. Instead, they will accumulate at the interface, where they represent a positive space charge. The effect of the field resulting from that space charge is to improve charge carrier balance. Firstly, it will impede the further injection of majority carriers (holes) from the anode, and secondly, it will enhance the injection of minority carriers (electrons) from the cathode. Also, excitons

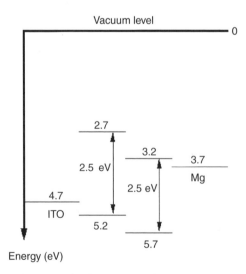

Figure 6.17 Energy level diagram of a fictitious double-layer device, using an ITO anode and a magnesium cathode

will form at the internal interface, far away from the electrodes. Cathodes in particular have been associated with exciton quenching (i.e., radiationless exciton decay); this is avoided by placing exciton formation in the centre of the device rather than close to the cathode.

As the Kodak Group used small molecules, bilayers could readily be manufactured by subsequent evaporation. This approach has been extended to sophisticated multilayer architectures, yielding brightnesses of over 10^5cdm^{-2} and external quantum efficiencies of over 7% to date.

With polymeric organic semiconductors, vapour deposition is not an option, devices have to be prepared by spin-casting instead. Multilayer architectures are harder to realize with spin-casting than with vapour deposition, because of the need for orthogonal solubilities (i.e., the solvent used for a particular polymer layer must not cause the dissolution of the preceding layer). To sidestep solubility problems, in principle a precursor route may be employed, where the first layer is prepared from a soluble precursor polymer that then is converted in situ into a conjugated and completely insoluble polymer. This has been successfully employed for hole-transporting PPV and electron-transporting CN-PPV double-layer polymer OLEDs. However, lengthy in situ thermal conversion under high vacuum is awkward, and the use of precursor polymers has generally fallen out of favour following the advent of soluble conjugated polymers.

A very favourable approach has recently emerged that combines the ease of injection into a double-layer device with the simplicity of solution processing. In that approach, a single layer of a blend of a hole-transporting and an electron-transporting conjugated polymer, namely poly(dioctyl fluorene) (PFO) and F8BT, is spin-cast in one single preparation step. As spin-casting implies the very rapid formation of a solid film from solution, the two polymers have little time to phase separate and a solid film may result where both polymers remain intimately mixed. Such a mixture has been termed a bulk heterojunction, and the preparation and morphology control of hole/electron transporting blends is the focus of much current research, mainly with a view to photovoltaic applications of organic semiconductors. Holes are injected and transported into the (majority component) PFO, but can be transferred easily to F8BT, as it has similar ionization potential. However, F8BT has poor hole mobility due to hole-specific traps. Instead, it has rather high electron affinity and displays comparatively good (albeit dispersive) electron transport. Thus electrons are mobile on the F8BT chain until they encounter a trapped hole. With some further device improvements, highly efficient ($>4 \text{cd A}^{-1}$) and low onset voltage ($\sim 3 \text{V}$) OLEDs have been prepared from such blends.

6.2.3.3 Exciton formation

When both hole and electron polarons have been injected into a device, and they drift towards each other under the applied voltage, they will combine into excitons that may emit light. The physics of this process is discussed in Section 6.1.4.2.

At first sight, it appears that exciton formation in multilayer architectures is hindered by the internal barrier encountered by carriers of either type at the HTL–ETL interface. However, this is generally not the case. Excitons in organic semiconductors generally display exciton binding energies E_b of 0.1–0.3eV. When a carrier has to overcome an internal barrier to form an exciton, this may require a certain amount of energy. However, on exciton formation, E_b is instantly refunded. Effectively, the internal barrier is reduced by E_b. Thus, majority carriers remain stuck at an internal barrier and redistribute the internal field in the favourable way discussed above, until a minority carrier arrives at the interface. As soon as a minority carrier is available, exciton formation is then helped by the effective barrier reduction E_b. High E_b also stabilizes excitons against dissociation and non-radiative decay.

In bulk heterojunction blends, one carrier has to transfer from one chain to another to form an exciton. This will be the type of carrier for which the energy level offset of the ionization potentials ($|\Delta I_p|$) or the electron affinities ($|\Delta E_a|$) is smaller. Let us call the smaller of these energies the transfer energy, E_t. Two very different scenarios emerge for the case $E_t < E_b$ as opposed to $E_t > E_b$. In the former case, formation of excitons from polarons will be favoured, while in the latter case, the dissociation of existing excitons into polarons will be preferred. In the case of F8/F8BT blends introduced previously, exciton formation is clearly favoured, and such blends are useful for OLED applications. In other hole/electron transport material blends, such as poly(alkyl thiophene)/ perylene tetracarboxyl diimide blends, exciton dissociation is favoured, which makes such blends attractive for use in photovoltaic devices. While measurements of $|\Delta I_p|$, $|\Delta E_a|$ and E_b with sufficient precision to predict exciton formation or dissociation are usually not available, there is a simple experimental approach to decide which is the case. If in a blend the fluorescence intensity is much reduced compared to the pure components, then excitons are separated efficiently due to the presence of the blend partner.

6.2.3.4 Optimizing OLED efficiency

The discussion so far outlines the strategy towards OLED devices with balanced carrier injection and quantitative exciton formation that can be driven at low voltage. The formidable challenge that remains is to maximize the amount of light generated from the excitons. It is obvious that we require a material with a high luminescence quantum yield. However, the formation of normally non-emissive triplet excitons presents an unwanted limit on OLED efficiency, and several approaches to overcome this have been explored.

Generally, η_{EL} and η_{PL} are related via:

$$\eta_{EL} = \frac{\sigma_S/\sigma_T}{\sigma_S/\sigma_T + 3}\eta_{PL}, \qquad (6.13)$$

where σ_S and σ_T are the polaron capture cross sections for singlet and triplet exciton formation, respectively. The naive assumption $\sigma_S = \sigma_T$ leads to $\eta_{EL} = \eta_{PL}/4$.

Enhanced singlet exciton formation

Comparisons of EL and PL quantum efficiencies have yielded contradictory results. Some experiments confirm that $\eta_{EL} = \eta_{PL}/4$, but other workers are adamant that they find high EL quantum efficiencies consistent with a singlet:triplet formation ratio of about 1:1, implying $\sigma_S \approx 3\sigma_T$. To determine singlet:triplet formation ratios directly, rather than inferring them from EL to PL efficiencies, a systematic magnetic resonance study on a number of organic semiconductors with band gaps in the visible range found that σ_S/σ_T was indeed generally larger than 1, and between 2 and 5 for different materials. $\sigma_S/\sigma_T \approx 2$ to 5 corresponds to η_{EL} of between $\sim 0.4\eta_{PL}$ and $\sim 0.6\eta_{PL}$ instead of $0.25\eta_{PL}$. Further work on an oligomer series has shown that σ_S/σ_T increases with conjugation length, which implies larger σ_S/σ_T for polymeric than for low molecular weight organic semiconductors. The marked violation of the naive 'one singlet to three triplets' rule for polymers (but not low molecular weight materials) may be explained as being the result of the relatively long quantum coherence in a polymeric organic semiconductor.

Electrophosphorescence

As alternative to the enhanced singlet formation cross section in polymers, the low molecular weight OLED community has developed the concept of harvesting triplets for light emission by using phosphorescence. In a typical electrophosphorescent device, a wide band gap host semiconductor is doped with a small percentage of a phosphorescent emitter. The excitation is transferred from the host to the guest via exciton energy transfer. Forrest and co-workers at Princeton have developed a range of green, yellow, orange and red organoiridium complexes, which are exemplified by the particularly efficient red phosphor $btp_2Ir(acac)$ (see the appendix). When doped into a wide band gap host, electrophosphorescence with better than 80% internal quantum efficiency and $60\,lm\,W^{-1}$ is observed.[4] Using pure $btp_2Ir(acac)$ without a host matrix had resulted in less efficient devices. In particular, electrophosphorescence is an attractive approach in the red. Due to the response characteristics of the human eye, red dyes must show very narrow emission peaks, otherwise colour purity will be compromised. Iridium-based phosphors display considerably narrower emission bands than typical fluorescent dyes, and are therefore particularly useful as red emitters. These phosphors also display relatively short triplet lifetimes ($4\,\mu s$), thus avoiding problems associated with triplet–triplet annihilation at high brightness, which had been encountered with longer-lifetime phosphors. However, electrophosphorescence is ambitious for blue emission due to the need for a high band gap host semiconductor.

Organolanthanides

Another approach to triplet harvesting is represented by the organolanthanide dyes. Organolanthanides are somewhat similar to organometallic phosphors, but the central metal atom is a lanthanide such as europium (Eu) or terbium (Tb). The red dye ADS053RE is a typical example. Organolanthanides owe their properties to the unique electronic structure of the lanthanides (or rare earth metals), reflected by their positioning in the periodic table. In a dye such as ADS053RE the organic ligand can absorb light (typically in the blue or near UV), or can be excited electrically. The exciton is then passed to the central lanthanide and excites an electron of the lanthanide 4f shell; notably this works for singlet and triplet excitons. The intramolecular excitation

---

[4] Luminous efficiency is measured in $lm\,W^{-1}$ and is a measure of the light obtained for a given power input. For comparison, a typical incandescent light bulb has a luminous efficiency of typically less than $20\,lm\,W^{-1}$.

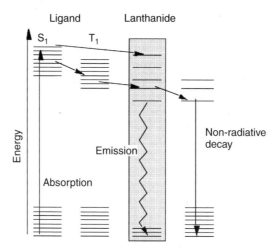

Figure 6.18   An excitation migrates through the molecular (ligand) and atomic (lanthanide) levels in an organolanthanide

transfer is shown schematically in Figure 6.18. Note that the observed emission comes from the radiative decay of the excited 4f state; it is an atomic transition, *not* a molecular transition. This is the marked difference between organolanthanides and conventional organometallic phosphors. Due to the localized and isolated nature of the excited state, organolanthanides do not suffer from triplet–triplet annihilation. Efficiencies well in excess of 25% can be achieved. Since emission comes from an atomic transition, the bands are extremely narrow (FWHM ≈ 10 nm), resulting in very pure colours (green from terbium, red from europium).

There are two major drawbacks for organolanthanide applications. Firstly, for electrical excitation, carriers have to be injected into a rather large band gap material (deep blue or near UV) even if the emitted light is red. Larger band gaps generally make good, well-balanced carrier injection harder and lead to reduced power efficiencies and higher onset voltages. Also, as in all transfer-based concepts, the generation of blue light is somewhat problematic. Secondly, the bonding between a lanthanide and an organic shell has a considerable ionic (as opposed to covalent) character. The organic shell acquires a partial negative charge, the lanthanide a partial positive charge. When an electron is injected to the ligand, it becomes rather unstable against degradation. Consequently, until now, even for encapsulated organolanthanide devices, device lifetimes have been poor.

Conjugated dendrimers

Another recent approach to improve the efficiency of light-emitting devices is the use of dendrons with a conjugated core surrounded by non-conjugated dendrimers. A schematic representation of the dendron concept is given in Figure 6.19. The conjugated core of the dendrimer can be either fluorescent or phosphorescent. The dendron concept seeks to combine the advantages of conjugated polymers and low molecular weight materials. Dendrimers can be processed from solution and form films in a manner

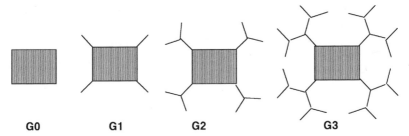

Figure 6.19 Schematic representation of zeroth- to third-generation conjugated dendrimers. Shaded: conjugated core. Thin lines: dendronic side groups (non-conjugated)

similar to polymers. However, due to the dendronic side groups, individual chromophores are shielded from each other. This avoids some of the problems encountered when conjugated polymers are being used; due to interchain interactions such as aggregation and excimer formation, quantum efficiency may be reduced, and excimer emission may compromise colour purity. The major drawback of the dendrimer approach is the much reduced charge carrier mobility due to the increasing separation between conjugated units. The mobility μ scales with the separation between conjugated groups, D, with a characteristic length R_0 according to $\mu \propto D^2 \exp(-D/R_0)$.

A group at St Andrews University has studied dendrimers with a core consisting of three distyrylbenzene groups grouped around a central nitrogen, and dendrimeric side groups consisting of meta-linked vinylene phenylene groups, up to third generation. OLEDs made from higher-generation dendrimers displayed narrow EL spectra that approached the solution PL spectra of the conjugated core, and quantum efficiencies rose steeply with dendrimer generation. This is the result of a successful isolation of the emissive core groups from each other. Carrier injection was not affected by dendrimer generation, but the carrier mobility decreased dramatically. For second- and third-generation dendrimers, which did display narrow spectra and improved efficiency, the mobility was of the order of only $10^{-8}\,\mathrm{cm^2V^{-1}s^{-1}}$.

Light outcoupling

After the efficient electrical generation of a photon, it must still leave the device to be observed by the viewer of the display. Practically, this outcoupling is often very inefficient, with often only 1 in 8 photons leaving the device. In other words, the *external quantum efficiency* η_{ext} is smaller than the internal quantum efficiency η_{int}. The main loss mechanism here is in-plane waveguiding. To solve this problem, devices have been designed that suppress waveguiding. With a photoresist-based technique, the St Andrews group manufactured a corrugated anode that scatters light out of in-plane modes, hence, out of the device and towards the observer. In this way, η_{ext} was improved twofold. A more sophisticated approach for improved outcoupling is to establish a resonant vertical cavity mode in a device. Such cavities have spectrally narrow modes and a strong directional emission characteristic. Thus, spectrally pure colours can be generated even from broadband emitters. The principle of EL from resonant cavities has been established, but device manufacture is difficult due to the need to incorporate a dielectric mirror into the device architecture. Resonant cavities have proven a powerful tool for the investigation of fundamental phenomena in quantum optics of organic semiconductors (strong coupling), but for practical devices probably will not be cost-efficient. They may, however, play an important role in future developments of organic injection lasers.

As a concluding remark, a single OLED device does not constitute a display that can communicate information, just as a single OFET does not represent a useful circuit. Consequently, device engineering issues are as prominent for OLED research as they are for OFETs, or even more so. Depending on the application, we may wish to have small or large, high-resolution, full-colour displays. This can only be achieved in pixellated displays, where individual pixels can be addressed independently. Currently, both *passive matrix* and *active matrix* displays are being developed.

6.2.4 Organic photovoltaics

The basic mechanism of photovoltaics is the reverse of electroluminescence. A solar photon is absorbed in a semiconductor, thus producing an exciton that has to be separated into electron and hole polarons, which migrate to opposite metal electrodes and discharge via an external load circuit, rather than recombining instantly under their mutual attraction. Obviously this exciton separation requires an internal field in the device that overcomes the electron–hole Coulomb attraction.

6.2.4.1 Basics of photovoltaic devices

Figure 6.20 shows the principle of a conventional (inorganic) photovoltaic device. A thin, semi-transparent layer of (in this case) a p-doped inorganic semiconductor is prepared on top of a thicker, n-doped layer. At the interface between the p layer and the n layer, a rectifying pn junction is formed. This results in the formation of a *depletion layer* with very few charge carriers, hence high resistivity, and a strong internal field. An exciton created by photon absorption near the pn junction may be torn apart by this field, and carriers will prefer to recombine via the external load R if its resistance is smaller than that of the depletion layer. Figure 6.21 illustrates the I/V characteristics of a rectifying junction in the dark and under illumination. V is taken positive if it points from p to n (forward bias), and negative otherwise. Under illumination, if the p and n

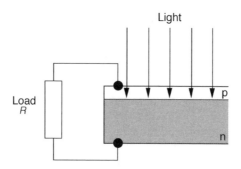

Figure 6.20 A photovoltaic diode based on inorganic semiconductors

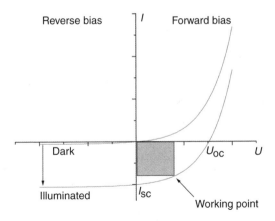

Figure 6.21   Characteristics of pn junctions with and without illumination

terminals are shorted ($R = 0$), then even at $V = 0$ a current will flow due to photo-generated carriers, and in the opposite direction as a current under forward bias. Under illumination the $I(V)$ characteristic will be moved downwards by this *short-circuit current*, I_{SC}.

I_{SC} is the maximum current the photovoltaic cell can deliver, and it will depend, among other factors, on the intensity of illumination. If the p and n terminals are unconnected ($R = \infty$), no current flows ($I = 0$) and the potential difference (voltage) between them will approach the intercept of the illuminated $I(V)$ characteristic with $I = 0$: the *open-circuit voltage* V_{OC}. V_{OC} is the highest voltage that the photovoltaic cell can deliver. Within the lower quadrant of the I/V diagram, between I_{SC} and V_{OC}, the photovoltaic device can provide electrical power $P = VI$ to an external load. However, at I_{SC}, we have $V = 0$ hence $P = 0$, and at V_{OC} we have $I = 0$ hence $P = 0$. To draw power from a photovoltaic cell, we have to sacrifice some bias and operate the cell at a working point $0 < V < V_{OC}$. The power drawn at V can be represented as the area of a rectangle, which is shown shaded. At one particular $V_{max} < V_{OC}$, this rectangle will have maximum area; at this voltage the power delivered by the cell is maximised. The *fill factor* (FF) is defined as

$$\text{FF} = \frac{P_{max}}{I_{SC} V_{OC}}, \tag{6.14}$$

which is often expressed as a percentage. FF is one of the most important characteristics of a photovoltaic cell, and large values of FF are obviously desirable. The efficiency of a photovoltaic cell is quantified by the *external quantum efficiency* (the number of carriers generated divided by the number of photons absorbed) and *power efficiency* η_p:

$$\eta_p = \frac{P_{out}}{P_{in}} = \left(\frac{I_{SC} V_{OC}}{LA}\right)\text{FF}, \tag{6.15}$$

where L is the light intensity and A is the device area.

The key challenge in producing efficient solar cells is to harvest as many as possible of the incoming photons. This is made difficult by the fact that only excitons created near the pn interface, or more precisely within one exciton diffusion radius (~5 nm) of the interface, can possibly be separated into carriers. An exciton produced further away from the interface will recombine, maybe under fluorescence, but will not provide a current. Since typical absorption coefficients of inorganic semiconductors imply light penetration depths of order 100 nm even at maximum absorption, only a small fraction can generate excitons within the region of interest.

Early organic photovoltaic devices closely mimicked the architecture of inorganic heterojunction devices. Since organic semiconductors are intrinsically p- or n-type semiconducting without doping, no internal field develops at a heterojunction. Instead, this has to be established by the use of metals with unlike work functions. However, these devices suffer from the same problem as their inorganic counterparts; namely, most excitons are created too far away from the junction for harvesting.

6.2.4.2 Organic bulk heterojunction diodes

The possibility to process organic semiconductors from solution is an asset that has been used to overcome the problem of short diffusion length. In the so-called *bulk heterojunction* photovoltaic devices, the electron donor and electron acceptor are spun from the same solution.

Typically, two unlike polymers, A and B, will not mix. A and B may share a common solvent and therefore may mix in solution, but in the solid state they will tend to separate into A and B islands. This incompatibility is a well-understood phenomenon for polymers in general, and has been the subject of much work in polymer physics. However, in the solid state at room temperature, conjugated polymers will typically be in a glassy state that does not allow for large-scale molecular rearrangements. How far phase separation will proceed during film preparation will therefore depend on the kinetics of the film formation process as compared to the kinetics of the phase separation process. In a fast film formation process such as spin-casting, phase separation may be arrested at an early stage, leading to a very fine texture, with A-rich and B-rich domain sizes of 100 nm or smaller. If both blend components are present in substantial amounts, this can lead to a *bicontinuous* structure, in which both components are *percolated*; that is, for each component, there is a path from the macroscopic faces of either side of the sample to the other which only consists of material of one component. Assuming that A and B are two semiconducting polymers, with A being an electron acceptor/transporter, and B an electron donor/hole transporter, then there is a conductive path for both electrons and holes through the whole film to reach their respective electrodes. This means that electrons and holes can be collected at opposite electrodes, in the same way as in discrete layer photovoltaic devices, thus fulfilling a necessary condition for photovoltaic applications. In the field of organic photovoltaics such a structure is sometimes called an *interpenetrating network* (not to be confused with a bicontinuous polymer mixture in which at least one of the components is cross-linked). Note that in a photoconductor system such as PVK/TNF, only holes have significant mobility while electrons are stuck. From a PVK/TNF blend, no photocurrent can be extracted, as the electrons left behind would rapidly build up a space charge field which would cancel any built-in field.

In a finely dispersed structure, the A–B interfacial area is much enhanced compared to discrete layer devices. Each A–B interface may act as a heterojunction to separate excitons into electrons and holes, so the entire layer represents a bulk heterojunction. When the length scale of the phase-separated structure is of the same order as an exciton diffusion length, then the bulk heterojunction approach will enable the separation of many more excitons than a discrete layer architecture. This is a key advantage of organic photovoltaics over inorganic photovoltaics.

A number of such bulk heterojunction photovoltaic devices have been reported in the literature. For example, a system based on PPV derivatives (MEH-PPV as hole transporter and CN-PPV as electron transporter) has achieved a rectification ratio of 1000 at ± 3.5 V, U_{OC} of 0.6 V and quantum efficiencies of up to 6%. A similar approach was based on MEH-PPV with a solubilized fullerene derivative, using a very similar material combination as had been studied for discrete layer devices. Note how the progress in materials chemistry that has made soluble C_{60} derivatives available has enabled the realization of a new device concept. In that work, optimum performance was found at rather high fullerene load (80%), indicating that percolation is usually easier to achieve for a polymeric component than for a non-polymeric component. Quantum efficiencies approaching 3% have been reported. Devices where both components were fluorene copolymers have also been demonstrated. An interesting and novel approach is the use of electron-transporting low molecular weight materials that can crystallize after film preparation in a matrix of a polymeric hole transporter. In these systems the thermally activated growth of relatively large electron-transporting crystals with high electron mobilities within a hole-transporting matrix improves quantum efficiencies to up to 10%.

To comment on the remaining problems of organic photovoltaics, lifetime issues are important as always: in particular, the presence of oxygen leads to problems with long-term efficiency, similar to those encountered in OLEDs. Another crucial point is that the systems investigated so far are not very well suited to harvesting solar photons. Absorption and photocurrent action spectra of the studied materials typically peak in the mid-visible range, not at 1.4 eV in the near-infrared band, where the sun emits most photons per unit wavelength interval. This lack of suitable material represents a challenge to materials chemistry.

6.3 CARBON NANOTUBES

6.3.1 Structure

Shortly after the discovery of buckminsterfullerene in 1990, a related form of carbon was discovered, the carbon nanotube (CNT). CNTs consist of a graphene sheet which has been rolled up into a cylinder. The tube is energetically stable relative to the unrolled sheet because the unrolled sheet has a row of dangling bonds at its edges. For the same reason, CNTs are usually found to have hemispherical caps at one or both ends, made up of carbon hexagons and pentagons, such that the tube is a completely closed surface with no incomplete bonds at all. CNTs have an extremely high aspect ratio; a single-walled tube is typically 1–20 nm in diameter, but can be up to 100 mm in length. Multiwall nanotubes also exist and consist of several concentric carbon cylinders. Figures 6.22 and 6.23 show examples of single-wall and multiwall CNTs, respectively.

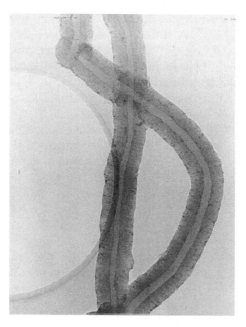

Figure 6.22   TEM image of single-wall nanotubes. The diameter of each tube is approximately 3 nm. Image courtesy of R. Brydson, Institute of Materials Research, University of Leeds

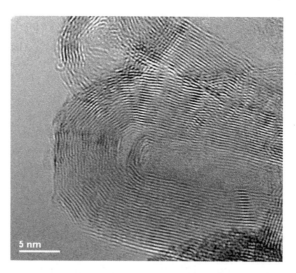

Figure 6.23   TEM image of multiwall nanotubes. Image courtesy of Z. Aslam, X. Li, B. Rand and R. Brydson, Institute of Materials Research, University of Leeds

Theoretically, there exist an infinite number of permutations of tube structure, depending on how the two opposite sides of a graphene sheet are joined together to form the nanotube. There are two high-symmetry configurations which are *non-chiral*: the so-called *armchair* and *zigzag* tubes. In an armchair tube, the tube axis is perpendicular to a line of C–C bonds in the graphene sheet, such that a thin ring cut from the tube will show a pattern of six-membered carbon hexagons joined to each other, vertex to vertex, by a C–C bond (Figure 6.24(a)). In a zigzag tube, the tube axis is parallel to a line of C–C bonds, such that a thin section cut from the tube shows a complete ring of carbon hexagons joined side to side as in Figure 6.24(b). All other tube configurations are *chiral*, with the tube axis offset relative to the two above cases, such that the chains of carbon hexagons can be traced in a spiral along the length of the tube. This is expressed mathematically by the *chiral vector*

$$\mathbf{C_h} = m\mathbf{a_1} + n\mathbf{a_2}, \qquad (6.16)$$

where $\mathbf{a_1}$ and $\mathbf{a_2}$ are unit vectors of the graphene sheet as shown in Figure 6.25. The chiral vector gives the direction normal to the tube axis on the graphene sheet. Thus, armchair tubes are all represented by chiral vectors with $m = n$, whereas zigzag tubes all have $n = 0$. The chiral vector is often abbreviated to $\mathbf{C_h} = (m, n)$.

The circumference of a zigzag tube $(m, 0)$ is thus given by $m|\mathbf{a_1}| = ma_C\sqrt{3}$ where the C–C bond length $a_C = 0.142$ nm. Although CNTs can be formed with a wide range of

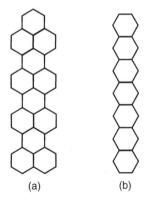

(a)   (b)

Figure 6.24 Strips cut from the circumference of (a) an armchair nanotube and (b) a zigzag nanotube, showing the C–C bonds

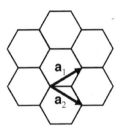

Figure 6.25 Portion of a graphene plane showing the unit vectors $\mathbf{a_1}$ and $\mathbf{a_2}$

diameters, the minimum tube diameter is governed by the structure of the endcap, since the smallest stable cap is half a C_{60} molecule.

6.3.2 Synthesis

The main synthesis methods for CNTs are laser ablation, arc discharge and CVD. In laser ablation, a graphite target is heated to 1200 °C in an argon atmosphere. Carbon is then evaporated from the target by a pulsed laser beam, and is collected on a cooled copper surface. The graphite target also contains small quantities of nickel and/or cobalt, which are also evaporated and deposited onto the copper surface; these act as catalysts for nucleation of the nanotubes. This method produces relatively wide (up to ~20 nm in diameter) and long (up to ~100 mm) nanotubes. The tubes often form in bundles or twisted ropes with a remarkably uniform distribution of diameters. However, the tangled nature of these bundles makes extraction and application of individual tubes very difficult. Tube growth is believed to occur via a *scooter* mechanism, in which a single catalyst atom travels, or scoots, around the end of an open tube, absorbing carbon from the argon atmosphere, and feeding this into the graphene sheet. Tube growth is terminated when the open end becomes saturated with catalyst atoms, at which point these detach from the tube, allowing a fullerene cap to form.

In the arc discharge method, a DC bias of 20–30 V is applied between two carbon electrodes in a helium atmosphere. Carbon atoms are ejected from the anode, and accumulate in the form of nanotubes on the cathode. The electrodes are typically 5–20 mm in diameter. As with laser evaporation, the anode includes small quantities of nickel, cobalt or iron, which are also deposited onto the cathode to act as a catalyst. Arc discharges tend to produce narrower and shorter tubes than those obtained from laser ablation (up to ~5 nm in diameter and around 1 mm long). Like laser ablation, arc discharges tend to produce bundles of nanotubes.

CVD synthesis of nanotubes involves decomposition of small hydrocarbon molecules (methane, ethene or acetylene) at temperatures between 500 and 1100 °C. The carbon atoms thus released are collected on a cooled substrate, which is pretreated with one of the above catalysts. CVD has the advantage that evaporation from a solid source is not involved, and promises to be the most amenable method for scaling up to large-scale production. Numerous variations on the CVD synthesis route have been explored, including PECVD for growth of well-aligned tubes (see below), and templated growth using porous silicon or lithographically defined arrays of catalyst particles.

In general, single-wall nanotubes cannot be grown in the absence of a catalyst. This implies that the catalyst atoms are essential in preventing the energetically favourable closure of the carbon cage by a process such as the scooting model described above. Multiwall nanotubes are better able to remain open-ended, as the exposed end of the concentric cylindrical structure is stabilized by extra carbon atoms which form bonds between adjacent graphene layers. These bonds can break and reform as the edge continues to grow. Single-wall nanotubes may also grow from the base upwards. In this model, nucleation on a catalyst particle produces a curved graphene sheet, which quickly forms a closed cap but remains open at its base. The open perimeter of the tube is continually fed by carbon atoms adsorbed onto the catalyst from the reaction chamber, and these extend the tube from the bottom upwards.

6.3.3 Electronic properties

The electronic properties of CNTs are closely related to those of two-dimensional graphite. A single graphene sheet is a two-dimensional conductor that has energy bands which are degenerate at the Brillouin zone boundary, therefore it exhibits metallic behaviour. When a graphene sheet is rolled to form a CNT, free electronic motion is allowed only along the tube axis; only discrete electronic states can be supported around the circumference of the tube. CNTs are therefore one-dimensional conductors (or quantum wires), and they can behave as metals or semiconductors, depending on their exact structure. All armchair tubes are metallic, and their energy bands can be obtained from the dispersion relation for a graphene sheet:

$$E(k_x, k_y) = \pm \Delta \left\{ 1 \pm 4 \cos\left(\frac{\sqrt{3}k_x a}{2}\right) \cos\left(\frac{k_y a}{2}\right) + 4\cos^2\left(\frac{k_y a}{2}\right) \right\}^{1/2}, \qquad (6.17)$$

where the lattice constant $a = |\mathbf{a}_1|$ and the energy parameter $\Delta = 3.033\,\text{eV}$, by imposing the circumferential confinement relation

$$k_x = \frac{2\pi p}{\sqrt{3}am} \qquad (p = 1, \ldots, 2m) \qquad (6.18)$$

on the wavevector k_x. Note here that k_x and k_y represent wavevector components in the two-dimensional reciprocal lattice of graphene which, like the real space lattice, also has hexagonal symmetry with the k_y and k_x axes directed parallel and perpendicular to a hexagon side, respectively.

For zigzag nanotubes, the appropriate confinement condition is

$$k_y = \frac{2\pi p}{am} \qquad (p = 1, \ldots, 2m). \qquad (6.19)$$

By substituting this condition into Equation (6.17), it can be seen that there is zero energy gap (metallic behaviour) at $k_x = 0$ whenever m is a multiple of three. In general, an (m, n) nanotube will be metallic if $m - n$ is a multiple of three. For all other tubes, an energy gap persists throughout the Brillouin zone, and the size of this gap varies inversely with the tube diameter. Thus, in a random collection of nanotube structures, one-third of the tubes will be metallic and two-thirds will be semiconducting.

The confinement conditions given by Equations (6.18) and (6.19) lead to a large number of energy subbands in the energy band structure. Figures 6.26 and 6.27 show energy bands for armchair and zigzag tubes, respectively. The energy bands in both cases have zero gradient at the zone centre which, for a one-dimensional conductor, gives rise to Van Hove singularities in the density of states, analogous to the case of

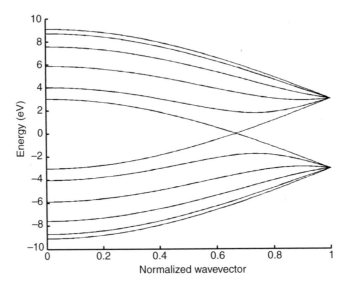

Figure 6.26   Electronic band structure for a (5, 5) armchair nanotube. The wavevector is normalized to the position of the Brillouin zone boundary, $k_y = p/a$

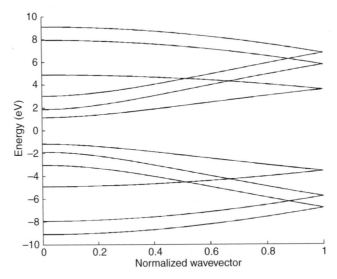

Figure 6.27   Electronic band structure for a (5, 0) zigzag nanotube. The wavevector is normalized to the position of the Brillouin boundary, $k_x = p/(a\sqrt{3})$

semiconductor quantum wires (Chapter 3). The armchair nanotube has two crossing bands, resulting in metallic behaviour, whereas the (5, 0) zigzag tube shown in Figure 6.27 has a clear band gap, hence it is a semiconductor. Figure 6.28 shows the band structure for a (6, 0) zigzag tube; this satisfies the condition described above for metallic behaviour, and the band gap reduces to zero at the Brillouin zone centre.

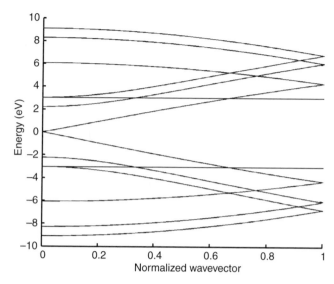

Figure 6.28 Electronic band structure for a (6, 0) zigzag nanotube. The wavevector normalization is as for Figure 6.26

Since the electronic states are confined in the directions perpendicular to the tube axis, electron momentum is only defined in the axial direction, therefore scattering (by phonons or impurities) is constrained to be purely in the forward direction (zero scattering angle) or purely in the reverse direction (backscattering or 180° scattering angle). This restriction greatly affects the transport properties of CNTs relative to bulk materials. At low temperatures, transport is ballistic, and the conductance becomes dependent on the length of the tube. For metallic nanotubes, the tube resistance is generally much less than the contact resistances between the tube and the connecting leads. In this configuration the tube exhibits Coulomb blockade behaviour (Section 3.6.4), with electron transfer into the tube suppressed by the presence of a single electron. The I/V curve contains steps, corresponding to an abrupt increase in current every time the bias is raised sufficiently to match the electrostatic charging energy required to place another electron on the tube. This behaviour persists up to temperatures of around 20 K in isolated single-wall nanotubes.

CNT analogues of electronic metal–semiconductor, semiconductor–semiconductor or metal–metal junctions can be formed by joining two tubes of different chiralities. The transition between the two tubes is accomplished by defects in the hexagonal carbon lattice, so a kink usually occurs at the junction.

6.3.4 Vibrational properties

Just as the electronic band structure of CNTs is related to that of a two-dimensional graphene sheet, so too are the vibrational modes or phonon band structure. Although extended vibrational waves can exist parallel to the tube axis, only certain mode configurations can be supported around the circumference of the tube, giving rise to a large number of closely spaced phonon subbands. The exact definition of the unit cell of a CNT depends on the chirality of the tube. In general, if the tube has a unit cell which contains $2N$ carbon atoms, then there will be $6N$ phonon modes. Four of these modes are acoustic, leaving $6N-4$ optical modes. Of the four acoustic modes, there are two transverse modes, in which the carbon atoms are displaced in the x and y directions, respectively, relative to the tube axis (z), and a longitudinal mode in which the displacement is parallel to the tube axis. The fourth acoustic mode is a twisting mode, so called because it involves a twisting motion of the tube about its axis, with the atomic displacements preserving the cylindrical topology. The optical modes include a radial breathing mode, in which all the atoms in a circle around the tube move in and out, in phase. This mode can be readily detected using Raman spectroscopy, and provides quite a sensitive measure of the tube diameter. Related to the radial breathing mode are a whole family of optical modes characterized by displacement waves around the tube circumference which have $1,2,3,\ldots$ nodes.

The phonon modes of CNTs give the dominant contribution to their thermal properties in the temperature range $\sim 10-400\,\mathrm{K}$. Molecular dynamics simulations of phonon modes in individual CNTs have predicted thermal conductivities of up to $4 \times 10^4\,\mathrm{W\ m^{-1}K^{-1}}$, significantly greater than those of both diamond and in-plane graphite. Such predictions raised great hopes that CNTs could be used for thermal management (component cooling) applications. However, it appears that the theoretically predicted values are difficult to realize in practice. Measurements on individual multiwall CNTs have yielded room temperature thermal conductivities of $\sim 3000\,\mathrm{W\ m^{-1}K^{-1}}$, but in samples containing large numbers of tubes, where a thermal connection cannot be made directly to both ends of each tube, the thermal conductivities are much lower. The thermal conductivity for a single tube is highly anisotropic; large values only prevail parallel to the tube axis, and it appears that the interfaces between separate tubes present large thermal resistances.

6.3.5 Mechanical properties

One aspect of CNTs which has particularly captured the imagination of many is their remarkable mechanical strength; popular reviews often quote statements such as 'a hundred times the strength of steel at one sixth of the weight'. The strength of CNTs is directly related to the C=C bond, which is one of the strongest of all chemical bonds, and to the relatively small number of crystalline defects present in the tubes. Evaluation of the tensile strength and the Young's modulus of CNTs is obviously difficult, due to their small size, and a range of methods have been employed, including TEM monitoring of tube vibrations, AFM manipulation and computer simulation. The measured tensile strength (the maximum load before breaking) of CNTs is typically around 50 GPa, compared with a theoretical maximum of 150 GPa for a completely defect-free tube. The compressive strength is slightly higher still, with measured values of around 100 GPa. In comparison, the tensile strength of steel is 2 GPa. The measured Young's modulus of CNTs is approximately 1 TPa, with simulation predicting a maximum value of over 5 TPa. Again, these values are considerably higher than for steel, which has a Young's modulus of 0.21 TPa. Under high tensile strain, dislocations form in which an adjacent pair of carbon hexagons distorts into a heptagon and joined pentagon. These 5/7 dislocations change the local chirality, hence the diameter of the tube. Multiple-wall CNTs show similar mechanical properties to single-wall tubes; however, it appears that normally only the outer tube can be coupled to the mechanical load, since the bonding between the different concentric layers is too weak to provide effective load transfer. Bundles of nanotubes show mechanical properties that are comparable, but somewhat weaker than those of single tubes, because of the difficulty of sharing the load equally between all the tubes. It is thought that twisting a bundle of tubes into a nanorope will increase mechanical strength by improving the mechanical coupling between the tubes.

Because of their high aspect ratio, CNTs are susceptible to buckling and bending when stress is applied perpendicular to the tube axis. However, unlike many materials, they are highly resilient to severe bending; rather than creating defects which crack the tube, the bending results in a rehybridization of the sp^2 bonds, a process which is reversible if the tube is straightened out again. Bending can be induced by AFM probing, and has been observed in TEM images. Collapse of the tube cross section has also been observed, resulting in flat carbon ribbons. Cross-sectional collapse is energetically more likely for large-diameter tubes (>3 nm).

6.3.6 Applications

6.3.6.1 Carbon nanotube transistors

In many respects, carbon nanotubes are extremely well suited for use as the conducting channel of a field effect transistor (FET). The high conductivity of CNTs gives high current handling and low power dissipation, while their strong charge confinement reduces the detrimental effects associated with reduction of transistor size – *short channel effects* – primarily, the loss of the ability of the gate contact to control charge flow between source and drain. Since CNTs are formed completely from covalent bonds, they do not suffer from electromigration effects under high-current drive, and since all the bonds in a CNT are properly connected, there are no exposed surfaces that require passivation. Consequently, there is no restriction on the choice of gate dielectric material: CNTs can be used with any one of a range of new high dielectric constant materials (high-k dielectrics) that are currently being investigated as an alternative to silicon dioxide. The conduction and valence bands in CNTs are symmetric, with both electrons and holes having the same effective mass; this means that CNT-based FETs are ideal, in principle, for complementary logic, since both n- and p-type transistors will operate at the same speed. Finally, the fact that semiconducting CNTs have a direct band gap raises the interesting possibility of integrated CNT-based optoelectronic systems.

A CNT FET can be formed by draping the nanotube between source and drain contacts as shown in Figure 6.29. A highly doped silicon layer underneath the CNT, and insulated from it by a thin oxide layer, acts as the gate contact of the FET. Without any intentional doping the CNT acts as a hole conductor, forming a

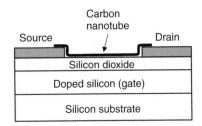

Figure 6.29   Schematic diagram of a carbon nanotube FET

p-channel device; n-channel devices can be formed by doping the CNT with an element such as potassium. The interfaces between the CNT and the source and drain contacts (usually formed from aluminium) are Schottky contacts; i.e., there is a potential barrier between the two materials. The barrier height can be reduced by increasing the tube diameter, which reduces its band gap and effective mass. The Schottky barriers lead to other interesting properties; for example, following oxygen exposure and annealling, a p-channel CNT FET can be converted to an n-channel device, since the electronic properties of the Schottky barriers are altered by oxygen adsorption on the source and drain contacts.

Top-gated CNTs have been fabricated by positioning the tube between source and drain contacts on an SiO_2 surface as before, and then growing a thin covering layer of SiO_2 before depositing a metal or polysilicon gate layer. It is also possible to fabricate side-gated CNT FETs by depositing a gold stripe perpendicular to the nanotube. Gold does not adhere to the nanotube, and therefore the resultant structure has metal contacts on either side of the tube, but with each separated from it by a small gap. A similar effect can be obtained with chromium.

CNTs may also be used as the basis of chemically sensitive FETs for sensing applications. For example, detection of NO_2 by a CNT FET has been demonstrated; the NO_2 molecules are adsorbed by the nanotube, creating additional free holes, which increase the tube conductivity. Further work is needed to produce a sensor whose response is reversible on removing the reagent from the atmosphere.

6.3.6.2 Field emission

Field emission – the electric field induced emission of electrons from a solid – is readily obtained in CNTs. Theoretically, the geometry of a CNT, comprising a long, thin cylinder terminated in a hemispherical cap, results in one of the highest attainable surface fields at its tip, relative to other attainable tip shapes. A CNT field emission device can then be made by applying a voltage between two electrodes, one of which is coated with an array of nanotubes.

An important commercial application of this property is in flat panel displays. Using transparent electrodes and an addressable array, pixels of field-emitting CNTs can be switched to excite red, green or blue phosphors and hence create a full-colour image. Compact (5″ diagonal) colour CNT displays have already been demonstrated, which have very high brightness compared to alternative display technologies. The electrons emitted from CNTs can be accelerated to such an extent that X-rays, rather than visible light, can be emitted from a target, and CNTs have already been used commercially in cold cathode X-ray sources for lead detection.

In these initial applications of CNTs for field emission, the tubes were deposited in a random array on the electrode surface. The random orientation of the tubes gives an inherent screening of the resultant electric field, relative to the ideal scenario in which all tubes would be aligned perpendicular to the electrode plane. Alignment of the nanotube array can be attained by careful control of the growth conditions using a PECVD technique. The screening effect can also be reduced by control of the lateral tube–tube spacing; the optimum spacing is approximately $2l$, where l is the average tube length, and this can be achieved by templated growth on isolated dots of a nickel catalyst. With such improvements in fabrication technology, other applications for field-emitting CNTs are under development, such as high-power microwave amplifiers and electron gun arrays for parallel electron beam lithography.

6.3.6.3 Mechanical reinforcement

Individual CNTs have excellent mechanical properties. However, the exploitation of these properties in bulk materials is a major challenge. One important approach is the formation of composite materials in which a small percentage of CNTs are incorporated into a metal or polymer host. This method has already been demonstrated for polypropylene and aluminium, with a doubling in tensile strength observed in both cases. Theoretical predictions show that the tensile strength of steel could be increased by a factor of 3 by the addition of just 10% by weight of CNTs.

6.3.6.4 Fuel cells

There has been great optimism that CNTs could be used as hydrogen storage vessels for application in non-fossil-fuel cells. Carbon is one of the most lightweight materials to remain solid at room temperature, and the graphene sheet presents a very high surface area to weight ratio. Graphite itself has not proved useful for hydrogen adsorption, since only a small proportion of the surface interacts with hydrogen molecules; however, it was envisaged that CNTs would have the added benefit of capillary forces which would draw hydrogen inside.

Simulations have shown that the adsorption of hydrogen by CNTs decreases rapidly with increasing temperature. A maximum storage capacity of 14% by weight is predicted at cryogenic temperatures and high pressures, based on the maximum number of covalent C–H bonds that can be formed inside the tube. This scenario is impractical since the hydrogen cannot be released unless the temperature is raised considerably. At room temperature, the predicted storage capacity drops to below 1% by weight. (As an indication of practical utility, the US Department of Energy has cited a value of 6.5% storage capacity as a minimum requirement for vehicular applications.) Experimentally, much higher storage capacities (over 60% by weight) have been reported, but these have not proved reproducible. In some cases the high reported values appear to be a consequence of contamination or adsorption of other gases present in the system. To this end, further progress requires, among other things, improvements in the purification of CNTs. Another key issue is to devise a storage process that is not only reversible, but also provides useful capacity at convenient operating temperatures.

APPENDIX: REFERENCE TABLE OF ORGANIC SEMICONDUCTORS

Low molecular weight organic semiconductors

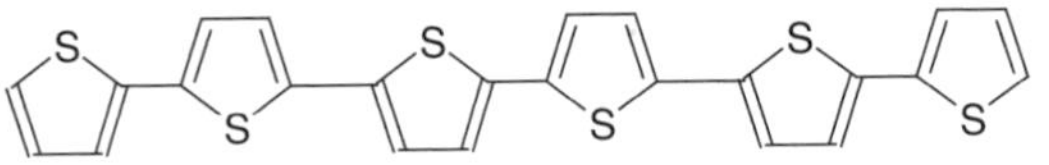

6T

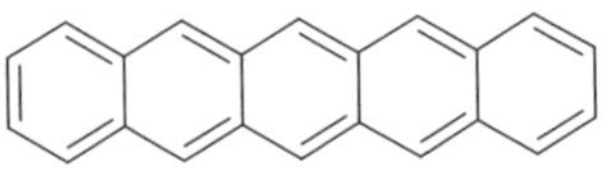

Pentacene

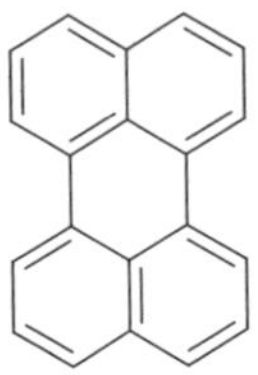

Perylene

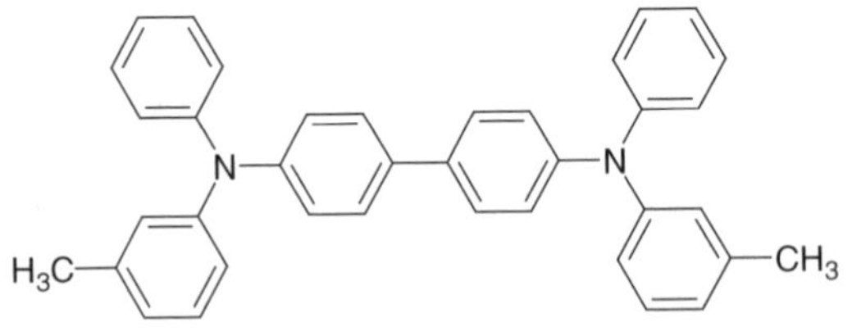

TPD

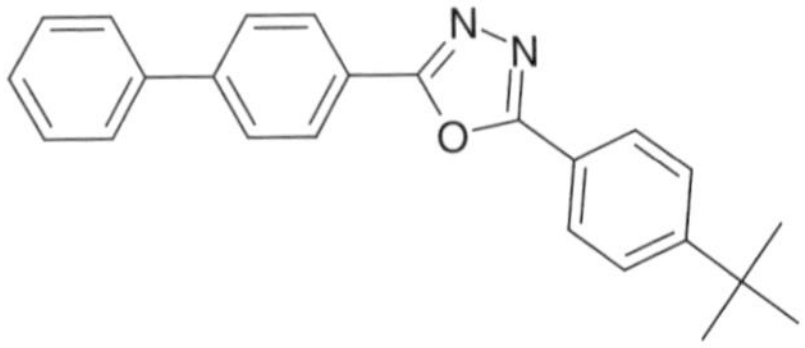

PBD

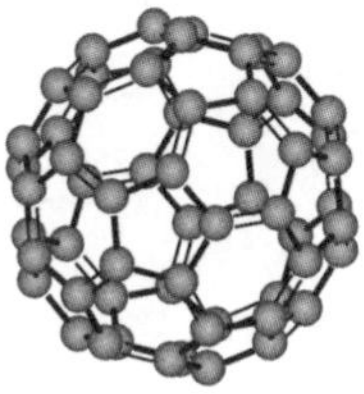

C_{60}

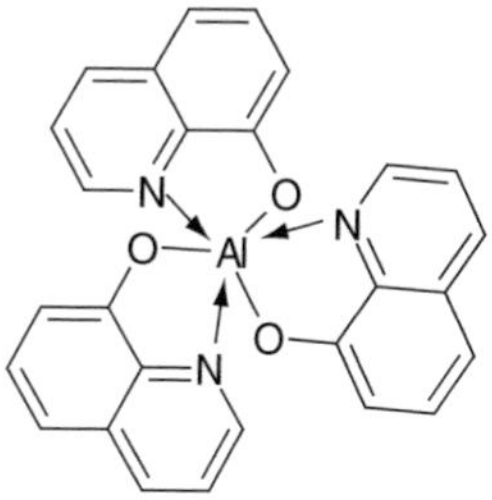

Alq_3

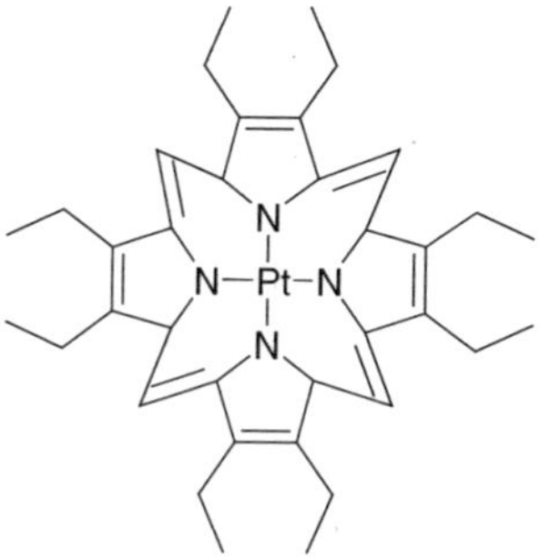

PtOEP

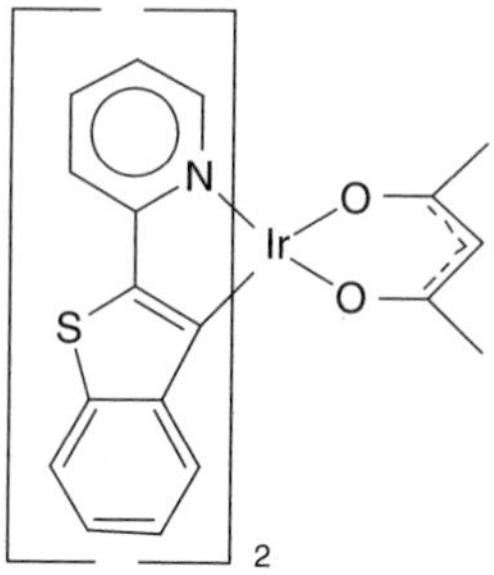

btpacac

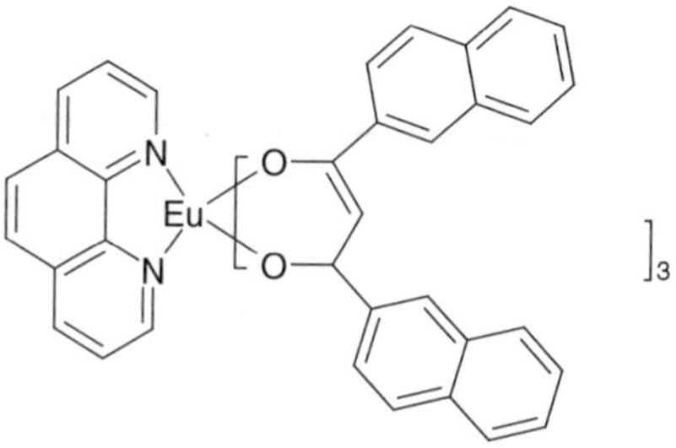

ADS053RE

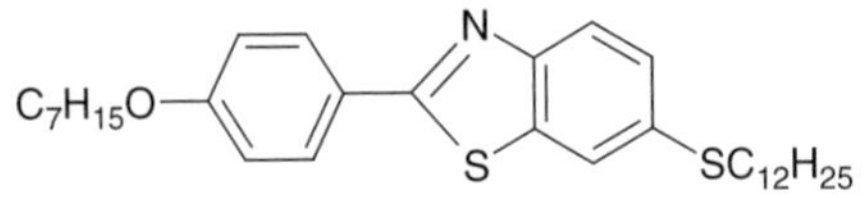

7O-PBT-S12

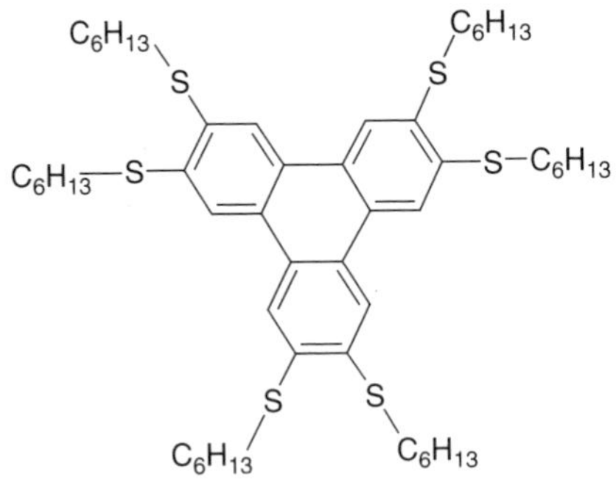

HHTT

Hexithiophene (6T), Pentacene, Perylene, and TPD (*N, N'*-bis-(*m*-tolyl)-*N, N'*diphenyl-1,1-biphenyl-4,4'-diamine) are hole-transporting and more or less strongly fluorescent organic semiconductors Perylene is more emissive, the others, less. 6T is one representative of the thiophene family of organic semiconductors, which are known for their fast hole mobilities and are often used in organic transistors. PBD (2-(biphenyl-4-yl)-5-(4-tert-butylphenyl)-1,3,4-oxadiazole) is an electron conductor. Both TPD and PBD have been used as carrier injection layers in multilayer device architectures. C_{60} is a material with very high electron affinity, and C_{60} derivatives have been used as electron acceptors in organic photovoltaic devices. However, electron transport in C_{60} is very sensitive to even traces of oxygen, which limits its practical potential. Alq_3 (tris(8-quinolinolato)aluminum(III)) is an organometallic Al chelate complex with efficient green electroluminescence and remarkable stability. Alq_3 was used as the emissive material in the first double-layer organic light-emitting device. PtOEP is a red phosphorescent porphyrine derivative. The central Pt atom facilitates spin–orbit coupling that allows light emission from triplet excitons. $btp_2Ir(acac)$ (bis(2-(2'-benzothienyl)-pyridinato-*N,C*-3') iridiumacetylacetonate) is a representative of a family of highly efficient phosphors that have been used successfully as triplet-harvesting emitters in efficient electrophosphorescent devices. ADS053 RE is a trade name for the red-emitting organolanthanide Tris(dinapthoylmethane)mono(phenanthroline)-europium(III). Organolanthanides transfer both singlet and triplet excitons to an excited atomic state of the central lanthanide, resulting in very narrow emission lines; i.e., spectrally pure colours (here the 612 nm red line of europium). 7O-PBT-S12 and HHTT are hole-transporting calamitic and discotic liquid crystals, respectively. Due to the stacking of conjugated cores in smectic (7O-PBT-S12) and some discotic (HHTT) liquid crystalline phases, both can display rather high charge carrier mobilities.

Polymeric organic semiconductors

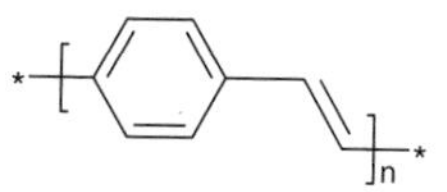

PPV

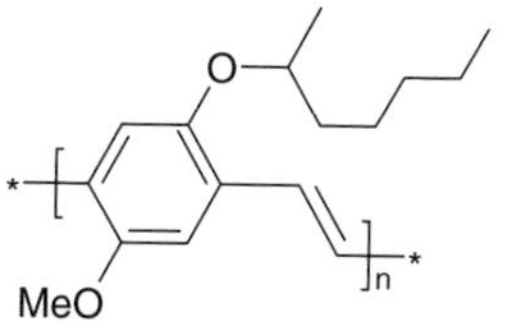

MEH-PPV

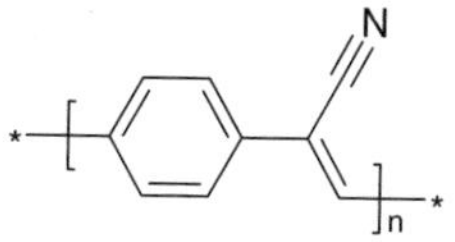

CN-PPV

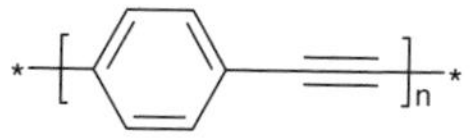

PPE

PPP

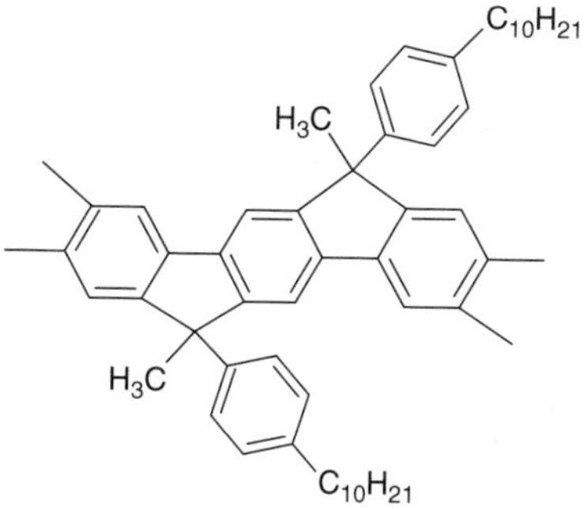

MeLPPP

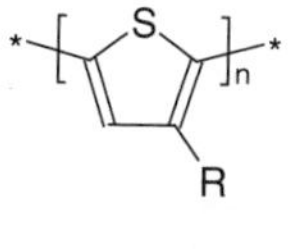

PAT

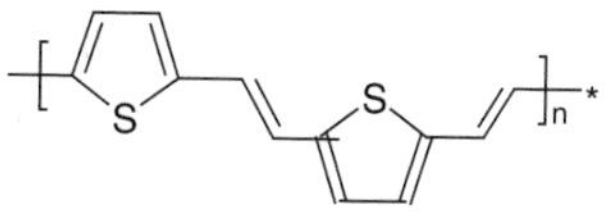

PTV

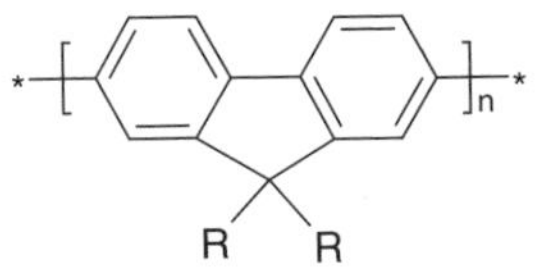

PF

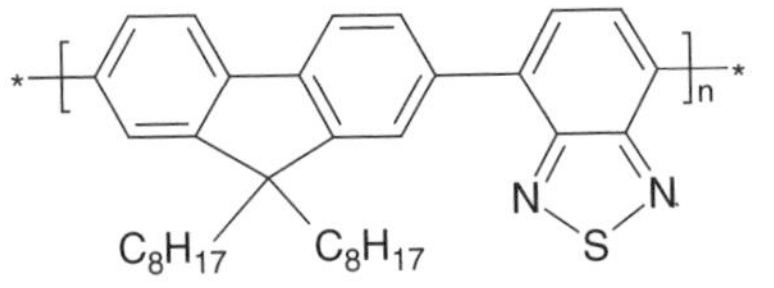

F8BT

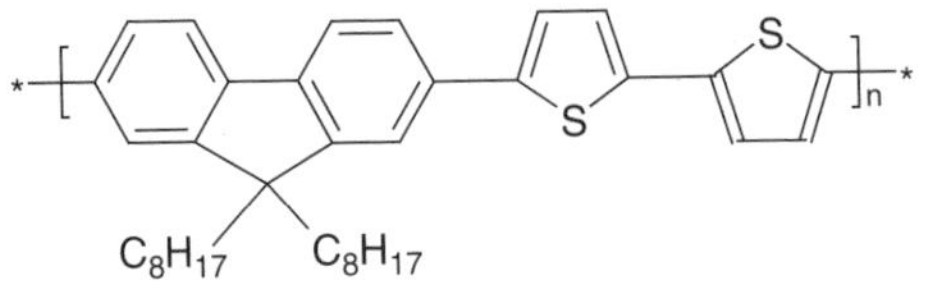

F8T2

Poly(p-phenylene vinylene) (PPV) played an outstanding role in the development of organic electroluminescence. MEH-PPV and Cyano-PPV (CN-PPV) are side chain substituted PPVs. Side chains promote solubility and can also change the band gap. Poly(phenylene ethynylene) (PPE) and poly(p-phenylene) (PPP) are variations on a similar theme. Methylated ladder-type PPP (MeLPPP) is similar to PPP, but with all backbone rings coplanar. PPVs, PPP, PPE and MeLPPP have been explored extensively in organic light-emitting devices. Poly(alkylated thiophene) (PAT) and poly(thienylene vinylene) (PTV) are less emissive, but have favourable properties in organic FETs. Polyfluorene (PF) is a blue emitter that has recently competed successfully with PPV as organic light-emitting material. Typically, PF is copolymerized rather than side chain substituted to modify its properties. F8BT is an electron transporter and efficient green emitter. F8T2 is a hole transporter that works well in transistors. PF, F8BT and F8T2 also display interesting liquid crystalline phases.

'Hybrid' materials

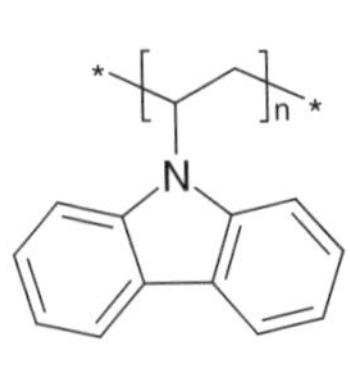

PVK

ST 638

sQP

oxTPD

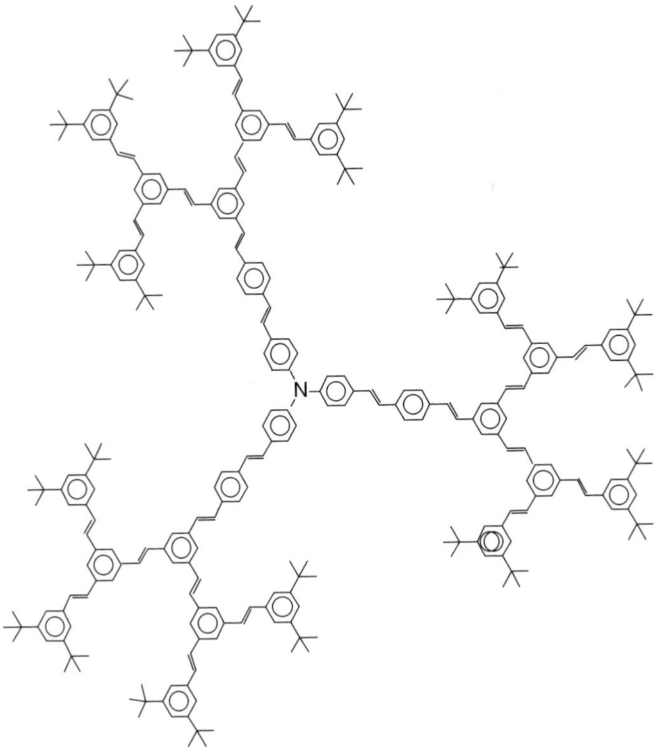

NDSB Dendron (G2)

Poly(vinyl carbazole) (PVK) is historically one of the first polymeric organic semiconductors. It is a polymeric material, with the film-forming and morphological properties typical of polymers. However, the semiconducting carbazole units dangle laterally from a non-conjugated backbone and are isolated from each other. The electronic properties of PVK are therefore very similar to those of low molecular weight carbazole. ST 638 is the trade name for 4,4',4''-tris(N-(1-naphthyl)-N-phenyl-amino)-triphenylamine. This is low molecular weight material, but due to its sterically hindered 'starburst' architecture it has a very high glass transition temperature and typically does not crystallize when spin-cast from solution, but forms a glassy film, like many polymers. The glassy morphology has considerable advantages for device applications; a tendency to crystallize is a major problem with hole-transporting small molecules such as TPD. The same structural theme was employed for the design of electron-transporting starburst-type phenylquinoxalines (not shown here). Another structural theme that can be used to suppress crystallization in non-polymeric materials is the use of spiro links between two (or more) p-phenylene units, here exemplified by a spiro-linked pair of quaterphenyls (sQP). Note the cross-shaped three-dimensional architecture of spiro compounds that is difficult to sketch on paper. oxTPD is clearly a low molecular weight compound, but via the oxetane functions that are attached with flexible spacers it can be cross-linked in situ with the help of a suitable (photo)initiator. The result is a highly cross-linked, inert hole-transporting film with no crystallisation tendency that has been used successfully in multiplayer devices. NDSB Dendron (G2) is a second-generation, nitrogen-cored distyryl benzene dendrimer. The core displays visible absorption and emission, the meta-linked dendronic side groups have a band gap in the UV, and for charge injection, transport and light emission they can be considered as inert.

Synthetic metals

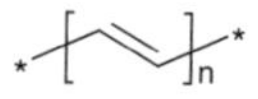

PA

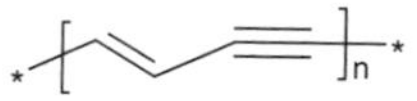

PDA

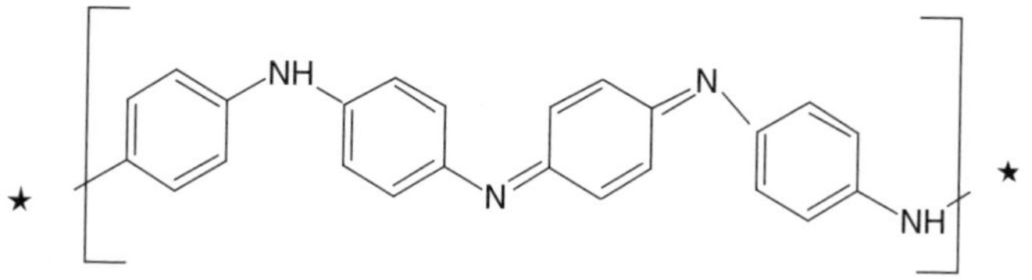

PAni

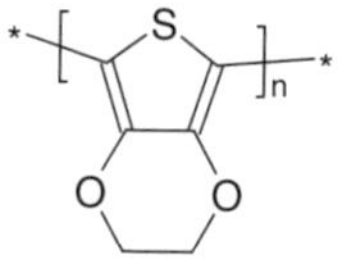

PEDOT

The distinction between organic semiconductors and synthetic metals is somewhat arbitrary, as the synthetic metals shown here are in the undoped state, when they display semiconducting rather than metallic or quasi-metallic properties. Metallic properties are induced by chemical doping with a substance that is either highly redox active, or an acid or a base. Poly(acetylene) (PA) is the classic example; the (chance) discovery that iodine-doped PA displays metallic conductivity was a milestone that earned the 2000 Nobel prize in chemistry. Poly(diacetylene) (PDA) has a widely tunable band gap if substituted with suitable side groups. Both PA and PDA are of historic, but no longer practical, interest. Poly(aniline) (PAni, here shown as emeraldine base) and poly (3,4-ethylenedioxythiophene) (PEDOT) are more modern developments. They are made metallic by acid rather than redox doping. Water-based PAni and PEDOT preparations are now commercially available. PEDOT that is acid-doped with poly(styrene sulfonic acid) (PSS), PEDOT/PSS, is now very popular in the OLED community to modify or replace the commonly used transparent ITO anodes.

BIBLIOGRAPHY

Books

An excellent text on the physics of solids is *Solid State Physics* by N. W. Ashcroft and N. D. Mermin, published by Holt, Rinehart and Winston, New York, 1977.

A good discussion of electronic transitions is given in *Physical Chemistry*, 5th edn, by P. W. Atkins, Oxford University Press, Oxford, 1994.

Much of interest, especially concerning space charge limited current, is contained in *Functional Organic and Polymeric Materials*, edited by T. H. Richardson and published by John Wiley & Sons, Ltd, Chichester, 2000.

Liquid crystalline polymers are discussed by A. M. Donald and A. H. Windle in their book *Liquid Crystalline Polymers*, Cambridge University Press, Cambridge, 1992.

Review articles

Tuning the bandgap in organic semiconductors: A. Kraft, A. C. Grimsdale and A. B. Holmes, *Angew. Chem. Int. Ed.*, **37**, 403 (1998); D. N. Batchelder, *Contemp. Phys.* **29**, 3 (1988).

PEDOT/PSS: L. Groenendaal, F. Jonas, D. Freitag, H. Pielartzik and J. R. Reynolds, *Adv. Mater.* **12**, 481 (2000).

Synthetic metals: A. B. Kaiser, *Adv. Mater.* **13**, 928 (2001).

7

Self-assembling nanostructured molecular materials and devices

7.1 INTRODUCTION

Soft materials are ubiquitous in everyday life, in soaps, paints, plastics and many other products. Their properties rely on the self-assembly of amphiphiles, colloids and polymers into a variety of mesophases, with nanoscale order of the constituent molecules. Nature also exploits self-organization of soft materials in many ways, to produce cell membranes, biopolymer fibres and viruses, to name just three. Mankind has recently begun to be able to design materials at the nanoscale, whether through atom-by-atom or molecule-by-molecule methods (top-down) or through self-organization (bottom-up). The latter endeavour encompasses much of soft nanotechnology, which is the focus of this chapter, as well as Chapters 6, 8 and 9. Strategies that take inspiration from nature are being followed, but alternative ab initio rational design methods which may one day improve on evolved structures (which often contain redundant elements) are also being pursued.

Self-organization of soft materials can be exploited to create a panoply of nanostructures for diverse applications. The richness of structures results from the weak ordering due to non-covalent interactions, and the consequent importance of thermal energy which enables phase transitions with differing degrees of order. The power of self-organization may be harnessed most usefully in a number of applications, including the preparation of nanoparticles, the templating of nanostructures, nanomotor design, the exploitation of biomineralization and the development of functionalized delivery vectors (for drugs, most importantly).

The intricate structures that can be formed by complex self-organizing molecules are truly amazing – for example helical and ferroelectric structures from non-chiral and

Nanoscale Science and Technology   Edited by R. W. Kelsall, I. W. Hamley and M. Geoghegan
© 2005 John Wiley & Sons, Ltd

non-polar molecules, biomimetic structures, artificial motors and muscles and protein or DNA recognition systems. The field of bionanotechnology is the subject of Chapter 9.

In the next few decades, the creation of a wealth of new materials and devices using soft nanotechnology will be witnessed. The impact of this on the quality of life of people everywhere will be immense. All this from a new materials chemistry, taking elements from colloid and polymer physical chemistry, sprinkling in a good dose of synthetic chemistry and biochemistry and sitting back to watch the explosion of novel research.

In this chapter the principles of self-assembly underpinning nanoscale structure formation in soft materials are elucidated. Examples of applications for nanoscale self-assembly are provided. In such a broad subject there are inevitably omissions. There is only a brief discussion of some aspects of bionanotechnology, since this chapter mainly emphasizes synthetic nanomaterials. Supramolecular chemistry is also not considered, although it is a powerful tool for the programmed self-organization of molecules and has been proposed as a means to create nanomachines (see the famous book by Drexler cited in the bibliography).

This chapter is organized as follows. Section 7.2 introduces the molecular components of self-organizing structures. Section 7.3 outlines the principles of self-assembly. Section 7.4 considers self-assembly routes to the preparation of nanoparticles as well as applications in nanotechnology; nano-objects are also briefly discussed. Section 7.5 summarizes templating methods for the fabrication of inorganic nanostructures. Section 7.6 covers nanotechnology applications of liquid crystal phases, both lyotropic and thermotropic. Section 7.7 contains a summary and outlook.

7.2 BUILDING BLOCKS

Self-assembling soft materials can be divided into synthetic and biological types. The chemistry of the constituent molecules or supramolecular aggregates are considered here, along with an introductory summary of the structures formed by these systems, which include polymers, colloids, liquid crystals, proteins, DNA and other biopolymers.

7.2.1 Synthetic

Polymers are long-chain molecules, usually organic. Biopolymers are considered separately in Section 7.2.2 and in Chapter 9. A wide variety of synthetic polymers can now be made by an extensive range of polymerization methods. Although conventional polymers spontaneously self-assemble into nanostructures, for example crystalline lamellae in crystalline polymers, the focus in this chapter will be on the engineered self-assembly of polymers into designed structures. A prime example is the microphase separation of block copolymers into a rich variety of nanostructures (Sections 7.6.2 and 7.6.3).

Surfactants are surface-active agents. Surfactant molecules are said to be amphiphilic; this means they contain both a hydrophilic (water-liking) tail group and a hydrophobic (water-hating) head group. The combination of these components leads to their preferential segregation to surfaces, where they can be active, such as in detergents.

Synthetic surfactants may have ionic head groups (as in cationic or anionic surfactants) or they may be non-ionic. The tail group is often a hydrophobic alkyl chain.

Lipids are biological amphiphiles. Many types of lipid such as phospholipids (containing a phosphate-based head group) have more than one hydrophobic tail. Amphiphiles aggregate into nanostructures in water to minimize the contact of the hydrophobic groups with H_2O molecules. A common nanostructure is a micelle, which can be spherical or cylindrical. These structures form with a hydrophobic core and a hydrophilic corona in order to avoid contact of the hydrophobic part of the molecule with water. Vesicles can also form; these are hollow spherical structures in which the shell is formed by layers of surfactant molecules. The lamellar phase comprises flat layers of amphiphilic molecules. Inverse micellar structures can be formed in 'oil' (i.e., an organic liquid) as the hydrophilic groups tend to segregate from the medium.

Colloids may be defined as microscopically heterogeneous systems where one component has dimensions in the range 1 nm to 1 µm, which at the lower end enters the nanoscale domain. This covers many types of material, including aerosols, foams and emulsions. This chapter mainly considers colloidal sols, which are dispersions of solid particles (often spherical latex beads) in a liquid.

Liquid crystals are materials with molecular order intermediate between that of a liquid and that of a crystal. Thermotropic liquid crystal phases are formed by organic molecules in the absence of solvent on heating from a low-temperature crystal phase. Lyotropic liquid crystal phases are formed by amphiphiles in solution, as discussed in Section 7.3.2. Molecules forming liquid crystal phases are termed mesogens. Thermotropic mesogens must be anisotropic. They may have a rod-like (calamitic) or disc-like (discotic) structure. Thermotropic liquid crystal phases are characterized by long-range molecular orientational order and in the case of smectic (layered) and columnar phases by various types of long-range translational order. There are many types of smectic and columnar phases; details can be found in *Introduction to Soft Matter* (Hamley) or *Introduction to Liquid Crystals: Chemistry and Physics* (Collings and Hird), both of which are cited in the bibliography for this chapter. The positional order of molecules in a nematic phase is short-range, as in a liquid, although there is long-range orientational order, characterized by the director (a unit vector along the direction of average orientation).

7.2.2 Biological

Structural proteins are commonly fibrous proteins such as keratin, collagen and elastin. Skin, bone, hair and silk all depend on such proteins for their structural properties.

Silk is produced by insects and arachnids to make structures such as webs, cocoons and nests. Silk from silkworm cocoons (of the moth *Bombyx mori*) has been used by mankind to make fabrics, because it has excellent mechanical properties, particularly its high tensile modulus. Spider silks also have outstanding strength, stiffness and toughness that, weight for weight, are unrivalled by synthetic fibres. The structures (several types have been recorded) all consist of silk based on antiparallel β-sheets of the fibrous protein fibroin. A β-sheet is a so-called secondary structure of a protein, formed by intermolecular hydrogen bonding between peptide chains (Section 7.3.1). Long stretches of the polypeptide chain consist of sequences (Gly-Ser-Gly-Ala-Gly-Ala), where the symbols indicate different amino acods. The Gly chains extend from one surface of the β-sheets and the Ser and Ala from the other, forming an alternating layered structure. The orientation of the chains along the β-sheet underpins the tensile strength of silk, while the weak forces between sheets ensure that silk fibres are flexible. Silk fibres have a complex hierarchical structure, in which a fibroin core is surrounded by a skin of the protein sericin. Within the core, termed *bave*, there are crystalline regions containing layered β-sheets and amorphous regions that may contain isolated β-sheets.

Collagen is the major component of connective tissues, found in all multicellular animals. The molecule has a triple-helix structure. In vivo, collagen is organized into covalently cross-linked fibrils. Denatured collagen is better known as gelatine. Denaturation refers to the destabilization of the secondary structure due to unfolding of the protein, commonly brought about by heating or by chemical means. Keratins are a group of fibrous proteins that form hair, wool, nails, horn and feathers. There are two major types of keratin: α-keratins in mammals and β-keratins in birds and reptiles. As the name suggests, the structure of α-keratins is based on the α-helix (actually a coil of α-helices). β-keratin proteins form a β-sheet in their native state. In the natural material, keratins are arranged into fibrillar structures. The α-helix is the other common secondary structure formed by intramolecular hydrogen bonding of peptide chains. Further details on protein secondary structures can be found in the texts listed in the bibliography.

Globular proteins are found in a range of substances, including enzymes, transporter proteins and receptor proteins. They may contain α helix and/or β-sheet secondary structures. Many common arrangements of these secondary structures occur in unrelated globular proteins and are termed motifs or domains; an example is the $\beta\alpha\beta$ motif. A recent model represents these structures in a 'periodic table' parameterized according to the number of layered structures (formed by packed α-helices or β-sheets) and the shape of the motif (flat, curled or barrel). The twist of β-sheets leads to a staggered arrangement for the secondary structures in the outer layers. These sheets can also curl, and a combination of curl and stagger can produce *barrels* (hydrogen-bonded cylinders). This periodic table provides a good match for over 90% of known single-domain structures. It has also been shown that, in many aspects of their physical chemistry, globular proteins behave like charged colloidal particles.

The double-helix structure of DNA is famous. Less well known is the formation of lyotropic liquid crystal phases by DNA fragments in solution. Short fragments behave like rods, and so the formation of liquid crystal phases is possible. On increasing the concentration (above 160 mg/ml for 50 nm DNA in physiological salt solutions), cholesteric and hexagonal columnar phases may be observed. Cholesteric phases are better known as chiral nematic phases, because this indicates that they are anisotropic fluids in which the local orientation follows a helix (the helical ordering refers to the director, not the individual molecules). Cholesteric phases (originally observed in a cholesterol derivative) are characterized by bright interference colours, as the pitch of the helix is usually close to the wavelength of light. At temperatures just below those at which the cholesteric phase exists, a 'blue phase' is sometimes observed. This phase is named for the colour arising from the double-twist cylinders that result from the packing of helices. These cylinders pack into various cubic structures, as discussed for example in the texts cited in the bibliography.

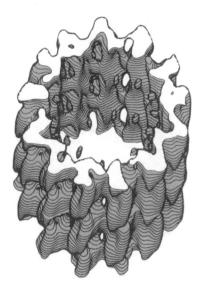

Figure 7.1 The 1.8 nm resolution X-ray structure of a microtubule. Reproduced from D. Voet and J. G. Voet, *Biochemistry*, Wiley, 1995, with permission

Microtubules are of interest in nanotechnology for a number of reasons. They are tubular structures formed from the protein tubulin (Figure 7.1), and could be used as nanochannels for the transport of liquid, or as 'struts' to support nanostructures. However, probably the main interest stems from the fact that they are a key component of one of the main motility systems of cells found in eukaryotes (the other being muscle). Eukaryotes are the cells of living organisms, except bacteria, and they contain a nucleus. The motion of cilia, which are the hair-like strands that undulate to sweep fluid across the surface of organs such as the respiratory tract, depends on the sliding of subfibres formed from microtubule arrays past one another. The whip-like structures responsible for motion in many types of cell called flagella (e.g., sperm cell tails) also move in this way. In contrast to the linear waving motion of cilia and eukaryote flagella, bacterial motion is impelled by rotation of flagella via a propeller-type structure that spans the bacterial membrane. As in muscles and cilia, the motion is driven by an ATPase (ATP is the molecule adenosine triphosphate) that acts as a transducer, converting the energy from ATP to ADP hydrolysis into mechanical energy (ADP is adenosine diphosphate). The detailed structure of the bacterial flagellum is complex, and further information can be found in a good biochemistry text. It remains to be seen whether nanomotors like those used in cilia or bacterial flagella will be incorporated directly into nanomachines or will inspire designs for artificial motors.

Viruses consist of nucleic acid molecules (RNA or DNA) encased in a protein coating. Virus capsids (protein shells) can be near spherical or rod-like (helical). Spherical viruses all have an icosahedral structure (a polyhedron with 20 triangular faces). Many common viruses, including rhinovirus (responsible for the common cold) and herpes simplex virus (Figure 7.2(a)) have this structure. The first virus to be discovered, tobacco mosaic virus, has a helical structure (Figure 7.2(b) and (c)), leading to a rod-shaped particle ~300 nm long and 18 nm in diameter.

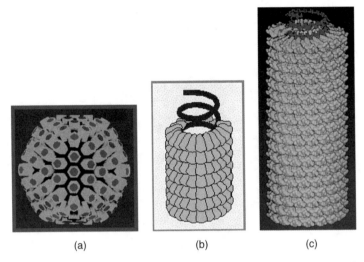

(a)   (b)   (c)

Figure 7.2 Models of (a) icosahedral structure of herpes simplex virus protein shell, (b, c) helical tobacco mosaic virus. Parts (a) and (b) reproduced from www.uct.ac.za/depts/mmi/stannard/ virarch.html; copyright L. M. Stannard. Part (c) reproduced from www.molbio.vanderbilt.edu/ fiber/ images.html. First published in K. Namba, D. L. D. Caspar and G. Stubbs, *Biophys. J.* **53**, 469 (1988)

7.3 PRINCIPLES OF SELF-ASSEMBLY

The term 'self-assembly' is used with a variety of meanings in different scientific contexts. For example, in Chapters 1 and 3, 'self-assembly' was used to describe the growth of semiconductor quantum dots using a particular MBE growth mode. In that context, 'self-assembly' referred to the spontaneous formation of the quantum dots as a consequence of the growth conditions (strain fields), without any need to define explicitly the dot size or shape. In the context of molecular materials, the term self-assembly is used to describe the reversible and co-operative assembly of predefined components into an ordered super-structure. Two categories of such self-assembly have been identified. Static self-assembly involves systems at equilibrium that do not dissipate energy. The formation of a structure may require energy, but once formed it is stable. In dynamic self-assembly, on the other hand, the formation of structures or patterns occurs when the system dissipates energy. Examples are patterns formed by reaction and diffusion processes in oscillating chemical reactions. The focus of this chapter is on materials that form static self-assembled structures, although a brief summary of possible routes to the fabrication of nanomotors involving dynamic self-assembly are outlined.

Self-assembly in soft materials relies on the fact that the fluctuations in the position and orientation of molecules or particles due to Brownian motion have energies comparable to thermal energy. Thermal energy has a dramatic influence on soft materials at the nanoscale as weak non-covalent bonds are broken and sometimes reformed. This enables the system to reach thermodynamic equilibrium, which is often a non-uniform state. Because of the relatively weak interactions between molecules, transitions between different structures can readily be driven by changes in conditions, such as temperature or pH. These external triggers that induce phase transitions could lead to a host of responsive materials, or coupled with an appropriate source of energy to nanomechanical systems. There is a diversity of phase transitions between different structures in soft materials, and examples are considered in subsequent sections.

7.3.1 Non-covalent interactions

For self-assembly to be possible in soft materials, it is evident that the forces between molecules must be much weaker than the covalent bonds that hold molecules together. Weak intermolecular interactions responsible for molecular ordering in soft materials include hydrogen bonds, coordination bonds in ligands and complexes, ionic interactions, dipolar interactions, van der Waals forces and hydrophobic interactions. These are now summarized.

The hydrophobic effect arises when a non-polar solute is inserted into water. The hydrophobic effect can be distinguished from hydrophobic interactions, which result from the association of two non-polar moieties in water. The hydrophobic effect is conventionally ascribed to the ordering of water molecules around an unassociated hydrophobic molecule. This leads to a reduction in entropy. This entropy loss can be offset when association of hydrophobic molecules into micelles occurs, because this leads to an increase in entropy as the 'structured water' is broken up. An enthalpy penalty for demixing of water and solute should also be outweighed by the entropy increase in order for the Gibbs free energy change for micellization to be negative. The structured water model is based on orientational ordering of water molecules around the inserted solute molecule. An alternative model proposes that the high free energy cost of inserting a non-polar solute into water is due to the difficulty of finding a cavity due to the small size of water molecules. However, it has been argued that the hydrophobic effect is more subtle, depending on *solute* size and shape as well.

Hydrogen bonding is particularly important in biological systems, where many protein structures in water are held together by hydrogen bonds. Of course, the existence of life as we know it depends on hydrogen bonds, which stabilize H_2O in the liquid form. In proteins, intramolecular hydrogen bonds between N—H groups and C=O groups that are four amino acid units apart underpin the formation of the α-helix structure. On the other hand, hydrogen bonds between neighbouring peptide chains lead to β-sheet formation. Similarly, collagen fibres contain triple-helical proteins held together by hydrogen bonding. The folding pattern of proteins is also based on internal hydrogen bonding The smaller the number of hydrogen bonds in the folded protein, the higher its free energy and the lower its stability. The reason that nature has evolved means to exploit hydrogen bonds in this way is due to the strength of the bond. Hydrogen bonds are weaker than covalent bonds (about $20 \, \text{kJ} \, \text{mol}^{-1}$ for hydrogen bonds compared to about $500 \, \text{kJ} \, \text{mol}^{-1}$ for covalent bonds), so superstructures can self-assemble without the need for chemical reactions to occur, yet the bonds are strong enough to hold the structures together once formed, since the energy is still larger than thermal energy ($2.4 \, \text{kJ} \, \text{mol}^{-1}$).

Molecular recognition between artificial receptors and their guests can be combined with self-organization to program the self-assembly of nanostructures. Many types of non-covalent interaction can be exploited in supramolecular chemistry, including hydrogen bonding, donor–acceptor binding and metal coordination complexation. A diversity of methods have been employed to create receptors for ionic and molecular guests. Another important example is the use of cyclodextrins as hosts for the delivery of drugs or pesticides.

Stabilizing colloidal dispersions against aggregation (known as coagulation when irreversible, flocculation when reversible) is important in everyday things such as foods or personal care products. Often the system is an oil-in-water dispersion that can be stabilized by adding interfacially active components such as amphiphiles or proteins. These segregate to the oil–water interface and stabilize emulsions by reducing interfacial tension, and the enhanced rigidity and elasticity of the membrane that forms also help to prevent coalescence. Colloidal sols found in paints and pastes also need to be stabilized for long shelf life; this can be achieved in several ways. First, for charged colloidal particles in an electrolyte medium, the balance between the repulsive electrostatic and attractive van der Waals contribution to the total potential energy can be adjusted, so that a barrier to aggregation is created. The phenomenon of charge stabilization can be analysed using the Derjaguin–Landau–Verwey–Overbeek (DLVO) theory, which is described in many colloid science textbooks. A second method to prevent aggregation is steric stabilization. Here long-chain molecules are attached to colloidal particles, creating a repulsive force as chains interpenetrate when the particles approach one another. The attached molecule may be chemisorbed (e.g., a long-chain fatty acid) or more commonly an adsorbed polymer. In contrast to charge stabilization, steric stabilization works in non-aqueous media and over a wide range of particle concentrations. The choice and concentration of polymer are critical in steric stabilization, since at low concentration, polymer chains can attach themselves to two (or more) particles, leading to so-called bridging flocculation. On the other hand, at higher polymer concentrations, if the polymer is non-adsorbing then it can lead to depletion flocculation, the mechanism for which was first recognized by Asakura and Oosawa. The polymers cannot penetrate the particles and are excluded from a depletion zone around them. When the particles are close together, the depletion zones overlap and the dispersal of polymers into the bulk solution is favoured entropically. An osmotic pressure of solvent from the gap between particles leads to an effective attraction between them, and hence flocculation. When the colloidal particle concentration is such that, on average, they are further apart than a polymer coil radius, and the polymer concentration is high, then depletion stabilization is possible. Forcing the particles together would require the 'demixing' of polymer from bulk solution. This increases the free energy so that the effective interaction between particles is repulsive.

7.3.2 Intermolecular packing

At high concentration, the packing of block copolymer or low molar mass amphiphilic molecules in solution leads to the formation of lyotropic liquid crystal phases, such as cubic-packed spherical micelles, hexagonal-packed cylindrical micelles, lamellae or bicontinuous cubic phases (Figure 7.3). The phase formed depends on the curvature of the surfactant–water interface. One approach to understanding lyotropic phase behaviour computes the free energy associated with curved interfaces. The curvature is analysed using differential geometry, neglecting details of molecular organization. In the second main model, the interfacial curvature is described by a molecular packing parameter. These two approaches will be described in turn.

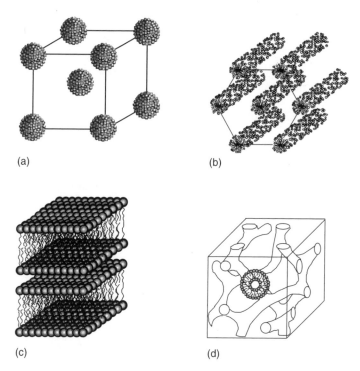

(a)

(b)

(c)

(d)

Figure 7.3 Lyotropic liquid crystal structures: (a) cubic-packed spherical micelles, (b) hexagonal-packed cylindrical micelles, (c) lamellar phase, (d) bicontinuous cubic structure. Here the amphiphilic molecules (not shown everywhere for clarity) form a bilayer film separating two continuous labyrinths of water. Reproduced from I. W. Hamley, *Introduction to Soft Matter*, Wiley, 2000, with permission

In the model for interfacial curvature of a continuous surfactant film, we use results from the differential geometry of surfaces. A surface can be described by two fundamental types of curvature at each point P in it: mean curvature and Gaussian curvature. Both can be defined in terms of the principal curvatures $c_1 = 1/R_1$ and $c_2 = 1/R_2$, where R_1 and R_2 are the radii of curvature. The mean curvature is $H = (c_1 + c_2)/2$, and the Gaussian curvature is defined as $K = c_1 c_2$.

Radii of curvature for a portion of a so-called saddle surface (a portion of a surfactant film in a bicontinuous cubic structure) are shown in Figure 7.4, although they can equally well be defined for other types of surface such as convex or concave surfaces found in micellar phases. To define the signs of the radii of curvature, the normal direction to the surface at a given point P must be specified. This is conventionally defined as positive if the surface points outwards at point P. In Figure 7.4 c_1 is negative and c_2 is positive. The mean and Gaussian curvatures of various surfactant aggregates are listed in Table 7.1.

It should be noted that end effects in elongated micelles due to capping by surfactant molecules that lead to an ellipsoidal or spherocylindrical (cylinder capped by hemispheres) structure are neglected. This will, however, change both mean and Gaussian curvatures to an extent that depends on the relative surface area of cap and tubular

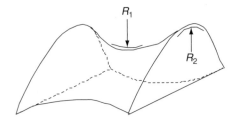

Figure 7.4 Principal radii of curvature of a saddle surface

Table 7.1 Mean and Gaussian interfacial curvature for common aggregate shapes. Here $R = R_1 = R_2$ denotes a radius of curvature

	Mean curvature $H=(c_1+c_2)/2$	Gaussian curvature $K=c_1 c_2$
Spherical micelles of vesicle (outer layer)	$1/R$	$1/R^2$
Cylindrical micelles	$1/2R$	0
Bicontinuous cubic phases	0 to $1/2R$	$-1/R^2$ to 0
Lamellae (planar bilayers)	0	0
Inverse bicontinuous cubic phases	$-1/2R$ to 0	$-1/R^2$ to 0
Inverse cylindrical micelles	$-1/2R$	0
Inverse spherical micelles or inner layer of vesicle	$-1/R$	$1/R^2$

parts. The elastic free energy density associated with curvature of a surface contains, for small deformations, the sum of contributions from mean and Gaussian curvatures. The interfacial curvature model is thus useful because it defines the elastic moduli κ and $\bar{\kappa}$ for mean and Gaussian curvature, respectively. These can be measured (e.g., by light scattering) and characterize the flexibility of surfactant films. Uncharged surfactant films typically have elastic energies $F_{el} \leq k_B T$; i.e., they are quite flexible.

An alternative approach to describing lyotropic mesophases in concentrated solution is based on the packing of molecules. The effective area of the head group, a, with respect to the length of the hydrophobic tail for a given molecular volume controls the interfacial curvature. The effective area of the head group (an effective molecular cross-sectional area) is governed by a balance between the hydrophobic force between surfactant tails which drives the association of molecules (hence reduces a) and the tendency of the head groups to maximize their contact with water (and thus increase a). The balance between these opposing forces leads to the optimal area per head group, a, for which the interaction energy is minimum.

Simple geometrical arguments can be used to define a packing parameter, the magnitude of which controls the preferred aggregate shape. For a spherical micelle, it can be shown that the following condition holds: $V/la \leq 1/3$, where V is the volume of a molecule and l is the length of an extended hydrophobic chain (which can easily be calculated). The term $N_s = V/la$ is called the surfactant packing parameter, or critical packing parameter. The surfactant parameter can be used to estimate the effective headgroup area, a, or vice versa. The surfactant parameter is concentration dependent, reflecting changes primarily in a (but to a lesser extent in V) on varying the amount of solvent.

Just as spherical micelles can be considered to be built from the packing of cones, corresponding to effective molecular volumes, other aggregate shapes can be considered to result from packing of truncated cones or cylinders.

The surfactant packing model and the interfacial curvature description are related. A decrease in the surfactant packing parameter corresponds to an increase in mean curvature. The packing parameter approach has also been used to account for the packing stabilities of more complex structures, such as the bicontinuous cubic phases. Here the packing unit is a wedge, which is an approximation to an element of a surface with saddle-type curvature (Figure 7.4). Then it is possible to allow for differences in Gaussian curvature between different structures, as well as mean curvature.

7.3.3 Biological self-assembly

Understanding the folding of proteins is one of the outstanding challenges of science, let alone biophysics and biochemistry. Although much progress has been made in modelling protein folding, there is no consensus on the best method. Most methods consider a protein folding energy landscape. The problem is that this is a rough surface, with many local minima, and it can often be hard to model the guiding forces that stabilize the native structure and cause the free energy to adopt a 'funnel' landscape. Many minimalist models are based on computer simulations of particles on a lattice, and they are always based on coarse-grained approaches. Fully atomistic models seem some way off. Some structural insights on protein conformational dynamics have emerged from steered molecular dynamic simulations in which Monte Carlo moves are used as well as molecular dynamics trajectories.

DNA will be an important component of many structures and devices in nanobiotechnology. DNA computing is an application currently attracting considerable attention. In one approach, single DNA strands are attached to a silicon chip. Computational operations can then be performed in which certain DNA strands couple to added DNA molecules. Multi-step computational problems can also be solved. Here the DNA strands encode all possible values of the variables. Complementary DNA strands are then added, and attach themselves to any strand that represents a solution to one step of the computation. Single strands remaining are removed. This process is repeated sequentially for each step, and the DNA that is left is read out, via polymerase chain reaction (PCR) amplification, to provide the solution, represented in binary form, where a given binary number corresponds to an eight-nucleotide sequence. The DNA-directed assembly of proteins, using oligonucleotides capped with the molecule streptavidin, is another exciting realm of applications. The method can be used to fabricate laterally patterned arrays of many types of macromolecules with the biotin organic group as an end group, since a strong complex is formed between this and streptavidin.

The charged nature of DNA has been exploited to bind metal ions that aggregate into nanoparticles of silver, for example; they are then used as seeds for further deposition of silver to produce nanowires. Positively charged C_{60} fullerene derivatives have also been condensed onto DNA. Similarly, CdS nanoparticles have been templated on the charged DNA backbone. Arrays of DNA-functionalized CdS have been assembled, layer by layer, on a gold electrode using a set of two populations of DNA-capped CdS nanoparticles and a soluble DNA analyte. The two oligonucleotides bound to CdS nano-particles are complementary to the ends of the target DNA. The construction of nanoscale geometric objects and frameworks using three- and four-arm synthetic DNA molecules has also been reported. The use of nanoparticle-tagged DNA solutions in gene sequence detection is discussed in Section 7.4.2.

Modified or artificial cells could find applications in bionanotechnology as nanoscale delivery agents or nanoreactors. A membrane in a cell wall fulfils a number of functions. It acts as a barrier to prevent the contents of a cell from dispersing and also to exclude external agents such as viruses. The membrane, however, does not have a purely passive role. It also enables the transport of ions and chemicals such as proteins, sugars and nucleic acids into and out of the cell via the membrane proteins. Membranes are important not only as the external cell wall, but also within the cell of eukaryotes, where they subdivide the cell into compartments with different functions.

A cell membrane is illustrated in Figure 7.5. It is built from a bilayer of lipids, usually phospholipids, associated with which are membrane proteins and organic macromole-cular sugar molecules called polysaccharides. The lipid bilayer is the structural founda-tion, and the proteins and polysaccharides provide chemical functionality. Proteins are associated with cell membranes in a variety of ways. Integral proteins are very tightly bound within the membrane. Some proteins are associated with a specific surface within the bilayer, such as the hydrophobic surface between hydrocarbon tails. Those spanning the membrane are known as transmembrane proteins. These are obviously important in transporting ions or molecules across the cell membrane. Integral proteins are amphi-philic; the two ends extending into the aqueous medium contain hydrophilic groups whereas the region within the bilayer is predominantly hydrophobic. Integral proteins are believed to form an α-helix in the transmembrane domain (Figure 7.5), due to hydrogen bonding in the polypeptide backbone. This reduces the exposure of polar $-OH$ and $-NH_2$ groups to the apolar environment of the bilayer. Because of the non-polar nature of the interior of lipid bilayers, they are impermeable to most ionic and

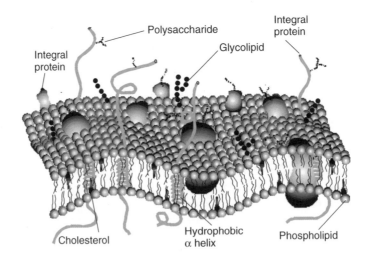

Figure 7.5 Schematic of a cell membrane. Adapted from D. Voet and J. G. Voet, *Biochemistry*, Wiley, 1995

polar molecules and indeed this is the basis of the barrier activity of lipid membranes. Integral proteins are bound in the lipid bilayer, and often act as channels for the transport of ions and molecules. These channels have to be highly selective to prevent undesirable material entering the cell and so are opened and closed as necessary. Membrane transport is also carried out by proteins that are not integral to the membrane. Transport proteins are then required to move ions, amino acids, sugars and nucleotides across the cell wall. They can either ferry these species across or form channels to transport them. An example of the latter is bee venom which contains the channel-forming protein mellitin. In contrast to integral proteins, peripheral proteins are not bound within the membrane but are associated to it either by hydrogen bonding or electrostatic interactions. Peripheral proteins often bind to the integral proteins.

Vesicles formed by lipids (termed liposomes) represent model systems for the cell membrane. The incorporation of channel-forming proteins (porins) into lipid bilayers has been studied for many years, and synthetic structural and functional mimics have been devised. It is straightforward to form vesicles from the lipid bilayers. Block copolymers form vesicles that can be polymerized, which has obvious advantages in encapsulation applications. The incorporation of channel-forming proteins into planar polymerized triblock copolymer membranes has been reported. This further extends the delivery and nanoreactor capabilities of the biomimetic structures. Recently, pH-swellable porous core–shell latexes have been developed. These are analogous to the pH-controlled pore opening of the protein shell of cowpea chlorotic mosaic virus. By appropriate surface functionalization, the recognition properties of bilayers can be enhanced, as required for many drug delivery applications. A model recognition system is the biotin–streptavidin complex, for which the free energy of binding is comparable to that of a covalent bond.

7.3.4 Nanomotors

A key element of any nanomachine is a nanomotor. A variety of approaches are being followed in the manufacture of nanomotors. The crudest is to make miniature versions of motors from the macroscopic world, however the ability to scale such structures downwards is limited by energy dissipation due to friction. Alternative strategies include attempts to mimic motors in biological systems, and the simpler 'motors' driven by chemical potential or concentration gradients, such as oscillating gels. Here we briefly discuss nanomotors based on soft matter.

Considering first biological motors as models for artifical motors, we can define two classes. In the first, proteins such as kinesin, dynein and myosin behave as linear slides. Among rotory motors, well-studied systems include the ATP synthesis complex, and bacterial flagellar motors, which are described in detail in Section 9.2.2.

Artificial motors exploit out-of-equilibrium chemical phenomena, such as a concentration gradient, as in ATP synthesis. Using this knowledge, it is possible to design simpler systems than those operating in nature. A minimal system can be constructed based on osmotic pumping using lipid vesicles in a gradient of solute concentration. The lipid bilayers act as osmotic membranes, allowing the passage of water molecules but not of solute molecules. Thus, when placed in a high osmotic pressure environment, the vesicles shrink and in a uniform solution do not move.

However, in a solute concentration gradient, a directional motion is imposed. Unidirectional motion can also be imparted to liquids confined in capillaries by a temperature gradient. An interesting concept to drive fluid motion in microcapillaries uses optical trapping of colloid particles, which can be manipulated to create pumps and valves. Although the scale of the particles is of order micrometres, it would be exciting if this could be extended to the nanoscale using smaller particles and shorter wavelength radiation. Other systems rely on the Marangoni effect. Due to dynamic surface tension fluctuations, surface-active molecules flow into higher surface tension regions (or away from low surface tension regions), to restore the original surface tension. This is the origin of the motion of camphor 'boats' that move freely on the surface of water. The origin of this motion was explained by Lord Rayleigh over a century ago, but the system has been revisited recently as a simple analogue of artifical motors.

A particularly attractive artifical motor system relies on oscillating chemical reactions to drive volume changes in polymer gels. The Belousov–Zhabotinsky (BZ) reaction was used to create an oscillating redox potential. This was then coupled to the most familiar polymer gel system exhibiting a volume phase transition, poly(N-isopropylacrylamide), or PNIPAM, in water. The PNIPAM was modified by covalent attachment of ruthenium tris(2,2'-bypyridine) units, which act as catalysts for the BZ reaction. Thus the oscillations in the BZ reaction were translated into periodic swelling and deswelling of the gel due to changes in the charge on the ruthenium complex.

7.4 SELF-ASSEMBLY METHODS TO PREPARE AND PATTERN NANOPARTICLES

7.4.1 Nanoparticles from micellar and vesicular polymerization

The fabrication of nanoparticles of controlled size, shape and functionality is a key challenge in nanotechnology. There are several established routes to nanoparticle preparation. Roughly spherical nanoparticles can be prepared by very fine milling (Section 5.3.5); this route is used to prepare iron oxide nanoparticles in ferrofluid dispersions or zinc oxide nanoparticles for use in sunscreens. So-called colloidal methods produce nanoparticles with much more uniform size and shape distribution than milling. Metal and metal oxide nanoparticles have been prepared using micellar nanoreactors where, for example, salts are selectively sequestered in the micellar core then reduced or oxidized. Such nanoparticles can be used in catalysis, separation media, biopolymer tagging and light-emitting semiconductor (CdS) quantum dots.

Recent work has shown that metal nanoparticles can be patterned at the surface using the self-organization of block copolymers. Two main routes have been exploited: nanoparticle formation within micelles in solution which may subsequently be deposited on a solid substrate, and direct patterning at the surface using selective wetting. Figure 7.6 shows examples of nanoparticle direct patterning by selective wetting at the surface of a diblock copolymer.

Nanocapsules; i.e., shell particles with a hollow interior, can be prepared by a number of routes, including the cross-linking of the shell of block copolymer vesicles.

(a)   (b)

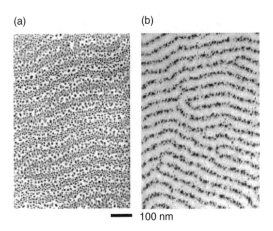

━━ 100 nm

Figure 7.6 Examples of nanoparticle and nanowire arrays templated by a stripe pattern formed at the surface of a polystyrene–poly(methyl methacrylate) diblock copolymer by vapour deposition of gold. The gold selectively wets polystyrene domains. Reproduced from W. A. Lopes and H. M. Jaeger, *Nature* **414**, 735 (2001)

An alternative approach has recently been developed, using polyelectrolyte multilayers assembled around a colloidal core that is subsequently dissolved. Biological particles, such as apoferritin and cowpea chlorotic mosaic virus, with hollow fillable interiors are described in Sections 7.4.6 and 7.3.3, respectively.

7.4.2 Functionalized nanoparticles

Functionalized nanoparticles will find numerous applications, for example in catalysis and as biolabels. Gold nanoparticles functionalized with proteins have been used as markers to detect biological molecules for some time. They may also be used to deliver DNA in a so-called gene-gun. Arrays of nanoparticles can be prepared via dip pen nanolithography (DPN), which is described in Section 9.1.2. For example, magnetic nanoparticles can be patterned into arrays, with potential applications in magnetic storage devices.

Functionalized nanoparticles are required for many biotechnological applications. Figure 7.7 shows a technique for detecting specific gene sequences that could be used in genetic screening. First the sequence of bases in the target DNA is identified. Then two sets of gold particles are prepared: the first set has attached DNA that binds to one end of the target DNA, and the second set carries DNA that binds to the other end. The nanoparticles are dispersed in water. When the target DNA is added, it binds both types of nanoparticle together, linking them to form an aggregate. The formation of this aggregate causes a shift in the light-scattering spectrum from the solution; i.e., a colour change in the solution that can easily be detected. This technique has recently been developed to allow the electrical detection of DNA. The principle is similar to that of the colour-change detection system, except one end of the target DNA binds to a short capture oligonucleotide attached to the surface of a microelectrode, and the other end

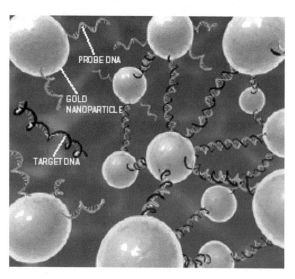

Figure 7.7 DNA-functionalized gold nanoparticle gene sequence detection system. Reproduced from *Scientific American*, September 2001, p. 63

binds to an oligonucleotide attached to Au nanoparticles. Binding of the target DNA causes Au nanoparticles to fill the gap between a pair of electrodes, an event that can be detected from capacitance or conductivity measurements. In practice the sensitivity of the device was enhanced by silver deposition on the nanoparticles. Arrays of electrode pairs were assembled to form DNA chip arrays in which each pair contained a different oligonucleotide capture strand.

7.4.3 Colloidal nanoparticle crystals

There is an immense interest in photonic band gap crystals because they can be used to confine photons or to modulate or control stimulated light emission or to construct lossless waveguides. A photonic band gap crystal or photonic crystal is a structure with a periodic variation in dielectric properties (Section 3.8.8). The propagation of electro-magnetic waves in such a crystal is analogous to that of electrons in semiconductors, in particular there exist band gaps that exclude the photon propagation modes in certain frequency intervals. In principle, three-dimensional crystals could have a complete bandgap; i.e., one for which photon propagation is prevented in all spatial directions; i.e., throughout the Brillouin zone, using the nomenclature of solid-state physics. The main focus on three-dimensional structures has been on the face-centred cubic (fcc) lattice because its Brillouin zone is most closely spherical, which might favour the formation of a complete photonic bandgap. However, it has been shown for an fcc crystal formed by colloidal spheres (opal structure) that, independent of the dielectric contrast, there is never a complete band gap. The inverse structure (spheres of air in a continuous solid medium) however holds promise, because calculations indicate the possibility of a complete 3D band gap. It has even been shown that by coating the air pores with nematic liquid crystal, a switchable photonic bandgap material can be achieved. Here the tunable localization of light or of waveguiding results from the electro-optic properties of the liquid crystal, where an electric field can be used to orient molecules in a particular direction with respect to the pore lattice.

To create a 3D photonic band gap, two conditions must be fulfilled. First, the colloidal particles must have low polydispersity (i.e., be almost the same size) in order to form a cubic crystal. Second, the number of defects in the cubic crystal must be minimized. Several strategies have been adopted to create macroscopic colloidal crystals. A common technique relies on sedimentation of particles under gravity. However, the resulting samples generally contain polycrystalline domains. Other approaches rely on surfaces to act as templates to induce order. For example, spin coating onto planar substrates can provide well-ordered monolayers, as can flow-induced ordering. A method that relies on so-called convective self-assembly has been used to create ordered crystals upon rapid evaporation of solvent. A related technique is the controlled withdrawal of a substrate from a colloidal solution, similar to Langmuir–Blodgett film deposition (Section 8.5.2.2), where lateral capillary forces at the meniscus induce crystallization of spheres, and if the meniscus is slowly swept across the substrate, well-ordered crystal films can be deposited. Convective flow prevents sedimentation and provides a continuous supply of particles to the moving meniscus. Actually, the controlled evaporation process alone is sufficient to produce films of controlled thickness that are well ordered up to the centimetre scale. An epitaxial mechanism has been employed, using a lithographically patterned polymer substrate to template crystal growth. Holes just large enough to hold one colloidal particle were created in a rectangular array. Controlled layer-by-layer growth starting from this template was then achieved by slow sedimentation of the silica spheres used. The formation of well-ordered crystal monolayer 'rafts' of charged colloid particles on the surface of oppositely charged surfactant vesicles has also been demonstrated.

As mentioned above inverse opal structures offer the greatest potential for photonic crystals. The most promising materials for the matrix seem to be certain wide band gap semiconductors, such as CdS and CdSe, because they have a high refractive index and are optically transparent in the visible and near-IR region. The preparation of porous metallic (gold) nanostructures within the interstices of a latex colloidal crystal has been demonstrated. Here a solution of gold nanoparticles fills the pores between colloidal particles, and the latex is subsequently removed by high-temperature furnace heating. A similar method has been used to fabricate inverse opal structures of titania. The same idea has been applied to form a nanoporous polycrystalline silica (deposited via low-pressure chemical vapour deposition); see Figure 7.8. In a related approach, silica spheres are coated with gold (to reinforce the colloidal crystal) then immersed in electroless deposition baths to deposit metal films within the porous template; the silica is removed in an HF rinse. These types of approach have been extended to a lost wax approach to prepare high-quality arrays of hollow colloidal particles (or filled particles) of various ceramic and polymer materials. Here a well-ordered silica colloidal crystal is taken and used as a template for polymerization in the interstices. If the pores are interconnected, the polymer forms a continuous porous matrix. By appropriate choice of polymer, either hollow or solid nanoparticles can be grown in it (hollow nanoparticles grow from the polymer matrix inwards, solid nanoparticles form within the voids). In this way it was possible to prepare colloidal crystals of solid or hollow TiO_2 particles,

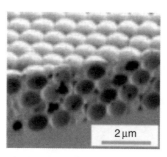

Figure 7.8 Cross-sectional scanning electron micrograph image of thin film inverse opal structure of polycrystalline silicon templated by 855 nm silica spheres. Reproduced from Y. A. Vlasov, X.-Z. Bo, J. C. Sturm and D. J. Norris, *Nature* **414**, 289 (2002)

as well as conductive polymer nanoparticles. An extension of colloidal polymerization techniques can be used to prepare defined waveguides. Crossed laser beams were used to polymerize polymer precursors within particular pores. By scanning the laser beams, a waveguide with a chosen path and shape can be fabricated.

The use of microgel particles of PNIPAM to form colloidal crystal arrays that selectively diffract light has been reported. Poly(*N*-isopropylacrylamide) in aqueous solution exhibits a volume phase transition at 32 °C; below 32 °C gels are hydrated and swollen but above 32 °C gels dehydrate and collapse. This transition has been used to vary the dimensions of PNIPAM microgel particles from 100 nm at 40 °C to 300 nm at 10 °C, a 27-fold volume change. This can be exploited to prepare a switchable selective diffraction array. Below the transition, the particles are swollen and only diffract light weakly; however, in the compact state, the diffracted intensity increases dramatically due to the enhanced contrast between particles and medium (the Bragg diffraction wavelength is unaffected). Wavelength-tunable arrays were fabricated by polymerizing PNIPAM in the presence of 99 nm polystyrene spheres. The embedded polystyrene spheres follow the swelling or shrinking of the PNIPAM hydrogel so that the wavelength of the Bragg diffraction can be tuned across the visible range of the spectrum.

7.4.4 Self-organizing inorganic nanoparticles

Within the past few years, there has been a surge of interest in composite materials consisting of a polymer filled with plate-like particles such as clay particles. Such fillers are extremely effective in modifying the properties of polymers, and orders of magnitude improvement in transport, mechanical and thermal properties have been reported. Examples of applications include low-permeability packaging for food and electronics, toughened automotive components, and heat- and flame-resistant materials. Polymer–clay nanocomposites have several unique features: First, they are lighter in weight than conventional filled polymers for the same mechanical performance. Second, their mechanical properties are potentially superior to fibre-reinforced polymers, because reinforcement from the inorganic layers occurs in two dimensions instead of one. Lastly,

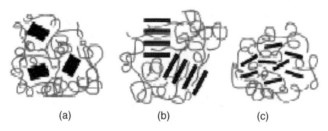

(a)　　　　　(b)　　　　　(c)

Figure 7.9 Possible structures for polymer–clay nanocomposites: (a) phase separated, (b) intercalated, (c) exfoliated. Reproduced from E. P. Giannelis, R. Krishnamoorti and E. Manias, *Adv. Polym. Sci.* **138**, 108 (1999)

they exhibit outstanding diffusional barrier properties without requiring a multipolymer layered design, allowing for recycling.

Clays are colloidal suspensions of plate-like mineral particles, with a large aspect ratio. Typically the particles are formed from silicate layers combined with layers of octahedrally coordinated aluminium or magnesium atoms. The layers lead to a lamellar phase for the clay in water. The aim in applications is to retain this structure in the polymer–clay nanocomposite; possible structures are illustrated schematically in Figure 7.9. Exfoliation and phase separation should be avoided and there is an immense literature (especially patent literature) on how to achieve this by chemical treatment of the clay particles, in particular by adsorption of organic molecules. The intercalated structure leads to enhanced barrier properties, due to the tortuous path for gas diffusion around the clay platelets.

Liquid crystal phases formed by mineral moieties have been known almost as long as organic liquid crystals. They have received renewed interest because of the ability to combine the properties of liquid crystals, in particular anisotropy and fluidity, with the electronic and structural properties of minerals. They may also be cheaper to produce than conventional liquid crystals, which require organic synthesis. Rod-like mineral systems that form nematic phases have been well studied. Sheet-forming mineral compounds that form smectic (layered) structures in solution are also known.

The colloidal behaviour of vanadium pentoxide (V_2O_5) has been investigated since the 1920s. Under appropriate conditions of pH, ribbon-like chains can be obtained via the condensation of V—OH bonds in a plane. Figure 7.10 shows a scanning electron micrograph of dried ribbons. A nematic liquid crystal forms in aqueous suspensions if the particle volume fraction, ϕ, exceeds 0.7%. A sol–gel transition occurs at $\phi = 1.2\%$, which divides the nematic domain into a nematic sol and a nematic gel. For $\phi > 5\%$, a biaxial nematic gel phase is formed. Suspensions of V_2O_5 can be aligned in electric and magnetic fields, similar to organic nematogens used in liquid crystal displays. Laponite and bentonite/montmorillonite clay particles also form nematic gels.

The previous section discussed formation of layered structures in intercalated suspensions of these types of clay. It can be argued that they are not lamellar or smectic phases since long-range order is not preserved upon swelling, where exfoliation occurs. Colloidal smectic phases have been observed for β-FeOOH, which forms Schiller layers (from the German for iridescence). The rod-like β-FeOOH particles form layers at the bottom of the flask. The spacing between the layers is comparable to the wavelength of light, hence the iridescence. A swollen liquid crystalline lamellar phase based on

Figure 7.10 Scanning electron micrograph of a dried V_2O_5 suspension. Reproduced with permission from J. Livage

extended solid-like sheets (rather than rod-like particles) has been rationally designed using the solid acid $H_3Sb_3P_2O_{14}$. In contrast, plate-like $Ni(OH)_2$ nanoparticles (91 nm radius, 12 nm thick) and $Al(OH)_3$ nanodiscs (radius 200 nm, thickness 14 nm) self-assemble into columnar mesophases. A nematic phase has also been observed for $Al(OH)_3$ nanoparticles. The formation of a smectic phase rather than a columnar phase is expected if the polydispersity in particle radius is large enough to prevent the efficient packing of columns. In fact, at very high volume fractions in $Al(OH)_3$ suspensions, evidence was obtained for a smectic phase, which can accommodate the polydispersity in radius (although a low polydispersity of particle thickness is required).

7.4.5 Liquid crystal nanodroplets

Figure 7.11 shows an array of block copolymer micelles containing liquid crystal solubilized in the micellar core. The self-assembly of the block copolymer micelles into a hexagonal close-packed arrangement is apparent. The long-range ordering of the structures could be improved as in other soft materials by using an alignment substrate or by annealing. The ability to pattern liquid crystal nanodroplets at the nanoscale is not required for conventional display applications (which do not require a resolution beyond that of visible light) but may find applications in phased array optics. Phased array optics is a method to reconstruct a three-dimensional image on a two-dimensional

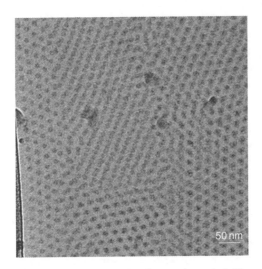

Figure 7.11 Transmission electron micrograph of quench-cooled film of poly(styrene oxide)-*b*-poly(ethylene oxide) block copolymer micelles containing liquid crystal solubilized in the poly(styrene oxide) core. Reproduced from I. W. Hamley, V. Castelletto, J. Fundin, M. Crothers, D. Attwood and Y. Talmon, *Colloid Polym. Sci.* **282**, 514 (2004)

surface. Optics allows this to be done if the phase and amplitude of the light waves from the virtual image are controlled. An array of switchable light sources 200 nm apart is sufficient to reconstruct any desired light wave pattern. It has been proposed that liquid crystals can be used as switchable birefringent phase shifters. However, no one has yet found the means to arrange the liquid crystal in nanometre-scale arrays. Patterning of liquid crystals in micelles or microemulsions is a promising way to achieve this.

7.4.6 Bionanoparticles

Viruses are natural nanoparticles which have evolved into a variety of shapes. A number of nanotech applications of viruses are now considered. First, they may be used as responsive delivery agents. Recent work has focused on the use of modified cowpea chlorotic mottle virus nanoparticles as biocompatible responsive delivery agents. At pH < 6.5 the virus adopts a compact spherical structure, however at pH > 6.5 the structure becomes porous, allowing the pH-controlled release of encapsulated drug molecules, for example. In non-responsive mode, viruses may be used as 'Trojan Horses' for the delivery of genes in transfection applications. Gene therapy is attracting immense attention as a means to treat diseases by modifying the expression of genetic material. Its premise is that disease can be prevented at the level of DNA molecules, thus compensating for abnormal genes. With an eleven-year history of clinical trials, and many more in progress, recent evidence that gene therapy may be efficacious in the treatment of medical conditions due to the deficiency of single genes has attracted worldwide attention.

Both viral and non-viral approaches have been used in clinical trials to treat illnesses such as cystic fibrosis and several forms of cancer. Viruses have evolved efficient ways of targeting cells, delivering genetic material and expressing it. However, inflammatory and immunological responses induced by viruses may limit their utility for repeated administration. Numerous systems have been studied for non-viral gene delivery, including synthetic polymers such as polylysine and poly(oxyethylene)-based block and graft copolymers or biologically derived liposomes or cationic lipids or the cationic polyelectrolyte poly(ethyleneimine) (PEI). PEI has a very high cationic charge density, making it useful for binding anionic DNA within the physiological pH range and forcing the DNA to form condensates small enough to be effectively transferred across the cell membrane. Furthermore, it has been shown that PEI enhances transgene expression when DNA/polymer complexes are injected into the cytoplasm.

Magnetotactic bacteria exploit magnetic nanoparticles to navigate from regions of oxygen-rich water (toxic to them) to nutrient-rich sediment. The bacteria contain grains of magnetite aligned in chains, as shown in Figure 7.12. The chain of crystals (and hence the bacterium) aligns along a magnetic field direction, which contains vertical and horizontal components (except at the poles). In the northern hemisphere, the bacteria move downwards by moving towards the north. In the southern hemisphere, the bacteria are south-seeking. The magnetic grains of magnetite in these bacteria contain single magnetic domains. Grains that are $<5\,\mathrm{nm}$ in size are magnetised to saturation, whereas grains that are larger than $10\,\mathrm{nm}$ contain several magnetic domains. Nanotechnologists can take inspiration from nature's use of chains of magnetic particles as navigational aids.

The use of chemically modified versions of the iron storage protein ferritin in high-density magnetic data storage devices is the focus of current commercialization efforts. Ferritin is a nearly spherical protein with an $8\,\mathrm{nm}$ diameter core of ferrihydrite ($5Fe_2O_3 \cdot 9H_2O$). The core can be removed by reductive dissolution to produce the shell protein 'apoferritin'. The core can then be 'refilled' by incubation with metal salts, and subsequent oxidation. In this way, the core can be filled with magnetite (Fe_3O_4), which unlike the native ferrihydrite is ferrimagnetic at room temperature; the resulting ferritin is called magnetoferritin.

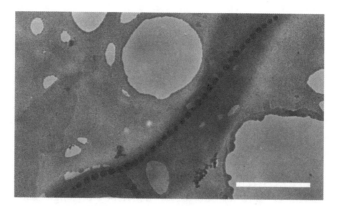

Figure 7.12 Chain of magnetite nanoparticles in a magnetotactic bacterium: scale bar $= 500\,\mathrm{nm}$. Reproduced from S. Mann, *Biomineralization: Principles and Concepts in Bioinorganic Materials Chemistry*, Oxford University Press, Oxford, 2001, with permission

7.4.7 Nano-objects

Nanoparticles with shapes other than simple spheres, shells or tubes have been prepared via soft material-mediated methods. The photoinduced conversion of silver nanospheres to silver nanoprisms has been reported. Photoinduced fragmentation of silver nanoparticles is believed to produce the single-crystal prism-shaped particles (whose faces correspond to planes of the crystal lattice). The growth habit of (nano)crystals can be controlled using organic agents such as surfactants (as well as through the degree of supersaturation or ionic strength), producing polyhedra with faces controlled by the growth rate of certain planes in the crystal unit cell. Nanoparticles of CdSe with rod, arrow, teardrop and tetrapod shapes may be fabricated by using surfactants to selectively control the growth of certain crystal faces. String and other superstructures of spherical nanoparticles may be prepared in the same way. Nature exploits soft materials to template the synthesis of hard nanostructures, as discussed further in Section 7.5.2, which includes examples of the intricate structures made by certain organisms. Self-assembled nanostructures may also be used to template the formation of helical nano-particles (using peptides in solution) or of string, necklace or vesicular structures formed by block copolymers in solution.

The self-assembly of rod-coil block copolymers can, for example, be used to make mushroom-shaped nano-objects that assemble into lamellar stacks which show polar ordering.

7.5 TEMPLATED NANOSTRUCTURES

7.5.1 Mesoporous silica

The self-assembly of surfactants can be exploited to template inorganic minerals, such as silica, alumina and titania. The resulting structures resemble those of zeolites, except that the pore size is larger for the surfactant-templated materials than the pore size resulting from channels between atoms in classical zeolite structures. In conventional zeolites, the pore size is typically up to 0.1 nm, whereas using amphiphile solutions it is possible to prepare an inorganic material with pores up to several tens of nanometres. Such materials are thus said to be mesoporous. They are of immense interest due to their potential applications as catalysts and molecular sieves. Just as the channels in conventional zeolites have the correct size for the catalytic conversion of methanol to petroleum, the pore size in surfactant-templated materials could catalyse reactions involving larger molecules.

It was initially believed that the templating process simply consisted of the formation of an inorganic 'cast' of a lyotropic liquid crystal phase. In other words, preformed surfactant aggregates were envisaged to act as nucleation and growth sites for the inorganic material. However, it now appears that the inorganic material plays an important role, and that the structuring occurs via a cooperative organization of inorganic and organic material. Considering, for example, the templating of silica, a common method is to mix a tetra-alkoxy silane and surfactant in an aqueous solution. Both ionic and non-ionic surfactants have been successfully used to template structures,

Figure 7.13 Hexagonal structure of calcined mesoporous silica, templated using an amphiphilic triblock copolymer. Reproduced from D. Zhao, J. Feng, Q. Huo, N. Melosh, G. H. Fredrickson, B. F. Chmelka and G. D. Stucky, *Science* **279**, 548 (1998)

as have amphiphilic block copolymers (these behave as giant surfactants, and enable larger pore sizes). The cooperative self-assembly process leads to a structure in which the silica forms a shell around amphiphilic aggregates, the latter being removed by calcination.

Figure 7.13 shows a hexagonal honeycomb pattern where the silica has been templated from a hexagonal-packed cylinder phase. Layered or bicontinuous structures have been prepared in a similar manner, by templating lamellar or bicontinuous phases, respectively. Similarly, highly monodisperse silica beads have been made by templating spherical micelles.

7.5.2 Biomineralization

Biomineralization involves the uptake and controlled deposition of inorganic moieties from the environment in biological systems. The main types of biominerals are the various forms of calcium carbonate (e.g., calcite and aragonite) and calcium phosphate. Calcium carbonate is the principal component of shells, which consist of an outer layer of large calcite crystals, and an inner region of layers of aragonite several hundreds of nanometres thick. Other marine organisms live within intricate exoskeletons formed from calcium carbonate. Examples include the so-called coccospheres (Figure 7.14). Calcium phosphate is the building material for bone and teeth, in the form of hydroxyapatite, which can be represented as $Ca_{10}(PO_4)_6(OH)_2$. Bone is formed by the organized mineralization of hydroxyapatite in a matrix of collagen fibrils and other proteins to form a porous structure. The mineral content controls the rigidity or elasticity of the bone. Tooth enamel also contains hydroxyapatite (more than in bone), and its ability to

Figure 7.14 Example of a coccosphere. scale bar $= 3\,\mu m$. Reproduced from S. Mann, *Biomineralization: Principles and Concepts in Bioinorganic Materials Chemistry*, Oxford University Press, Oxford, 2001

withstand abrasion results from a complex structure where ribbon-like crystals are interwoven into an inorganic fabric. A great deal of research activity is currently focused on the construction of artificial bone for replacement joints, and as scaffolds for tissue engineering. However, the porous macrostructure of bone is outside the nanodomain, and so this fascinating subject is not considered further here.

Unicellar organisms called radiolarians and diatoms produce their beautiful microskeletons (Figure 7.15) from amorphous silica. Lamellar aluminophosphates can also be

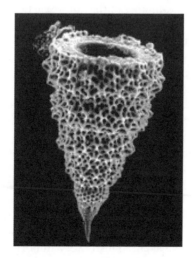

Figure 7.15 Examples of radiolarian microskeletons of length $330\,\mu m$ (left) and $220\,\mu m$ (right). Reproduced from www.ucmp.berkeley.edu/protista/radiolaria/radmm.html

templated to create patterns that mimic diatom and radiolarian microskeletons. Their nanoscale features are formed by the templated self-assembly of minerals via biological structures. In particular, the lace-like structures are formed from vesicles, packed together at the cell wall. The vesicles are arranged in a thin foam-like film, and biomineralization occurs in the continuous matrix.

7.5.3 Nanostructures templated by block copolymer self-assembly

Nanolithography using block copolymers is the subject of Section 8.7. The patterning of inorganic nanoparticles using block copolymer micelles adsorbed onto solute substrates is another exciting application of block copolymer self-assembly, as discussed in Section 7.4.1.

7.6 LIQUID CRYSTAL MESOPHASES

7.6.1 Micelles and vesicles

Micelles and vesicles formed by surfactants and block copolymers are widely used in systems as diverse as personal care products, agrochemicals and pharmaceuticals to solubilize fragrances, pesticides and herbicides, or drugs. Usually the aim is to solubilize organic compounds in the core of micelles in aqueous media.

The primary nanotechnology applications of micelles and vesicles result from their use as templates to synthesize nanoparticles with a multitude of structures and functionalities. Core cross-linking reactions to form organic nanoparticles containing functionalized coatings (tailored through the choice of corona chain) have also been used. In particular, cross-linking of the non-toxic biodegradable polylactide core of micelles with an end-functionalized poly(ethylene glycol) corona leads to sterically stabilized and biocompatible nanoparticles for drug delivery applications. Another approach is to cross-link the shell and remove the core, for example by ozone etching. Similarly, cross-linking the shell of a vesicle leads to hollow nanoparticles that can be used to encapsulate compounds.

As mentioned in Section 7.4.1, micelles can also be used as media for the production of inorganic nanoparticles. The synthesis of metal nanoparticles in aqueous block copolymer micelles has recently attracted a great deal of attention. Metal ions or complexes that are insoluble in water are sequestered in the micellar core. The block copolymer micelles containing the metal compounds then act as nanoreactors where, upon reduction, nucleation and growth of metal nanoparticles occurs. Applications of such metal nanoparticles are extensive, including catalysis, electro-optical materials (quantum dots) and in patterning of semiconductors. Using block copolymer micelles, it is possible to control the size of the particles by changing the copolymer composition and molecular weight. This is very important for the synthesis of magnetic nanoparticles, to ensure they are large enough to exceed the superparamagnetic limit but small enough to comprise a single domain (Section 4.1.4).

7.6.2 Lamellar phase

The lamellar phase (known as the smectic phase for low molar mass liquid crystals) is found in diverse systems, ranging from surfactants in solution to clays to block copolymers. The layered structures in clays and polymer–clay nanocomposites were discussed in Section 7.4.4. Here the focus is on recent examples of high-tech applications for lamellar phases in block copolymers.

Non-centrosymmetric structures can possess a macroscopic electric polarization, hence piezo- and pyroelectricity, as well as second-order non-linear optical activity. The fabrication of non-centrosymmetric stacks of block copolymer lamellae has been demonstrated in blends of ABC triblock and AC diblock copolymers. The structure is illustrated schematically in Figure 7.16. It is favoured over others (macrophase separated, random lamellar, centrosymmetric lamellar stack) if the asymmetry in aA and cC contact energies is large enough.

It has been proposed to exploit lamellar block copolymer structures to self-assemble all-polymer solid-state batteries, by using a triblock copolymer where the three blocks correspond to the anode, electrolyte and cathode. This has the advantage that leakage of toxic liquid electrolyte is avoided, and furthermore the processing is straightforward (e.g., spin coating of thin films). Similar applications of lamellar block copolymers in nanocapacitors and nanotransistors have also been envisaged.

Lamellar block copolymer nanostructures can be used as selective one-dimensional dielectric reflectors if the layer thickness is large enough (close to the wavelength of light) and the refractive index difference between blocks is large enough. Polystyrene–polyisoprene diblocks swollen with the corresponding homopolymers, for example, exhibit a limited angular range stop band at visible frequencies with potential applications in photonics; for instance in waveguiding.

Rod-coil diblocks can form a range of lamellar structures, as demonstrated by transmission electron microscopy images of polystyrene–poly(hexyl isocyanate) diblocks, which form wavy lamellar, zigzag and arrowhead morphologies. Distinct structures result because the rod block can tilt with respect to the layers, and the tilt can alternate between domains. The coupling of liquid crystal ordering to that of block copolymers extends considerably the range of nanostructures available to the nanotechnologist.

Inspired by a similar concept, it has been shown that ordering on multiple nano-length scales can be achieved in complexes of diblock copolymers and the amphiphilic long-chain alcohol pentadecylphenol (PDP). Hydrogen bonding of the alcohol to the

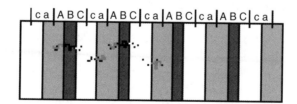

Figure 7.16 Schematic of a non-centrosymmetric lamellar structure observed in a blend of ABC triblock and AC diblock copolymers. Reproduced from T. Goldacker, V. Abetz, R. Stadler, I. Erukhimovich and L. Leibler, *Nature* **398**, 137 (1999)

—NH group in poly(4-vinylpyridine) (P4VP) produced a comb-like block, whereas no hydrogen bonding occurred to the coil-like polystyrene block. The usual ordered structures were observed due to microphase separation in the melt of the diblock, however, in addition, mesogenic ordering was observed within the P4VP–PDP phase due to formation of a lamellar structure below the liquid crystal–isotropic phase transition for the PDP. Since the lamellar–isotropic phase transition for the PDP–P4VP lamellae occurs below that for the PS–P4VP block copolymer, it is possible to switch off the lamellar ordering on one length scale independent of the other. It was shown that this transition was accompanied by a large change in the electrical conductivity (P4VP being a semiconducting side chain conjugated polymer). The potential to create switchable nanoscale structures with ordering in two and three dimensions has obvious implications in other applications such as alignment layers in liquid crystal displays, nanoscale sensors and optical waveguides.

7.6.3 ABC triblock structures

The phase behaviour of ABC triblocks is much richer than that of AB diblocks because there are two independent compositional order parameters and three Flory–Huggins interaction parameters (this expresses the energy of interaction between segments for the three different block combinations), the subtle interplay of which gives a varied morphospace. Figures 7.17 and 7.18 show examples of the intricate morphologies that are observed. A remarkable structure consisting of helices of minority polybutadiene domain wrapped around polystyrene cylinders in a PMMA matrix has even been reported.

State-of-the-art self-consistent mean field theory calculations have been used to predict a number of intricate nanostructures at the surface of ABC triblock copolymers (Figure 7.19). It should be noted that the patterns in Figure 7.19 are simulated in a 2D system. Due to confinement and surface energy effects, such morphologies may not be realizable at the surface of a bulk sample, however they could be accessed by sectioning of a glassy bulk sample. Potential exploitation of surface structures formed by ABC triblock copolymers can be envisaged when domains are selectively doped with metal or semiconductor; for example, Figure 7.6 shows patterning using a diblock copolymer. Applications include nanowire arrays for addressing nanoscale electronic devices or three-colour arrays for high-resolution displays.

7.6.4 Smectic and nematic liquid crystals

Conventional methods of fabricating liquid crystal displays are not usually regarded as nanotechnology. Present-day displays are based on nematics sandwiched in thin films between electrode-coated glass substrates. The supertwist nematic (STN) is used in low-cost small displays, for example in watches and calculators. Its principle of operation relies on the Fréedericks transition, in which the orientation of molecules (on average defined by a director) is switched by application of an electric field. In the STN, the orientation of the director in the off state is induced to follow a helix by the use of perpendicularly oriented grooves in the glass plates used to sandwich the liquid crystal.

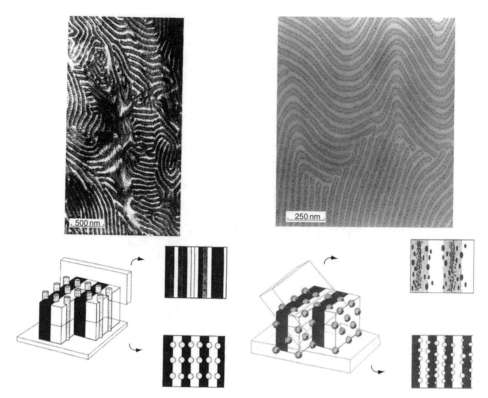

Figure 7.17 Examples of morphologies observed for polystyrene-*b*-polybutadiene-*b*-polymethyl-methacrylate triblock copolymers with a minority midblock component. Left: cylinders at a lamellar interface, Right: spheres at a lamellar interface ("ball at the wall" morphology). The upper images are transmission electron micrographs. The lower figures are schematic diagrams. Reproduced from R. Stadler, C. Auschra, J. Beckmann, U. Krappe, I. Voigt-Martin and L. Leibler, *Macromolecules* **28**, 3080 (1995)

In the on state, the molecules align along the electric field lines, perpendicular to the glass plates. TFTs operate on the same principle, but each element of the display is individually addressed using a thin film transistor.

The fabrication of a liquid crystal display on a single substrate, which could ultimately lead to flexible or paintable displays has recently been demonstrated based on an array of encapsulated liquid crystal cells. Stratified polymer structures self-assemble through phase separation of a photopolymerizable prepolymer and a nematic liquid crystal. Horizontal stratification creates the walls of the cells and vertical stratification (using a different wavelength of UV) produces lids. At present, the technique has been used to fabricate micron-sized polymer cells, although extension to the nanoscale using harder radiation should be feasible.

Usually a nematic phase is cloudy due to light scattering from fluctuating micron-sized domains with different orientations (creating refractive index variations, since the refractive index of liquid crystal phases is anisotropic). Nanometre-scale phase-separated structures formed in a lyotropic structure of surfactant micelles in a liquid crystal matrix have been shown to lead to a transparent nematic phase. In the nanoemulsion, the

Figure 7.18 'Knitting pattern' morphology observed by TEM on a polystyrene-*b*-poly(ethylene-co-butylene)-*b*-poly(methylmethacrylate) triblock copolymer (stained with RuO$_4$). Reproduced from U. Breiner, U. Krappe, E. L. Thomas and R. Stadler, *Macromolecules* **31**, 135 (1998). Also used as cover of *Physics Today*, February 1999

droplets of surfactant disrupt the long-range orientational order of the nematic phase, leading to optical isotropy and transparency, although the local nematic ordering is retained. Mixing spherical colloidal particles with liquid crystal likewise produces a phase-separated structure as colloidal particles are expelled from nematic droplets

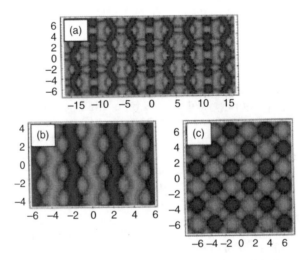

Figure 7.19 Examples of predicted morphologies for linear ABC triblock copolymers, from self-consistent mean field calculations. Reproduced from Y. Bohbot-Raviv and Z.-G. Wang, *Phys. Rev. Lett.* **85**, 3428 (2000)

below the isotropic–nematic phase transition temperature. The particles are expelled because the trapping of defects in the nematic phase by colloidal particles has too high an energy penalty. The colloid particles therefore separate into an interconnected network (the struts of which are several nanometres thick). The result is a waxy soft solid with a high storage modulus.

7.6.5 Discotic liquid crystals

Columnar phases formed by discotic liquid crystals such as those based on triphenylenes form one-dimensional conductors, due to the overlap of π^* orbitals of the aromatic moieties which are surrounded by a hydrocarbon insulator coating. Each column thus acts as a nanowire, and applications in molecular electronics have resulted, in particular in gas sensors. They could also be used in molecular electronic devices, for example in electroluminescent displays or in three-dimensional integrated circuits.

7.7 SUMMARY AND OUTLOOK

Self-assembly is responsible for nanostructure formation in colloidal, amphiphilic, polymeric and biomolecular materials. In this chapter, the principles of self-assembly in synthetic and biological systems were first considered. Then selected examples of self-assembly routes to the production of nanostructures and nanodevices were presented. A key theme is that self-assembly in soft materials (synthetic and biological) can be used to template nanostructures in inorganic matter, either in bulk or at a surface. The range of structures that can be fabricated in equilibrium depends (following the Gibbs phase rule) on the number of components in the system. In the case of ABC triblock copolymers this leads to a large number of possible nanostructures with different symmetries. An additional complexity in phase behaviour results from the coupling of distinct types of order, for example orientational order of liquid crystals with translationally ordered block copolymer nanostructures.

Out-of-equilibrium processes can also be exploited, for example in nanoscale motors or actuators. Actually, out-of-equilibrium structures may also be useful, since they could be captured when templating a hard material. It has to be kept in mind that the rich structural diversity and access to out-of-equilibrium structures are both different aspects of the weak ordering due to non-covalent interactions that characterizes soft materials.

Many developments are under way to exploit self-assembling soft materials in nanotechnology. The first commercial nanostructures are likely to be nanoparticles fabricated in micellar or vesicular nanoreactors, and mesoporous templated materials for catalysts and separation media. Uses of more intricate structures such as those formed by ABC triblock copolymers are still some way off. Downstream applications of biomineralization (in prostheses, artifical bone and teeth) are less distant. The development of drug delivery systems using functionalized nanoparticles is also the subject of intense research activity at present. This is only a flavour of the many different approaches being investigated. The use of block copolymer films in nanolithography (Section 8.6.2) and to pattern nanoparticles into regular arrays are the focus of much attention.

Arguably the most important nanodevice is the nanomotor, and self-assembly routes to the production of simple oscillating 'motors' have already been developed. To fabricate directional motors with a renewable energy source, inspiration is being taken from nature, where ATP synthesis underpins distinct linear and rotary motors. This is discussed in more detail in Chapter 9. Supramolecular chemistry also has much to offer here, although this is outside the scope of the present chapter. Other nanodevices will contain passive nanostructures which can be built using self-assembly, examples including waveguides and optical filters. Nanowires and ferroelectric piezo- and pyro-electric structures can also be produced. Self-assembled nanocapacitors and nanotransistors can also be envisaged, although as yet there has been little research in this area. Using a combination of self-assembled nanostructure elements from the broad palette available, together with a suitable power source (e.g., nanostructured polymer solid-state battery) a customized nanodevice could readily be put together. The prime limitation is that certain non-periodic structures require atomic or molecular manipulation, outside the realm of self-assembly.

BIBLIOGRAPHY

Supramolecular chemistry

J.-M. Lehn, *Supramolecular Chemistry: Concepts and Perspectives* (VCH, Weinheim, 1995).

Molecular machines

K. E. Drexler, *Engines of Creation* (Fourth Estate, London, 1990).
E. Regis, *Nano. The Emerging Science of Nanotechnology: Remaking the World – Molecule by Molecule* (Little-Brown, Boston, 1995).
J. P. Sauvage, (ed.), *Structure and Bonding*, Vol. 99 (Springer-Verlag, Heidelberg, 2001).

Self-assembly

G. M. Whitesides and B. Grzybowski, Self-assembly at all scales. *Science* **295**, 2418 (2002).

Bionanotechnology

V. Balzani, A. Credi, F. M. Raymo, and J. F. Stoddart, Artificial molecular machines. *Angew. Chem., Intl Edn Engl.* **39**, 3349 (2000).
C. M. Niemeyer and C. A. Mirkin, *Nanobiotechnology: Concepts, Applications and Perspectives*, (Wiley-VCH, Weinheim, 2004).

Biochemistry

D. Voet and J. G. Voet, *Biochemistry* (Wiley, New York, 1995).

Soft matter (general)

I. W. Hamley, *Introduction to Soft Matter* (Wiley, Chichester, 2000).
R. A. L. Jones, *Soft Condensed Matter* (Oxford University Press, Oxford, 2002).
T. A. Witten and P. Pincus, *Structured Fluids* (Oxford University Press, Oxford, 2004).

Nanotechnology with soft materials

I. W. Hamley, Nanotechnology with soft materials. *Angew. Chem.*, **42**, 1692 (2003).

Liquid crystals

P. G. de Gennes and J. Prost, *The Physics of Liquid Crystals* (Oxford University Press, Oxford, 1993).
P. J. Collings and M. Hird, *Introduction to Liquid Crystals: Chemistry and Physics* (Taylor and Francis, London, 1997).

Colloids

D. J. Shaw, *Introduction to Colloid and Surface Chemistry* (Butterworth-Heinemann, Oxford, 1992).
R. J. Hunter, *Foundations of Colloid Science* (Oxford University Press, Oxford, 2001).

Polymers

J. M. G. Cowie, *Polymers: Chemistry and Physics of Modern Materials* (Blackie, London, 1991).
R. J. Young and P. A. Lovell, *Introduction to Polymers* (Chapman and Hall, London, 1991).
L. H. Sperling, *Introduction to Physical Polymer Science* (Wiley, New York, 1992).
D. I. Bower, *An Introduction to Polymer Physics* (Cambridge University Press, Cambridge, 2002).
M. Rubinstein and R. H. Colby, *Polymer Physics* (Oxford University Press, Oxford, 2003).

Block copolymers

I. W. Hamley, *The Physics of Block Copolymers* (Oxford University Press, Oxford, 1998).
I. W. Hamley (ed.), *Developments in Block Copolymer Science and Technology* (Wiley, Chichester, 2004).

Surfactants

D. F. Evans and H. Wennerström, *The Colloidal Domain: Where Physics, Chemistry, Biology and Technology Meet* (Wiley, New York, 1999).

B. Jönsson, B. Lindman, K. Holmberg and B. Kronberg, *Surfactants and Polymers in Aqueous Solution* (Wiley, Chichester, 1998).

8

Macromolecules at interfaces and structured organic films

8.1 MACROMOLECULES AT INTERFACES

The previous chapter introduced us to the importance of soft matter in nanotechnology, focusing on the importance of self-assembly as a parallel, scalable route to creating highly controlled and ordered structures on the nanoscale. In this chapter we pursue the topic of how the properties of organic materials, and in particular polymers, can be exploited to make functional nanoscale devices, focusing now on their use in ultrathin films and single-molecule layers.

This area has its roots in a number of classical areas of materials science and chemical engineering. The exploitation of macromolecules at interfaces has historically been a central concern of colloid science, and already forms the basis of major industries. Examples include:

(i) the use of interfacially grafted and adsorbed polymers to control colloid stability in industries such as food, pharmaceutical formulations and personal care products;

(ii) control of interactions between submicron polymer particles to optimise the process of film formation in water-based varnishes and paints;

(iii) submicron polymer coatings on plastic film to add functionality by controlling surface properties such as adhesion and printability;

(iv) control of the wetting properties of surfaces by treatment with self-assembled monolayers, such as in the surface treatment of glass fibres for use in composite materials.

As these technologies become refined and our understanding and control of them at the nanoscale becomes more precise, new areas exploiting the properties of

Nanoscale Science and Technology   Edited by R. W. Kelsall, I. W. Hamley and M. Geoghegan
© 2005 John Wiley & Sons, Ltd

macromolecules at interfaces are becoming important. Areas in which rapid progress is being made include:

- DNA microarrays, more colloquially known as gene chips, are already a big industry. These devices, using an array of different macromolecules tethered to a planar surface, permit the rapid screening of gene expression, and are likely to be the first of a variety of new highly parallel approaches to systems biology and chemical analysis.

- Other approaches to miniaturising techniques in chemistry and biochemistry, including microfluidics, and nanodevices for microreactors and for sorting and separating molecules and biomolecules.

- Plastic electronics, in which the self-assembling properties of polymers in thin films are used to make fully optimised electronic and optoelectronic devices by simple and cheap processing routes.

- The precise control of interactions between synthetic materials and human cells, for tissue engineering and other biomedical purposes.

In the future we anticipate that exploitation of the properties of self-assembly and responsiveness to the environment that are characteristic of macromolecules at interfaces will be used to make wholly new types of functional nanodevices. Some of these will be synthetic devices that mimic some of the operating principles of cell biology, exploiting the principles of conformational change and molecular recognition in nanomachines such as molecular motors and selective valves and pumps. Other areas of application will be found in truly molecular-scale electronic devices, in which much of the nanofabrication of the molecular components of circuits is achieved by self-assembly.

The study of the properties of macromolecules at interfaces represents a fascinating interaction between fundamental surface physics and applied chemistry. In asking why surfaces and interfaces are so important for macromolecular systems, it is worth making a fundamental comparison between the surface of a simple polymer, like polystyrene, and the surface of an elemental material such as silicon.

Silicon has an open diamond structure with each atom separated by $0.54\,\mathrm{nm}$ whereas polystyrene has a random coil structure, with a segment length of $0.67\,\mathrm{nm}$. However, when considering the effect of a surface it is perhaps better to consider the size of the polymer chain, which scales as the square root of the number of monomers in the chain. As an example, polystyrene of molecular weight $100\,000\,\mathrm{g\,mol^{-1}}$ has $\sim$960 monomers and a chain size of $\sim$8.5 nm. To a first approximation, the effect of a surface can be considered relevant over a distance 15 times greater in polystyrene than in silicon. In macroscopic situations this may not always be significant but on length scales of several nanometres, it is particularly important because virtually every chain in a polystyrene nanoparticle will have some contact with the surface. This will not be true of silicon particles of similar size. From this point of view, polymers at surfaces are considerably more influential than small-molecule systems.

The interaction of uncharged materials with each other at the nanoscale is largely controlled by long-range (van der Waals) forces. The forces are long-range because their extent of interaction covers distances of up to 100 nm. Macroscopically, we consider these van der Waals interactions to be surface forces. Nanoscopically, these surface forces are the size of the materials themselves. Therefore, whatever the material, surfaces are very important in nanotechnology because it is via these surface forces that materials interact with each other.

8.2 THE PRINCIPLES OF INTERFACE SCIENCE

8.2.1 Surface and interface energies

The presence of an interface between a material and the vacuum (i.e., its surface) represents a major perturbation, and the importance of this perturbation necessarily increases as the dimensions of a system approach the nanoscale. The size of the perturbation is measured by the *surface energy*. The surface energy – to be thermo-dynamically precise, we should talk about a *surface free energy*, which is equivalent to a *surface tension* – tells us how much the system energy increases if new surface is created from bulk material. The equivalence of surface free energy and surface tension is easily demonstrated if one thinks about the energy required to create a new surface from, say, a liquid trapped in a square wire frame of length l. The energy required to extend the frame in any one direction by a distance x will be $2lx\gamma$. This is work done against the surface tension, and so the surface free energy per unit area, γ, must be equal to the surface tension. The factor 2 is a simple statement that there are two surfaces in the liquid film. The magnitude of surface energies varies from around $30\,\mathrm{mJ\,m^{-2}}$ for hydrocarbon-based materials such as polymers, to $70\,\mathrm{mJ\,m^{-2}}$ for water, to nearly $0.5\,\mathrm{J\,m^{-2}}$ for mercury. Most materials have surface energies that are lower than water, which means they will form a coating on water. The ability to coat water is extremely useful in creating thin organic coatings by, for example, the *Langmuir–Blodgett* technique (Section 8.5.2.2).

The origin of the surface energy lies in the cohesive force that binds together condensed matter of any kind, either solid or liquid. A crude but useful way of thinking about surface energy is to imagine the atoms or molecules in a solid or liquid to be held together by bonds of energy ε. If each molecule occupies a volume v_0, and each molecule has bonds to, on average, z other molecules, then the total cohesive energy per unit volume of the solid or liquid is given by $L = z\varepsilon/2v_0$. To a first approximation we can identify this with the latent heat. If we now create some new surface, we have to 'cut' some bonds – at a surface a molecule will interact with fewer molecules than it does in the bulk. If each molecule at the surface on average interacts with z' other molecules, then the surface energy $\gamma = z'\varepsilon/2v_0^{2/3}$. From this we can show that the surface energy is directly proportional to the cohesive energy. Solids or liquids that have strong cohesive forces holding them together have large values of the surface energy.

An interface between two different solids, between two different liquids, or between a solid and a liquid also has an energy associated with it, assuming the pairs of materials are not mutually soluble. The values of these *interfacial energies* are generally up to an order of magnitude or so less than surface energies; to make an interface between two materials A and B one has to cut A–A bonds and B–B bonds, but at the interface some of this energy cost is recovered by the formation of new A–B bonds. If the energy gained by making new A–B bonds actually exceeds the energy cost of breaking A–A bonds and B–B bonds then the materials must be mutually soluble.

The cohesive forces binding atoms and molecules together can have different origins. The simplest situation to consider is that of two atoms or molecules that are alike. In this case the forces interacting between them are always attractive and are known as van der Waals interactions. These van der Waals interactions have an interaction energy proportional to r^{-6}. At very small r (typically less than 0.5 nm) electron clouds begin to overlap and the Pauli exclusion principle forces electrons into more energetic orbitals. This means there is a repulsive term in the interaction potential between atoms and molecules, which is commonly represented as a potential proportional to r^{-12} (though there is nothing fundamental about this functional form, and many other short-ranged functions have also been proposed): see also Section 1.4.2.5.

If one knows the form of the potential between any two *atoms*, then to find out how two larger *bodies* of material interact, an obvious (though not rigorously correct) approach is to sum the interactions between every pair of atoms in the two bodies. In practice this is done by integration. The simplest case is when we have a single atom or molecule interacting with a surface that is sufficiently thick and sufficiently extensive in area that we can consider the surface to be a *semi-infinite medium*. If the medium has density ρ and the atom is distance D away from its surface, then we consider the interaction between the atom and an annulus of the material of radius x and depth z. We then integrate these interactions throughout the depth of the medium and over all annuli:

$$W(D) = -2\pi C\rho \int_{D}^{\infty} dz \int_{0}^{\infty} \frac{x\,dx}{(x^2 + z^2)^{6/2}} = \frac{-\pi C\rho}{6D^3}. \tag{8.1}$$

It is easy to alter this equation for other forms of the attractive potential. If we replace the atom or molecule by a surface of the same material, the energy of interaction of the two surfaces is best given per unit area:

$$W(D) = \frac{-\pi C\rho}{6} \int_{D}^{\infty} \frac{\rho\,dz}{z^3} = \frac{-\pi C\rho^2}{12D^2}. \tag{8.2}$$

This equation represents the energy required to separate to infinity two surfaces that are initially a distance D apart. If the original distance can be thought of as the separation of atoms in a bulk material, we can identify this with the surface energy. In fact, the energy required to separate two surfaces must be 2γ per unit area. This can be equated to the attractive potential between two surfaces derived above:

$$2\gamma = \frac{-\pi C\rho^2}{12D^2}. \tag{8.3}$$

Usually this equation is rewritten

$$\gamma = \frac{A_H}{24\pi D^2},$$ (8.4)

where $A_H = -\pi^2 C \rho^2$ is known as the *Hamaker constant*. For two different materials the Hamaker constant may be rewritten $A_H = -\pi^2 C \rho_1 \rho_2$. This simple result is very important in considering the stability of films on surfaces. It should be noted that such van der Waals forces depend on the nature of any intervening medium, and also that the above calculation neglects any other form of interaction between the components.

It is clear that in any system consisting of a dispersion of nanosized droplets or particles, surface and interfacial energies provide a major driving force for change. The smaller the particle, the larger the proportional importance of the surface or interface energy becomes. There are two particularly important ways in which a system can reduce its overall interfacial energy: aggregation and adsorption.

In the phenomenon of aggregation, nanoscale particles reduce their area of effective interface by joining together. If the particles are liquid, two such particles will lose their identity on coming together and will *coalesce* to create a single, larger particle. Solid particles are unable to deform on contact and instead *flocculate* to produce more or less open networks of joined together particles.

For adsorption to take place, a third material must be present (usually in solution in a liquid phase). Suppose this material, which we call C, has the property that the sum of the energies of the A–C interface and the B–C interface is less than the energy of the A–B interface. Then if a thin layer of this material is interposed between the two materials, it can have the effect of reducing the interfacial energy. The most common way of achieving this is by using a molecule with an *amphiphilic* character. For example, if A is a hydrocarbon (perhaps submicron polymer particles of the kind found in emulsion paint) and B is water, then if the molecule C is of mixed character, with part of the molecule being hydrophobic and part hydrophilic, then it is likely to be effective at reducing the interfacial energy.

The phenomenon of adsorption can, according to circumstances, be beneficial or deleterious in its effects.

8.3 THE ANALYSIS OF WET INTERFACES

Some general methods of analysing surfaces and interfaces have been introduced in Chapter 2. The specific problems of analysing interfaces involving polymers are that they are often wet, in the sense that we are interested in a solid–liquid interface rather than a solid–vacuum interface. This means we need tools that, rather than being surface-specific, interrogate buried interfaces, and we need techniques that do not need to operate in ultra-high vacuum conditions. Some key techniques that fulfil some or all of these condition include:

- *Light reflection techniques*: ellipsometry and surface plasmon resonance can provide sensitive measures of the amounts of material that adsorb at a solid–liquid interface.

- *Neutron reflectometry*: the smaller wavelength of neutrons compared to light allows one to probe the structure of an adsorbed or grafted layer in more detail, with subnanometre resolution. It is necessary to label the species that is being probed with deuterium. This is discussed in more detail in Section 2.8.4

- *Atomic force microscopy*: see Section 2.5.2.

8.4 MODIFYING INTERFACES

Interfaces have an associated energy, and because the surface-to-volume ratio of a particle is much greater when the particle is small, such small particles are always highly unstable unless their interfaces have been modified to prevent aggregation. Classically, this is the province of *colloid science*. When nanoscale particles are dispersed in water, steps are required to prevent the particles aggregating. Such *colloidal stabilisation* is usually achieved by modifying the surfaces of the particles to create an energy barrier which prevents the particles from sticking to each other. This can be done by arranging for the surfaces of the particles to be charged, or by coating them with a molecular layer of material. This molecular layer, often made up of polymers, may be physically or chemically attached to the surface.

Physical attachment has the advantage that it is simple and often requires nothing more complex in processing terms than exposing the surface to be modified to a solution of the modifying agent. Its disadvantages are that the molecules are not firmly attached to the surface, and with a change of conditions it is possible for them to leave the surface once again. Chemical attachment avoids this difficulty, and typically such layers are considerably more robust than physically attached layers, but of course the consequences of this are the more complex processes required to attach the molecules.

8.4.1 Adsorption and surfactancy

It is a general principle that if one has a two-component mixture near a surface, the composition of the mixture near that surface will be altered in a way which lowers the energy of the surface. This is the principle on which surfactants work. In a solution of surfactants in water, surfactant molecules migrate spontaneously to the surface of the water and in the process lower its surface tension, often by a very significant amount. There is a fundamental thermodynamic relation, the *Gibbs adsorption isotherm*, between the excess amount of surfactant at the surface, the degree to which the surface energy is lowered, and the thermodynamics of the surfactant–water interaction.

The most familiar class of surfactants are *amphiphilic* molecules – molecules with two separate parts, one hydrophobic and one hydrophilic. As discussed in the previous chapter, these materials are the basis of soaps and detergents. In addition to lowering the energy of a water surface, they will usually lower the interfacial energy between water and various solids as well. This is the basis of detergency, in which fat or dirt particles are coated by surfactants to be lifted off a substrate.

8.4.2 Polymer adsorption

Polymers are more effective at adsorbing at interfaces than chemically similar small molecules, and having adsorbed at such interfaces they are more effective than analogous small molecules at keeping interfaces apart. This makes them extremely valuable for colloidal stabilisation, and they are used to stabilise nanoscale dispersions of liquid and solid particles in a wide variety of existing industrial and domestic processes.

Why are polymers so effective at adsorbing to surfaces? The simplest way of thinking about this is to realise that adsorption describes an equilibrium between molecules attached to the surface and molecules free to move around in the solution. An adsorbing species will lower the energy of the surface by becoming attached there, but the cost of this is a loss of entropy that arises because the molecule is localised there. Compare a small molecule and a chemically identical polymer consisting of N subunits. If we move the small molecule and the polymer from the solution to the surface, we lose the same amount of entropy but we gain a factor of N more energy from the polymer, simply because it has N subunits, each of which lowers the energy of the surface as much as the small molecule.

The shape that a polymer adopts at a surface is conventionally thought of as consisting of three different elements: *trains* of consecutive polymer segments all stuck to the surface, *loops* of polymer segments leaving the surface but anchored to it by trains at each end, and *tails* of polymer segments attached at one end only. Statistical mechanical theories predict the relative amount of loops, trains and tails, as well as their distributions of sizes. We should picture a surface coated with adsorbed polymer as presenting a rather fluffy appearance, with a polymer coating which has rather low density but which protrudes a distance of up to tens of nanometres into the surrounding fluid (Figure 8.1).

The interaction of two nanoparticles coated with such a fluffy layer is rather complicated, though generally it is repulsive. The biggest effect is osmotic pressure; as we bring two polymer-coated surfaces together, the polymer segment concentration between the surfaces rises. This leads to an osmotic force pushing the surfaces apart. However, it is possible, if the density of polymers coating the surfaces is relatively low, for polymers attached to one surface to find free areas on the other surface to stick to as well, leading to the formation of polymer bridges between the two surfaces. This can lead to a net attraction between the two surfaces.

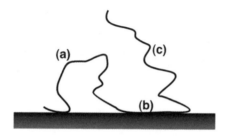

Figure 8.1 An adsorbed polymer may consist of (a) loops, (b) trains and (c) tails

8.4.3 The chemistry of grafting

Adsorption processes, whether of small molecule surfactants or of polymers, are effective ways of modifying a solid surface or a solid–liquid interface, but because the molecules are only physically attached to the surface the process can often be reversed, and with a change of conditions the adsorbed molecules may well become detached. More permanent molecular layers can be created by covalently bonding short to medium-sized organic molecules by one end to the surface. In some cases, such permanently bound layers can be formed with a very complete degree of surface coverage and a high degree of order within the layer; these systems are known as *self-assembled monolayers* (SAMs). Two classes of SAM-forming systems have been particularly important: alkylthiol molecules bound to a gold surface, and alkoxysilane or chlorosilane molecules bound to a surface of silicon oxide.

Self-assembled monolayers have a variety of uses. Broadly speaking, they form a convenient way to modify the wettability of surfaces, they can control the adhesion between a surface and a bulk phase in contact with it, they can control the interaction of a surface with cells (Chapter 9), and they can be used to introduce controlled chemical functionality at a surface. In the latter case, further reactions can be carried out exploiting that functionality; in particular, this route can be used to immobilise macromolecules at the interface. A very important example of this is in DNA chips. A DNA chip, more properly a DNA microarray, consists of an array of different oligonucleotides immobilised on a flat surface; they allow the rapid, parallel detection of specific nucleotide sequences, and as such have revolutionised the study of gene expression and promise major advances in genetic diagnosis in medicine. Although DNA microarrays have so far made the biggest impact, other types of microarrays, involving small molecules, proteins or polysaccharides, are likely to grow in importance in the future, as the power of this type of massively parallel experimentation becomes apparent.

The lateral patterning methods collectively described as *soft lithography* also rely on the facile formation of self-assembled monolayers, and their subsequent modification. Most usually, these methods exploit the reaction between gold surfaces and alkylthiols.

8.4.3.1 Gold/thiol chemistry

Perhaps the most reliable way of making a high-quality self-assembled monolayer uses alkanethiols attached to a gold surface. The required alkanethiol, RSH, is simply applied from solution to the gold surface; it is believed that it binds as a thiolate, RS^-, to a well-defined site in the gold (111) surface. As a result of a fortuitous commensurability between the separation of the binding sites and the effective cross-section of the alkyl chain, the chains form a densely packed array (Figure 8.2).

The dense packing of the chains in an alkylthiol monolayer means that the gold surface is effectively completely screened from the environment. This means that the chemical character of the surface can be easily changed; by using alkylthiols with different terminal functional groups, a wide variety of chemical functionalities can be introduced at the surface. Thus surface properties such as contact angle can be changed over a very wide range of values. The ease with which functional groups can be

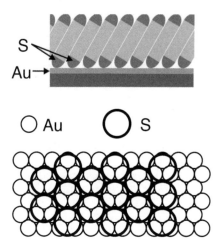

Figure 8.2   The attachment of alkanethiols to gold. Adapted from T. P. Sullivan and W. T. S. Huck, Reactions on monolayers: organic synthesis in two dimensions, *Eur. Journal Org. Chem.* **1** 17–29 (2003)

introduced at the surface also allows one to carry out further chemical reactions, as discussed below. The only disadvantages of gold/thiol self-assembled monolayers are their somewhat limited thermal stability and the need for a gold surface.

8.4.3.2 Silane coupling agents

Possibly the most well-established technique for creating a single molecular layer relies on the use of so-called silane coupling agents to create covalent bonds between the surface of glass, silica or the native oxide of silicon and small organic molecules.

There are two basic starting points. The first uses chlorodimethyl long-chain alkylsilanes or alkyltrichlorosilanes, which readily react with silanol groups (Si—OH) on the surface of silicon, silicon dioxide or glass to yield a robust covalent linkage between that surface and the hydrocarbon group. The second uses trialkoxy(alkyl)silanes, which are somewhat less reactive but which produce the same end result. The use of trialkoxy or trichloro compounds tends to produce more robust monolayers than the corresponding mono compounds, because having made one bond to the surface, each molecule has available two more reactive groups with which it can form bonds to its neighbours, resulting in a tough two-dimensional network. The downside is that in some reaction conditions alkyltrichlorosilanes in particular can begin to react with each other in solution, resulting in large aggregates which subsequently become attached to the surface. The resulting surface layers are much thicker and rougher than true monolayers.

Perhaps the most well-known of the silane coupling agents is aminopropyltriethoxysilane (AMPTES), which is a very widely used route to putting amine groups on a surface. This is the first step in many procedures for attaching proteins, DNA or other molecules to a glass surface.

The alkyl group in AMPTES is probably too short to pack in a well-ordered self-assembled monolayer. Monolayers with a degree of two-dimensional order comparable to that obtained with gold/thiol SAMS are achievable with molecules such as octadecyl-tricholorosilane (OTS). OTS provides a very effective way of rendering a glass, silica or silicon surface strongly hydrophobic.

In this context it is worth mentioning one other approach to making silicon surfaces hydrophobic. In a process widely used in the semiconductor industry for 'priming' silicon wafers prior to coating with photoresist, hexamethyldisilazane (HMDS, $Me_3SiNHSiMe_3$), is applied from vapour. The reaction

$$2Si-O-H + (CH_3)_3Si-NH-Si(CH_3)_3 \rightarrow 2Si-O-Si(CH_3)_3 + NH_3,$$

is different to the reaction that produces the coating of silicon dioxide by trialkoxyalkyl-silanes or trichloroalkylsilanes, but the end result is rather similar.

8.4.3.3 Macromolecular layers

In many cases the creation of a self-assembled layer at a surface is simply the first step towards introducing a more sophisticated functionality at that surface. In many applications one needs to graft a layer of macromolecules at that surface. Such a grafted layer is often referred to in polymer and colloid science as a *polymer brush*. This is a process that has been extensively used in colloid science to stabilise particles against aggregation. An application that can possibly be considered as a special case of colloid stabilisation is the use of grafted poly(ethylene oxide) to protect a surface against non-specific protein adsorption. An application that depends on very specific interactions between grafted macromolecules and molecules from the environment is exemplified by DNA micro-arrays, which rely on the covalent attachment of strands of DNA to surfaces.

We can distinguish between two broad approaches to attaching macromolecules to surfaces. In the grafting-to approach, preformed macromolecules are attached to the surface either chemically or physically. For physical grafting, one needs a polymer with a sticky end-group; for example, in water a block copolymer with a long water-soluble block, such as poly(ethylene oxide), with a shorter hydrophobic block on the end will form a grafted layer on a hydrophobic surface, with the end-block strongly and selectively adsorbing to the surface. Covalent attachment, which has obvious advantages of stability over physical attachment, is achieved by reacting a functional group on the end of the polymer chain to a complementary group on the surface. Surfaces with amine functionality have been extensively exploited in this way.

In the grafting-from approach, on the other hand, polymerisation of a macromolecule is initiated at the surface, growing the grafted chain in situ. Both approaches have advantages and disadvantages. Grafting-to generally allows one to use a much wider variety of synthesis routes, so the range of polymer architectures that can be used is generally much wider. But grafting-from usually produces layers with a considerably greater grafting density. In the original developments of the grafting-from technique, radical polymerisation was carried out from surface-bound initiators, resulting in dense and thick layers of highly polydisperse polymers. More recently, however, controlled radical polymerisations such as atom transfer radical polymerisation have been developed which allow the potential for much greater control of the architecture of the surface-grafted polymers.

8.4.4 Physical properties of grafted polymer layers

The questions that arise about the structure of self-assembled monolayers generally concern the perfection of their two-dimensional packing. In the ideal case, an alkanethiol or alkyltrichlorosilane with an alkyl chain of length between 12 and 20 carbon atoms will form a highly ordered, densely packed two-dimensional layer. In contrast, the disposition of long, flexible, macromolecules grafted to interfaces is determined not by considerations of packing and order, but by entropy. The freedom of grafted polymer molecules to adopt many different conformations, and the fact that in an aqueous medium, a solvent or a melt, chain conformations are constantly changing under the influence of Brownian motion is crucial in determining the properties of the grafted layer. In particular, the remarkable efficiency with which end-grafted chains can stabilise colloids against aggregation and prevent non-specific protein adsorption to surfaces is bound up with the condition that the configurational entropy of the grafted chains must be maximised.

More exotic applications of polymer brushes may rely on reversible behaviour; this is the ability of the polymer brush to change its conformation in response to changes in its environment. Depending on the pH, salt concentration, or temperature of the surrounding medium, a brush layer may swell or collapse. Such switchable behaviour is very useful if, for example, one wishes to control wettability or adhesion by changing these parameters. One brush layer may achieve this goal but sometimes two brush layers (a mixed or binary brush) can also be used. Here two different polymers are attached to the substrate, and their behaviour is complementary; i.e., when one brush polymer collapses, the other swells. This has been demonstrated experimentally by exposing a binary brush layer to different solvent vapours.

To understand the properties of grafted polymer layers, polymer brushes, it is helpful to compare one brush polymer with an untethered chain in a good solvent. The simplest model of a free polymer is the freely jointed chain; i.e., each monomer can move around at any angle compared to the next. This is physically not true, because bond angles are fixed. However, over a certain number of monomers, information about the bond angle is lost; the corresponding distance is known as a *persistence length*, and as a result we can define an effective statistical step length, in terms of which the long-range properties of a real polymer with short-range interactions can be mapped onto an equivalent freely jointed chain. The (end-to-end) length of the chain in a neutral, or *theta*, solvent is given by the random walk result $\langle R^2 \rangle^{1/2} = Na^{1/2}$, where a is the step size. This is exactly the same result as the mean free path traversed before an electron undergoes the spin-flip scattering of Equation (4.30). The random walk enables the polymer chain to maximise its *conformational entropy*. However, the freely jointed chain model, even with an effective step length, does not account for long-range interactions. For a solvent whose interaction with a polymer segment is indistinguishable from the interaction between two polymer segments, an *athermal solvent*, the effect of these long-range interactions is to expand the chain. The fact that two polymer segments cannot physically be in the same place at the same time leads to an effective repulsive interaction, known as the *excluded volume* interaction.

Most solvents are not neutral, and in addition to the excluded volume interaction we need to consider the effect of interactions between solvent molecules and chain segments. In a poor solvent, polymer segment–segment interactions are more favourable than polymer–solvent interactions, and the chain tends to collapse; conversely, in a good solvent the chain will tend to expand to maximise segment–solvent contacts.

The effect of both excluded volume and polymer–segment interactions can be expressed by monomer repulsion energy with the following form:

$$E_{\text{repulsive}} \approx k_B T v(T) c^2 R^3 = k_B T v(T) N^2 / R^3 \tag{8.5}$$

where T is the absolute temperature, $v(T)$ is a temperature-dependent constant and c is a concentration. The concentration can be replaced by $1/R^3$, which simplifies the equation. Note that we ignore numerical prefactors because they contain no important physical information; this is known as a *scaling theory*. A chain cannot extend indefinitely, and so there must be an attractive energy countering this repulsion. This attractive energy may be obtained by considering the chain to be a spring obeying Hooke's law. This elastic term is then given by

$$E_{\text{elastic}} \approx k_B T \frac{R^2}{Na^2}, \tag{8.6}$$

where Na^2 is the chain's unstretched (random walk) length. The length of a polymer chain in a good solvent is thus a compromise between these two results, which may be added and minimised with respect to R. The final result gives the *Flory radius*, $R_F \propto N^{3/5}$. The constant v is known as *excluded volume*. This result is valid for dilute and semi-dilute polymer solutions. In molten polymers (i.e., without solvent), the chain has a random walk conformation ($R \propto N^{1/2}$) because there a monomer cannot differentiate between monomers of its own chain and monomers of other chains.

The conformation of polymer brushes is dependent upon two parameters: the number of monomers in the chain, N, and the distance between polymer chains, d. The distance between brushes can be replaced by a *grafting density*, $\sigma = a^2/d^2$, where a is the polymer segment size. On length scales smaller than d, chains may be treated as if they were in a good solvent. We then suggest that the brush can be made up of n blobs, each containing $N_b = N/n$ monomers (Figure 8.3). The better the solvent, the more extended the brush but this does not affect the arguments that we shall describe below. In poor solvents, brushes tend to collapse to exclude solvent. The worse the solvent, the more collapsed the brush until the equivalent of a dry polymer brush is obtained. In a poor solvent, the height of a brush is proportional to the number of monomers in the brush and some function of grafting density, which depends on solvent quality.

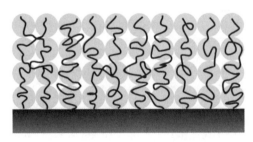

Figure 8.3 Brushes can be thought of as several blobs, where the number of monomers N_b in each of the n blobs is related to the number of monomers N in the polymer brush by $N = N_b n$; $n = 4$ in this sketch

Figure 8.4 In dilute solution the chains are not close enough to interact with each other. They can then adopt a 'mushroom' conformation

We shall only consider brushes in good solvents. We divide our consideration of polymer brushes into three sections: semi-dilute solutions, concentrated solutions and molten brushes. The case of the isolated brush (dilute solutions) is less interesting because the brush may be treated as a single chain in a good solvent. For this reason, such a brush is commonly known as a mushroom (Figure 8.4).

8.4.4.1 Semi-dilute solutions

In a semi-dilute solution we can use a scaling model to understand the physical properties of polymer chains. We know that $\sigma = a^2/d^2$ and $d = aN_b^{3/5}$, and can use these equations to calculate the height of the brush, which must be given by $h = Nd/N_b$. From substitution we obtain

$$h = Na\sigma^{1/3}. \tag{8.7}$$

In semi-dilute solutions, the brush height is directly proportional to the number of monomers in the brush.

8.4.4.2 Concentrated solutions

A concentrated solution may be considered to be one in which the brush chains overlap and strongly interact. This strong interaction occurs when $\sigma > a^2/\langle R^2 \rangle$, which means that the projected area of a tethered chain on the substrate is smaller than the cross-sectional area of an untethered chain in the solvent. We need to include this interaction in a model for the brush height. We achieve this by considering excluded volume, which is a means of describing the affinity of a chain for a solvent rather than itself or other chains of the same species. In this respect, we apply exactly the same argument for a concentrated brush layer as for a single chain in a good solvent:

$$E_{\text{conc}} \approx k_{\text{B}}T\left(\frac{h^2}{Na^2} + \frac{v(T)N^2\sigma}{ha^2}\right). \tag{8.8}$$

In this case we have replaced the $1/R^3$ term in the repulsive component of Equation (8.5) by σ/ha^2, which may be understood if one writes $\sigma/ha^2 = 1/hd^2$, which may be taken as a concentration; i.e., $1/R^3$. If we minimise the above equation with respect to h and equate the result to zero, we obtain $h \propto N(v(T)\sigma)^{1/3}$, which is the same grafting density and chain length dependence as in the semi-dilute regime. This does not mean that the physics is the same: in this concentrated regime the origin of the extension of the brush is the osmotic pressure difference between brush and solvent.

8.4.4.3 Molten brushes

A molten brush in this case is one where the solvent is replaced by a melt of the same polymer. Unfortunately, it is not one that can be considered by using simple arguments. We have mentioned that a polymer in the molten state has a random walk configuration, which would imply that $h \propto aN^{1/2}$ with no dependence on grafting density. This cannot be true, because it is not possible to indefinitely graft chains into ever smaller areas. Another objection is that by the nature of the random walk result, the brush layer would be very well defined. This cannot be true, because such a well-defined layer would have a very large entropic penalty in the free energy of the system (sharp interfaces have a large energy penalty). The interface between the brush and molten polymer must be diffuse, and it is calculated by using self-consistent mean field theory. This is a much more complex calculation and involves starting with a suitable test chain that interacts with other chains via a spatially varying free energy (or chemical potential). An iterative process of guessing the energy associated with a given test chain as a function of the total mean field interaction takes place before the best (self-consistent) result is achieved. Readers familiar with atomic physics will recognise these ideas from the Hartree–Fock method for determining the electronic orbitals for many-electron atoms.

8.4.5 Nanostructured organic coatings by soft lithography and other techniques

The utility of the above methods of coating surfaces is greatly increased if we have the power to apply the chemistry only to specific regions of the surface. To achieve this we require certain lithographic or patterning techniques. We shall address the ideas behind creating patterns on surfaces using self-assembly in Section 8.7 and here we shall consider lithographic methods. The standard lithographic method involves the use of self-assembled monolayers and a poly(dimethyl siloxane) (PDMS) mask. We can assume that the PDMS mask already contains the information on the required structure. This mask is coated with a self-assembled monolayer (SAM), which is imprinted on the surface of interest. The SAM can then be used as a template for further chemistry. This form of lithography (microcontact printing) is discussed in Section 1.4.1.2. Another important form of lithography is photolithography, also discussed in Chapter 1. Here UV radiation illuminates a mask. The mask is a glass slide covered with a chrome pattern. The UV radiation passes through the unpatterned part of the glass onto a photoresist. A negative photoresist is produced when the UV photopolymerises or cross-links the exposed

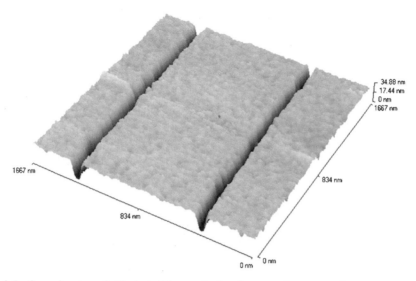

Figure 8.5 Scanning near-field photolithography has been used to create these nanometre-scale trenches in a gold layer. UV light has been passed through a SNOM tip to oxidise an alkanethiolate self-assembled monolayer. The exposed region and the underlying gold were removed by a wet chemical etch. This image was taken using atomic force microscopy. Image reproduced courtesy of Dr Shuqing Sun and Professor Graham Leggett

region. The remainder of the (unexposed) film can be dissolved away. A positive resist is created when the UV radiation destroys the exposed region.

One limitation of these techniques of lithography is the length scale of the pattern that may be achieved. The mask for microcontact printing contains patterns of several microns in size (hence microcontact); smaller sizes are difficult to achieve in PDMS (PDMS is an excellent material because it is deformable and easily cross-linked from a mould to give it its elastic qualities). Photolithography is also generally limited to micron-scale structures because of the wavelength of light; the technique is *diffraction limited*. However, as mentioned in Chapter 2, the diffraction limitation can be circumvented by using a SNOM. In the case of photolithography the UV light is passed through the SNOM tip, enabling the creation of structures of some 50 nm or even smaller; an example is shown in Figure 8.5. Again, the structure created can be used as a template for the chemistry described above.

Another interesting use of the scanning probe techniques in this area is the dip pen nanolithography method in which an AFM tip can be coated with a material, and then moved around a surface depositing this material (Section 7.4). A great advantage of this technique compared to many other forms of lithography is that an array of AFM tips may be used to produce the same pattern simultaneously.

8.5 MAKING THIN ORGANIC FILMS

One of the easiest ways to make a nanoscale structure involves making thin polymer coatings. Coatings of submicron thickness are routine even on the cheapest commodity polymers, and the process of spin-coating, an important part of fabricating semiconductor nanostructures, can be effectively carried out using equipment costing only a few dollars. Making *structured* films on nanometre length scales is a little more difficult, but nonetheless a number of fairly straightforward methods exist for making *layered* films; that is, films with structure at nanometre length scales in the direction perpendicular to the film surface. It is more difficult to pattern films laterally and we discuss this in the next section.

What all these methods have in common, and what sets them apart from the methods used to make magnetic multilayers and semiconductor heterostructures, is that they are processes that are carried out in *ambient conditions*, rather than in high or ultra-high vacuum. This means that the capital cost of the plant needed to make these structures is usually very low, and one can countenance high-volume, low-cost, applications.

8.5.1 Spin-coating of polymers and colloids

An extremely good way of making a solid film is to start with a liquid film. Liquids have the remarkable property of being *self-levelling*; the action of surface tension keeps a liquid surface smooth and flat. If the liquid is a polymer solution, then if we can remove the solvent in a way that maintains the flatness of the original liquid film, we end up with a polymer film whose thickness can be closely controlled and which in favourable circumstances is smooth down to the nanometre scale.

A relatively thick layer of a polymer solution can be made by simply coating a surface with the liquid, then scraping off the excess with a blade held at a controlled distance from the substrate. This process, known for historical reasons as *doctor-blading*, can produce smooth and uniform films with thicknesses in the range of tens of microns. But to obtain uniform films with thicknesses less than one micron, the technique of choice is *spin-coating*.

In spin-coating, a sample is flooded with a polymer solution and then rapidly spun round (a spin speed of 2000 revolutions per minute is typical). The hydrodynamics of the situation leads to a rather uniform thickness of solution being left on the substrate, the thickness of which depends on the viscosity of the polymer solution and the spin speed. This film then thins as the solvent evaporates.

The ideal outcome of a spin-coating process is a film that is uniform in thickness to within a few percent over an area of many square centimetres, with a root mean square surface roughness of around 1 nm, and a thickness that can be precisely controlled within the range 10–1000 nm. To achieve this, one needs a solvent with the following properties. It should be not too volatile; if significant evaporation occurs during the initial film thinning phase then hydrodynamic instabilities can lead to non-uniformities in the final film. It should remain a good solvent for the polymer in question at all concentrations, to avoid problems that arise from premature precipitation of the polymer before the spinning process is complete. Finally, the solution should wet the substrate well, to avoid film break-up during spinning.

The use of this kind of solution-based coating process is not restricted to simple solutions of polymers; it can also be used to apply materials that are in an essentially colloidal form, or materials that are soluble precursors which react to form an insoluble final product. Photonic crystals are an example of colloidal system (Section 7.4.3).

The most common polymer-based colloidal suspensions are the water-borne latexes that are the basis of emulsion paint. Such colloidal suspensions when applied in a thin film will in some circumstances form a uniform thin coating. This process has a number of stages; as the water evaporates, the colloidal particles pack together in a dense array (if the colloidal particles are spheres with a narrow size distribution, the most efficient packing will be a colloidal crystal). If their physical state permits it, the particles will then deform, and diffusion of polymer molecules will take place at their interface, until the integrity of the individual particles is completely effaced and a smooth film is obtained. In industrial systems the last stages are often helped by the inclusion of a volatile plasticiser which helps the polymers in the particles to flow and interdiffuse. Some chemical cross-linking will generally take place in order to increase the robustness of the final films.

Polymer latexes are generally in the size range 100 nm to 1 µm, and as such cannot be used to make nanoscale polymer films. Smaller colloidal particles include so-called *microgels*, in which one or a few chains are internally cross-linked to make polymer particles on the length scale of 10 nm or less. A polymer colloid of growing commercial and technological importance is formed when the conducting polymer poly(3,4-ethylene dioxythiophene) PEDOT is synthesised in the presence of polystyrene sulfonate (PSS). It seems that this material forms, in aqueous solution, a colloidal dispersion with some similarity to a microgel, in which small clusters of PEDOT and PSS chains are linked by physical cross-links. This material is important because it provides a practical route for spin-coating a conducting layer, opening up the possibility of low-cost all-polymer optoelectronic devices (Chapter 6).

Inorganic thin films can also be made by spin-coating, using the sol–gel chemistry described in Section 1.4.2.6. These systems, sometime known as *spin-on* glasses, are based on solutions of silicon and metal alkoxides which slowly hydrolyse to form first polymers, and then networks of inorganic oxides. The resulting films are composed of a swollen gel, which can be baked to form a dense, sintered inorganic glass.

8.5.2 Making organic multilayers

Nanostructured polymer multilayers are not quite as easy to make as single thin films, but a number of effective approaches exist which exploit the idea of self-assembly in more or less pure forms to make multilayers using relatively simple apparatus.

In some circumstances, spin-coating can be adapted to making multilayer coatings by simply successively spin-coating multiple layers on the same substrate. The requirement for this is that, having laid down one polymer, the solvent used to deposit the next polymer is not a solvent for the first one. Even in cases where the solvent for the second polymer is a non-solvent for the first polymer, so that the first polymer is not physically removed, one should still expect some intermixing of the polymers at the interface due to the solvent-induced swelling of the first polymer near its surface. In practice the range of polymers that can be made into multilayers by this route is relatively restricted.

More powerful routes to making multilayers rely on the phenomenon of *adsorption*. A polymer molecule adsorbed from a dilute polymer solution at an interface generally forms a rather loose layer about one molecule thick; if the interface is removed from the solution, this adsorbed layer will stay on the substrate and collapse to form a thin, dense layer whose precise thickness is controlled by the nature of the polymer–substrate interaction and the thermodynamics of the polymer solution. If the final substrate is dipped sequentially into a series of different polymer solutions, we have the technique of *layer-by-layer deposition*. If, on the other hand, the monomolecular layer is created at an air–water interface, and such layers are picked up sequentially onto a solid substrate, we have the classical *Langmuir–Blodgett* (LB) technique.

8.5.2.1 Layer-by-layer deposition

In layer-by-layer adsorption, one needs to ensure that if polymer A is deposited first on the substrate, the next polymer to be deposited will adsorb at the polymer A – water interface. The original way of ensuring that this is so relies on making multilayers in which each layer is an alternately charged polyelectrolyte. So, starting with a substrate that has an overall positive charge in water, one would dip this into a solution of a polymer with a net negative charge. On removing the substrate, we would have a monomolecular layer with a negative charge, which can then be dipped into a solution of a positively charged polymer. Attracted by the opposite charge, this polymer layer will strongly adsorb on top of the previously deposited layer. This cycle of deposition of alternately charged polymer layers can be continued to build up a multilayer (Figure 8.6).

Using a typical pair of polyelectrolytes such as sodium poly(styrene sulfonate) and poly(allylamine hydrochloride), it is possible by this method to build up multilayers consisting of as many as 60 pairs of layers, with each pair of layers having a thickness of a few nanometres. This layer thickness can be controlled by variables such as the concentration of polyelectrolytes and salts in the adsorbing solutions. X-ray and

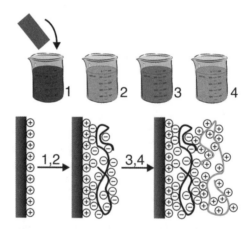

Figure 8.6 Layer-by-layer deposition of charged polymer films. In the first instance, a positively charged substrate is dipped into a polyanion solution (1) and then washed (2). Only one negatively charged layer is deposited on the substrate because of the Coulombic repulsion between like charges. This process can then be repeated with a polycation solution (3) followed by further washing (4). The process can be repeated until numerous layers have been deposited. Counterions are not shown in the figure. Adapted with permission from G. Decher, *Science* **277** 1232–1237 (1997). Copyright 1997 AAAS

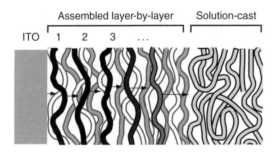

Figure 8.7 The layer-by-layer deposition method has been demonstrated in a light-emitting diode. Layer-by-layer deposition is used to deposit the anionic complex of PEDOT (darkest chains) and PSS (light chains) on an indium tin oxide (ITO) electrode. A cationic poly (*p*-xylylene-α-tetrahydrothiophenium) layer (grey chains) is then deposited. The doping density of the PEDOT is reduced in each subsequent layer (the chains are shown less dark). A light-emitting polymer layer can then be spin cast onto the graded interface. Reproduced with permission from P. K. H. Ho, J.-S. Kim, J. H. Burroughes, H. Becker, S. F. Y. Li, T. M. Brown, F. Cacialli and R. H. Friend, *Nature* **404**, 481–484 (2000). Copyright 2000 Nature Publishing Group

neutron reflectivity studies of the structure of the multilayers reveals that the interfaces between each layer are rather broad; in fact the interface width is comparable to the overall layer thickness. These films should be thought of as *fuzzy multilayers*.

A strength of the layer-by-layer approach is that it can be extended to incorporate many other species besides simple polyelectrolytes. Examples of such species include colloidal particles such as gold colloids and clay platelets, biological macromolecules such as proteins and DNA, dendrimers and C_{60}, and nanoparticles such as cadmium telluride.

The layer-by-layer approach has been demonstrated in practical devices. We recall the blend of PEDOT and PSS, which has applications as a hole injection layer in semiconducting polymer devices. The interface between the active layer and the PEDOT/PSS injection layer is sharp (because the layers are deposited separately) and the associated energy is likely to hinder carrier motion. By depositing PEDOT/PSS complexes in a layer-by-layer approach, a graded interface can be created (Figure 8.7). This gradient is achieved by chemically reducing the doping density in the PEDOT/PSS complex in each subsequent layer. The reduction of the doping density causes an increase of the ionisation potential of the PEDOT and should be modified to ensure a match to the active layer (in this case a light-emitting diode), which is deposited onto the graded interface. Of course, the doping may be varied to allow different active layers to be used. We return to the theme of hole injection layers from PEDOT/PSS blends in Section 8.8.1.

8.5.2.2 Langmuir–Blodgett techniques

The Langmuir–Blodgett technique is another means of layer-by-layer adsorption, but it does not rely on electrostatic attraction to provide a stable, well-defined layer. Another practical difference between the LB technique and the layer-by-layer adsorption method is that multilayers produced by the LB technique are not fuzzy but rather have an interface which can be on the subnanometre length scale. The deposited layers must fulfil certain requirements, which will be discussed below; the technique is not general, but neither is it limited to polymers, and small-molecule multilayers may equally well be deposited.

The first layer is deposited by simply spreading a few drops of solution onto a liquid surface (usually water because of its low vapour pressure and high surface tension). The solvent evaporates, leaving a monolayer of the solute on the fluid surface. Here it is best to use a solvent with a high vapour pressure. The resultant surface is usually not completely covered, and barriers are used to move the molecules together to form a monolayer. These barriers are used to reduce the total surface area of liquid until it is equal to that of the coating layer (Figure 8.8). The surface pressure of the monolayer can be measured and this gives a measure of the interaction of the molecules on the surface. As such, they can be treated as if they are the different phases of matter as the barriers push the molecules together, beginning with a gas of non-interacting particles, through forming a liquid-like layer of interacting molecules, before, at high pressures, the strength of interaction is great enough that the molecules form a 'solid' ordered surface phase. This solid phase can have highly crystalline properties over very large length scales, depending on the material deposited. Fatty acids are a class of material that in general provide very high quality monolayers with long-range order.

The trick to forming multilayered structures via LB deposition is to use a molecule with ends that respond differently to the solvent. For example, a monolayer can be achieved by pushing the substrate into the LB film, and allowing it to adsorb (Figure 8.9(a)). In this case the adsorbed side of the film will be hydrophobic because it is the hydrophobic component that is at the air surface before introducing the substrate. We denote the hydrophobic end by A and the hydrophilic end by B. Bringing the substrate up again will allow the other end (B) to be deposited first on the substrate. In this way multilayers can be deposited. Whether the structure is (substrate upwards) AB–BA–AB as in repeating this procedure, or otherwise, depends on the deposition method used. For example, continually using downward movement of the substrate for the deposition would result in a multilayer with a structure AB–AB–AB, and so on; permutations are illustrated in Figure 8.9(b).

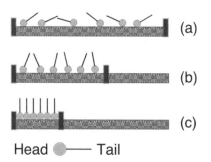

Figure 8.8   During Langmuir–Blodgett film deposition the coating layer orients itself such that the hydrophilic head group of the coating molecule is in contact with the water surface. The tail is hydrophobic. After solvent evaporation the molecules are dispersed on the surface; this is the gas phase (a). By moving the barriers, the molecules interact to form a liquid-like layer (b). Eventually the molecules cannot be moved any closer, and a solid, often highly crystalline layer emerges (c). Reproduced with permission from L. T. Jones, PhD thesis, University of Sheffield, 2003

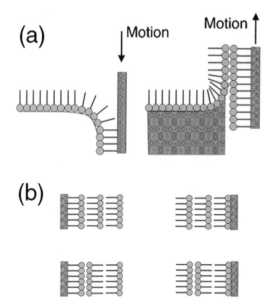

Figure 8.9 LB deposition can be used to create multilayers. (a) The nature of the multilayer can often be varied by controlling whether the substrate is pulled from or dipped onto the monolayer film. (b) Using this method, four different forms of multilayer can be created with head groups opposing each other or in contact with each other. Reproduced with permission from L. T. Jones, PhD thesis, University of Sheffield, 2003

Although LB film deposition is a cumbersome technique, it provides excellent control over film thickness and over macroscopic length scales. Although the technique is perhaps best known for producing multilayers of controlled properties (e.g., thickness, refractive index and polarisation), it is also very often useful simply to have a discrete number of layers in a uniform film, which is not possible using more rapid means of film creation such as spin-casting.

8.6 SURFACE EFFECTS ON PHASE SEPARATION

Here we shall consider how the surface can be used to trigger specific patterns in polymers. This patterning may occur laterally or perpendicularly, with both possibilities being useful in their own right. We first consider order forming perpendicular to the surface, giving rise to lamellar structures.

8.6.1 Polymer blends

Many materials are used in laminate form, which gives them a rigidity and strength that might not be very easily achieved otherwise. A difficulty with creating such structures is that the preparation of multilayers is usually a multi-step process involving different stages such as coextrusion, high-pressure or high-temperature treatment.

Phase separation is an extremely good example of how bulk behaviour is modified by the presence of a surface. Before we consider the effect of the surface, it is necessary to grasp a broad understanding of the mechanism in the bulk. One form of bulk phase separation was covered in Chapter 7 during the discussion of the hydrophobic effect. In the hydrophobic effect, the phase separation is a competition between different entropic effects; there might be ordering around a hydrophobic aggregate but this has a smaller energy cost as ordering around many individual molecules. A more general view ignores the precise nature of what makes a mixture immiscible; hydrophobicity might cause individual molecules to avoid contact in a mixture, but there might be other reasons, such as packing considerations. To consider a very simple formulation, we consider the Gibbs free energy to consist of an entropic term (favouring the mixing of the two components) and an enthalpic term (favouring phase separation). The physics behind the enthalpic component need not be under-stood, and in polymer mixtures it is represented by a term called the Flory–Huggins interaction parameter.

The competition between entropy and enthalpy leads to a dominant wavelength (called a spinodal wavelength) during the early stages of the phase separation. The energy cost of having separate phases is contained within the free energy. More inter-faces means that the enthalpic component of the free energy is dominant over the entropic component. Fewer interfaces (i.e., larger phases) means that entropy still plays an important role in the final structure. Given that the number of phase-separated domains of each phase per unit volume must be constant (because the free energy per unit volume must be constant), there will be a constant distance between these interfaces or between the different phases. The consequence of this behaviour is that the phenomenon of spinodal decomposition results in ordered structures in three dimensions. These structures may be either bicontinuous (two continuous phases) or one phase dissolved in another.

In the event of a surface being present, the phase separation will be modified. Because the lower-energy phase of the phase-separated mixture will preferentially segregate to a surface, the phase-separated morphology must be altered by the presence of a surface. In practice the surface triggers the direction of the phase separation because the symmetry of the mixture has been broken by the presence of a surface. This can give rise to a lamellar structure (Figure 8.10(a)). Although multilayered structures are possible, experimentally it is very difficult to realise more than two or three layers close to the surface because deeper in the film, thermal noise will force the lamellar structure to break up. Figure 8.10(b) and (c) show examples of multilayered structures in polymer films. The formation of stratified films in this fashion is known as *surface-directed spinodal decomposition* and has been studied in various polymer blends. Strictly, spinodal decomposition is an initial growth process; although the formation of the structure shown in Figure 8.10(b) and (c) would have been initiated by spinodal decomposition, late-stage coarsening effects will have been signifi-cant. The principles will still apply to other mixtures but polymers have shown the most striking behaviour, mainly because it is easier to achieve larger length scales with polymers, due to their greater size.

We know from the above discussion that phase separation in the bulk has a constant length scale. In polymer blend films, the length scale of this layering is close to that in bulk mixtures. However, in thin films (typically $\sim$100 nm or less) this spinodal length scale will be reduced with decreasing film thickness. For films as thin as, or slightly

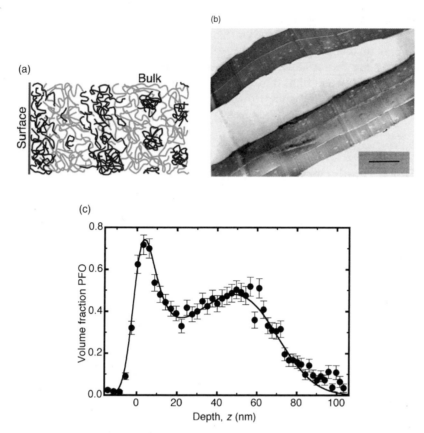

Figure 8.10 (a) Surface-directed spinodal decomposition is shown schematically. The dark chains segregate preferentially to the surface due to their lower surface energy. The phase separation is directed parallel to the surface and results in a layered structure. Deeper into the film, thermal noise breaks up the layered structure to leave an isotropic phase-separated structure. (b) An electron micrograph of two cross sections of a film of polystyrene and polybutadiene embedded in a resin (white matrix). Note the different layers in this film. Darker regions are polybutadiene-rich (the polybutadiene had been stained with OsO_4 to provide contrast with the polystyrene). The craters at the surface of these sections contained polystyrene that dissolved in the resin. The scale bar is 1 μm. (c) These ion beam data show the structure perpendicular to the surface in a blend of poly(9,9-dioctyl fluorene) (PFO) and poly(9,9-dioctyl fluorene-*alt*-benzothiadiazole) (F8BT); their chemical structures are shown in Chapter 6. The ordinates represent the fractional volume occupied by PFO. In this case there is a surface layer, rich in PFO, followed by a depletion layer, rich in F8BT, before bulk structure exists. Note that ion beam experiments have a finite resolution of $\sim$10 nm, so the exact volume fraction at the surface ($z = 0$) is washed out. Part (b) is reproduced from M. Geoghegan, R. A. L. Jones, R. S. Payne, P. Sakellariou, A. S. Clough and J. Penfold, *Polymer* **35**, 2019–2027 (1994). Part (c) is adapted from J. Chappell, D. G. Lidzey, P. C. Jukes, A. M. Higgins, R. L. Thompson, S. O'Connor, I. Grizzi, R. Fletcher, J. O'Brien, M. Geoghegan and R. A. L. Jones, *Nature Materials* **2**, 616–621 (2003). Copyright 2003 Nature Publishing Group

thinner than, the bulk spinodal length, the length scale of the phase separation will decrease, but for very thin films, the lamellar structure will not be stable at all, and the film will break up laterally. Of course the length scale where this becomes significant will depend on parameters such as temperature and polymer molecular weight, but as a rule of thumb one can say that films thinner than 100 nm are not likely to be stratified. When very thin films phase separate, therefore, larger length scales are likely to be observed as lateral domains (in the plane of the film).

8.6.2 Block copolymers

We know from the previous chapter how block copolymers pack in the bulk. We have also seen how curvature and interfaces are important parameters in this packing. However, in a film the surface inhibits some bulk conformations, because the surface tends to prefer orientations with one component preferentially adsorbed. An example of a morphology that would be stable on a surface is a lamellar A–B diblock copolymer structure. However, such a film would not necessarily be self-levelling, as we have mentioned to be the case for liquid films. We assume that the copolymer is symmetric, and has lamellar spacing l. If the surface and substrate favour the A block, then the film will have a stable uniform lamellar structure provided its thickness is an integer multiple of $2l$. However, if the two surfaces favour different components then the film has a uniform lamellar structure if its thickness is given by $2(n + \frac{1}{2})l$, where n is an integer. For any other thickness, the film has to orient itself into steps of different thicknesses so that the relevant criterion is satisfied. We might want to create a structure of lamellae lying perpendicular to the surface, and this is somewhat more challenging. One way to achieve this is to produce an energetically neutral surface. A random copolymer-coated surface of the appropriate composition of A and B monomers is one way to achieve such a surface, because the net copolymer–surface interaction will be the same for both of the blocks. Figure 8.11 illustrates the various ways that block copolymers may be oriented when in contact with a surface. Another way to recreate a lamellar block copolymer pattern perpendicular to the substrate is via an epitaxial route (Section 8.7.1). Rough substrates can also inhibit ordering parallel to the substrate because of the amount of bend required in the copolymer layer. In this case it is energetically favourable for the polymers to order perpendicularly to the surface.

An effective route to the creation of perpendicular lamellae on a given surface is to use external fields to align the block copolymer. Even copolymers consisting of polymers with a very small polarisability can be aligned using electric fields. Although enormous electric fields ($\sim 10^7$ V m^{-1}) are needed to cause rearrangement in materials with small polarisabilities, this is easily manageable with very thin films. In the first experiments, utilizing diblock copolymers of polystyrene and poly(methyl methacrylate) (PMMA), films of ~ 20 μm thickness were aligned in this way.

Block copolymer structures can be used as a template for nanocontact printing, for example. To achieve such structures one component of the block copolymer needs to be fixed before being etched in plasma. The etch selectively removes the unfixed part of the copolymer film, whereas the fixed part will remain; this is known as reactive ion etching,

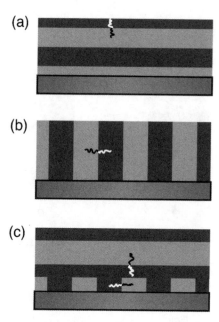

Figure 8.11 Orientation of symmetric diblock copolymers on a surface. (a) Generally symmetric diblock copolymers will form lamellae parallel to the surface. (b) Lamellae perpendicular to the surface is a higher energy state and normally requires a surface neutral to both blocks (other possibilities are discussed in the text). (c) Both possibilities may coexist, with lamellae both perpendicular and parallel to the surface. This combined geometry has the advantage that the film thickness requirements in the text need not be met

and can be used to create both positive and negative masks (Figure 8.12). This might be by cross-linking, or by using a stain. Staining is common in electron microscopy to create contrast where no heavy elements exist (Figure 8.10(b) contains an example of a stained polymer). Oxides of metals such as ruthenium or osmium react with double bonds present in the material and create a cross-link containing the heavy metal. Such a structure would be stable under plasma etching. Instead of plasma etching, UV radiation may be used. For example, in the case of the polystyrene–PMMA diblock copolymers discussed in the previous paragraph, UV radiation destroys the PMMA but fixes (cross-links) the PS. The PMMA can then be dissolved away. An application of this for creating magnetic nanowires is discussed in the next section.

We have described how to make lamellar blocks perpendicular to a substrate surface, but lamellae parallel to the substrate are also useful. We know that this is readily achievable with symmetric diblock copolymers, but such structures can also be achieved using asymmetric block copolymers by the addition of a suitable homopolymer. Figure 8.13 shows how a diblock copolymer, which forms spheres in the bulk, can be made to orient itself into lamellae by the addition of a homopolymer that is miscible with the shorter of the two blocks.

Pattern formation in block copolymers has many parallels in crystallinity; indeed block copolymers form crystal structures, and like small-molecule crystals, block

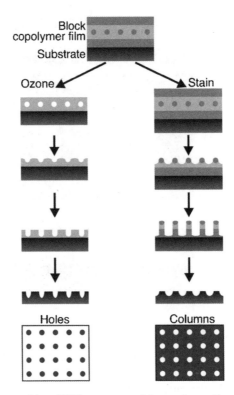

Figure 8.12 Reactive ion etching (RIE) creates positive and negative masks from thin films of block copolymers of polystyrene and polybutadiene which form polybutadiene spheres in the bulk. Here the polybutadiene is the darker component and wets both interfaces. By exposing the film to ozone, the polybutadiene can be degraded and removed, but by staining the polybutadiene with osmium, it becomes fixed and more resistant to an ion etch than polystyrene. Depending on the preparation route, the RIE will remove more or less polymer where the spheres initially were located, leaving holes or columns in the substrate. Adapted with permission from M. Park, C. Harrison, P. M. Chaikin, R. A. Register and D. H. Adamson, *Science* **276**, 1401–1404 (1997). Copyright 1997, AAAS

copolymer ordering is only perfect over small length scales. Part of the problem lies with preparation techniques, any impurities in the polymer can give rise to defects. The issue of polydispersity (spread in molecular weight distribution) is also relevant here, but relatively monodisperse block copolymers can order despite the effects of polydispersity because the chains in each block still have random conformations. A different major problem comes from ordering initiating at different points, which gives rise to grain boundaries. As a general rule, ordering in block copolymers is at best of order a few hundred nanometres. There are efforts to overcome the length-scale difficulty using, for example, solvent vapours to plasticise the copolymers, which enables them to order. A more practical disadvantage of using block copolymers is that they are expensive to synthesise, which may outweigh the cost benefits of using a self-assembly process.

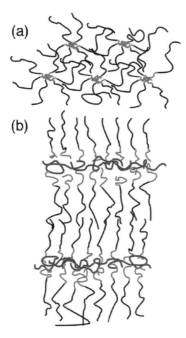

Figure 8.13 (a) A block copolymer A–B in which the B block is shorter than the A block will form spheres of the B block in the bulk. (b) A lamellar structure can be formed by the addition of a homopolymer that is miscible with the B block. This homopolymer may be the same as the B block, but the only criterion is that it be miscible with the B block

8.7 NANOPATTERNING SURFACES BY SELF-ASSEMBLY

Although there are many different methods of creating patterns on surfaces, most of those discussed so far are periodic. The goal for patterning will be achieved when methods of producing non-repeating or more complex structures are achieved. A good example of the need for such structures is in the creation of integrated circuits. Other examples of the need for complex structure on a substrate include micro- and nano-fluidics. However, uniform patterning could also play a role, such as in structures relating to data storage, where complex pathways are much less important than compressing very many bits in a single structure. Similarly, protein and DNA arrays, which might be used for gene sequencing, and chemical sensing devices require only ordered structures as the basic templates.

Sections 1.4 and 8.4 briefly explain how lithographic means can be used to create templates. The use of stamps is a well-accepted route to creating a pattern on a surface. However, such soft lithography does have some disadvantages. There may well be only one master stamp. This stamp will be used to create perhaps a hundred production stamps, which in turn will be used for the lithography. Clearly such a serial process is inefficient, and possibly quite expensive for large-scale production. Any refinement of the stamp, such as the addition of chemical heterogeneity, will also add time and cost to the process, which could well be prohibitive. Finally, the creation of the smallest structures for the stamp will be diffraction limited at some point.

There are a variety of non-lithographic means towards pattern formation, which have inherent advantages. The major disadvantage of non-lithographic processes is that they tend to be rather specific and we have rather limited control over the parameters influencing the desired structure. In one example, we shall show how dewetting can be used to create long-range patterns. However, not all films will dewet and specific substrate preparation may be necessary in order to tune the polymer–substrate inter-action to create the desired structure. This is a general truth about self-assembly; the physics gives us a choice of results over which we have limited control. Therefore, to gain the widest possible choice about the use of non-lithographic printing, we shall need to know all of the many possible routes to creating the structures. The future for soft lithography will probably require a marriage of top-down approaches and self-assembly.

The self-assembly of block copolymers in an electric field has been used to create a series of upright block copolymer cylinders which could be treated with UV radiation. The radiation could be used to destroy the cylinders (PMMA) but fix (by cross-linking) the matrix (polystyrene). Electrodeposition could then be used to inject cobalt into the cylindrical voids. The end result is cobalt nanowires in a polystyrene matrix. The success of such an experiment is due to self-assembly and provides an excellent example of the sort of technologies that could develop from these ideas. However, the nanowires created in this way are not flexible and bends and turns are not built into the system. There is only one length scale involved, hence incorporating the structure into any workable device would require considerable engineering of the other components so this technology would be successfully assimilated into the final structure. It is quite possible that an integration of nanowires and a lithographic template (as described above) could produce more useful structures, as well as perhaps having even better block copolymer ordering.

Another interesting example of the use of both lithography and self-assembly con-cerns inkjet printing as the starting point for creating a thin film transistor circuit. This experiment, more than most, illustrates how microcontact printing and self-assembly can be used together to make usable devices. It is possible to use inkjet printing to spread a conducting polymer onto a surface, but unfortunately the droplets tend to spread with accidental short circuits as a consequence. A photolithographed film can be used to stop this spreading by providing regions of different wettability across the surface. Blends of PSS and PEDOT can be used as source and drain electrodes of a transistor, which can then be inkjet printed onto the substrate; shorting will be pre-vented by the repulsion provided by a polyimide line on the otherwise hydrophilic substrate. A film of semiconducting polymer, F8T2 (Chapter 6), is then spin-coated onto the resultant structure, providing broad and complete coverage. A dielectric layer is spin-coated onto the F8T2 before a PSS/PEDOT gate is deposited on top by inkjet printing. This describes a perfectly workable device, but the device quality is actually improved by self-assembly in the F8T2 layer. Annealing the F8T2 at high temperature (in its nematic liquid crystalline phase) enables it to align parallel to the polyimide structure. The alignment remains after the structure is quenched down to room tem-perature. The major benefit of this alignment is greater current flow due to fast intrachain charge transport.

8.7.1 Patterns produced on heterogeneous substrates

A rather obvious way to produce patterned structures is to have a pattern to begin with. The film will simply replicate that pattern. This is, after all, the basis of lithography. However, there are subtleties that should be understood in these processes. We begin our discussion of heterogeneous substrates with the simplest example of the use of self-assembled monolayers to provide a basis for pattern replication. Experiments using block copolymer templates are useful to show why pattern replication does not always occur, and we shall then close this section by discussing the importance of topography.

Earlier in this chapter we discussed how to produce a lamellar surface structure using surface-directed spinodal decomposition. In thin films (i.e., films significantly thinner than the bulk spinodal wavelength) a lamellar structure is unstable and patterning takes place in the plane of the film. There is likely to be preferential segregation of one component to the surface or substrate (or both) but the dominant effect will be lateral phase separation, where again the dominant length scale involved is related to the spinodal wavelength. Suppose we now attempt to deposit such a structure on a micro- or nanopatterned surface by spin-coating. Figure 8.14 shows how a polymer blend

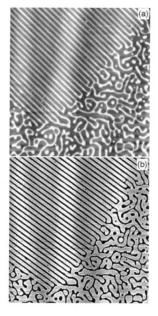

Figure 8.14 A polymer blend phase separates to replicate the structure of a self-assembled monolayer. An immiscible blend of polystyrene and poly(vinyl pyridine) is shown on a substrate patterned with stripes of gold and the SAM; the image was taken using atomic force microscopy and the periodicity of the pattern is 2.4 µm. (a) The poly(vinyl pyridine) adsorbs onto the gold and the polystyrene adsorbs onto the SAM. Away from the pattern, the phase separation proceeds as expected on a pure substrate. (b) The poly(vinyl pyridine) has been removed by dissolution of the sample in ethanol (a non-solvent for polystyrene). The removal of the poly(vinyl pyridine) demonstrates that the polystyrene segregates to the SAM. Reproduced with permission from M. Böltau, S. Walheim, J. Mlynek, G. Krausch and U. Steiner, *Nature* **391**, 877–879 (1998). Copyright 1998 Nature Publishing Group

structure is modified by spin-coating onto a lateral gold pattern, which has been created by microcontact printing of a SAM onto a silicon substrate. Notice how the blend phase separation follows the SAM structure, replicating its wavelength. Outside the SAM structure, the pattern replication fails and the structure takes on its own somewhat larger wavelength. This means that the phase-separated morphology will take on a shorter wavelength than would be expected on a neutral unpatterned substrate because the energy cost associated with having a greater areal density of interface is paid for by the energy gain due to the preferential attraction of one of the components (here polystyrene) to the gold pattern.

The use of block copolymers to reproduce nanoscale patterns is also useful as a means to control the behaviour of other polymer mixtures. Phase separation in polymer blends takes place over rather large length scales, given initially by the bulk spinodal wavelength and more practically by various coarsening processes. However, if the polymer blend is on a (patterned or ordered) copolymer surface, spatial variations in the surface energy will control the final morphology, reducing the length scale of the phase separation. The use of ordered block copolymers as a substrate is slightly different from using random copolymers because a random copolymer-coated surface will essentially result in a surface with a particular value of the surface energy, whereas a block copolymer can provide a spatial variation in the surface energy. The use of an A–B copolymer template for an A/B polymer blend would probably be the most useful variation on this theme, but the copolymer can still be effective in changing the behaviour of other blends provided there are sufficient variations in surface energy. Figure 8.15 shows an example of the use of block copolymer patterns to control the length scale of phase separation.

Suppose we can create a self-assembled monolayer that is an exact match for a block copolymer lamellar length scale. In this case we can align a lamellar block copolymer structure perpendicular to a substrate rather than parallel. This substrate replication (epitaxy) has the advantage of ensuring long-range order with the block copolymer films; there are fewer dislocations (Figure 8.16). We know that dislocations and grain boundaries can occur in block copolymers but the energy cost of a dislocation is not great, because block copolymers can easily reorient themselves around dislocations. Grain boundaries occur because nucleation of lamellae starts at different points. Self-replication of a self-assembled monolayer by a block copolymer eliminates grain boundaries because *translational order* is inherent in the structure due to the SAM underlayer. The number of dislocations are minimised because the energy cost of their presence is increased; a dislocation will force part of the polymer to lie on an unfavourable part of the substrate pattern.

8.7.2 Topographically patterned surfaces

Any topographical variations of the surface of a substrate have so far been ignored (e.g., surfaces with a chemical topography) or treated as boundaries (e.g., certain examples utilizing lithography). However, in the case of slight variations of substrate topography there are important and significant changes that need to be considered and understood. Wetting of a rough surface is a relatively new subject, and relatively few experiments have

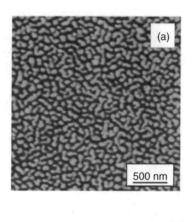
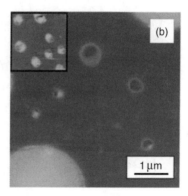

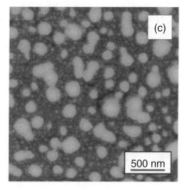

Figure 8.15 Block copolymers can alter the phase separation behaviour of polymer blends by providing a substrate with a spatial surface energy variation. (a) AFM image of a triblock copolymer film on a silicon wafer. The triblock copolymer contains blocks of polystyrene, poly(vinyl pyridine), and poly(*tert*-butyl methacrylate). (b) AFM image of an immiscible blend of polystyrene and poly (*tert*-butyl methacrylate) on silicon. (c) The same blend on the triblock copolymer. Note the much smaller length scale exhibited in (c) than in (b). Reproduced with permission from K. Fukunaga, H. Elbs and G. Krausch, *Langmuir* **16**, 3474–3477 (2000) Copyright 2000 American Chemical Society

been reported. However, the ability to control the contact angle of a fluid by varying substrate topography would be beneficial in producing water (or other) repellent surfaces, as well as micro- and nanofluidic applications. Experiments and theory have shown that a sharp pattern on a surface exaggerates the contact angle, making wettable surfaces even more wettable and non-wettable surfaces even more repellent. The exaggeration of contact angle to make, say, hydrophobic surfaces even more hydrophobic is largely due to the droplet not being able to conform to its equilibrium contact angle on a microscopic scale, forcing the droplet to an even greater contact angle (Figure 8.17).

A film coated on a substrate by some solvent evaporation technique will usually have a relatively flat surface, with the only contribution towards surface roughness being thermally induced capillary waves. However, in thin films (a few nanometres thick) the film's surface will convey some aspects of the substrate topography, so if

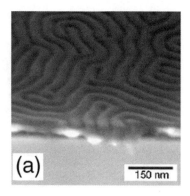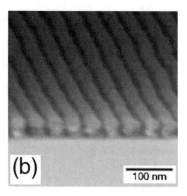

Figure 8.16 Cross-sectional scanning electron microscopy images of a diblock copolymer of polystyrene and PMMA on (a) an unpatterned substrate and (b) a nanoscopically patterned substrate whose pattern replicates the length scale of the copolymer microphase separation. In (b) the polystyrene and PMMA wet different parts of the pattern, causing perpendicular lamellae with impressive long-range order. Reproduced with permission from S. O. Kim, H. H. Solak, M. P. Stoykovich, N. J. Ferrier, J. J. de Pablo and P. F. Nealey, *Nature* **424**, 411–414 (2003) Copyright 2003 Nature Publishing Group

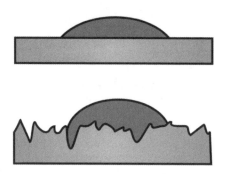

Figure 8.17 A droplet on a rough surface cannot adopt its true contact angle on the microscopic scale, which forces a greater contact angle than it would have on a flat surface of the same material

the substrate is not perfectly flat then it is possible that some imperfections will be replicated in the film surface. Such imperfections can have undesirable effects because it has been shown that ultrathin polymer films can dewet wettable surfaces. This may be because, in films thinner than the polymer chain size, a polymer may be able to adopt its ideal Gaussian (random walk) conformation only if it dewets; i.e., if it forms droplets of a size thicker than the polymer chain. It is nevertheless possible that this dewetting can be used as a means for pattern formation. Figure 8.18 shows a model topography imposed on a silicon substrate. Here the silicon surface is of a sawtooth structure and a polystyrene film, somewhat thinner than the size of a polymer chain in the bulk, is spin-coated onto it. The height of the sawtooth is only a few nanometres, comparable to the size of the polymer film, which forms nanochannels

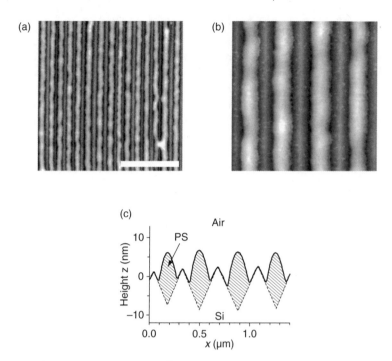

Figure 8.18 A thin ($\sim$5 nm) polystyrene film is spin coated onto a corrugated silicon substrate and heated to 150 °C. Despite the small height and large width of the sawtooth structure on the silicon, the polymer dewets to the troughs in the structure, forming channels. (a) AFM image of the structure. The scale bar is 2 μm. (b) Smaller area (1.5 μm × 1.5 μm) scan. (c) Line average scan of the data in (b). The position of the silicon substrate underneath the polymer is obtained by interpolation. Note that these AFM images have an exaggerated aspect ratio. Reproduced with permission from N. Rehse, C. Wang, M. Hund, M. Geoghegan, R. Magerle and G. Krausch, *Eur. Phys. J. E* **4**, 69–76 (2001) Copyright 2001 EDP Sciences

when heated above the glass transition (so the polymer can flow). This phenomenon may be exploited to much greater effect if surface chemistry is utilized. Glancing angle evaporation of gold onto the silicon sawtooth structure will result in gold on the peaks with pure (untouched) silicon troughs. The gold may then be used as the basis for thiol chemistry, giving a functional polymer surface (Section 8.4).

8.7.3 Patterns produced by thin film dewetting

Dewetting of films from a surface can have a variety of origins, but the most common of these are for metastable and unstable films in which surface energies and long-range forces play the major role. Metastable films will break up by the thermal nucleation of holes in the film or by the presence of impurities, which will nucleate a hole. Dewetting in metastable films is a random process, so ordered structures will not be possible by this mechanism. However, most unstable films can be used to create a dominant length scale because a spinodal mechanism is responsible for film rupture. If a film is thermodynamically unstable on its substrate, it will break up. Many small droplets will create a large surface area, whereas few large droplets will not satisfy the thermodynamic driving force for dewetting. The result is that the compromise between limiting the amount of interface created, while enabling dewetting to proceed yields a wavelength, the *spinodal wavelength*, characteristic of the dewetting process. For more unstable films, this spinodal wavelength decreases, reflecting the increased role of the immiscibility between substrate and film. The thermodynamic quantity responsible for the dewetting is usually long-range (van der Waals) forces. The strength of these forces is characterised by the Hamaker constant; Figure 8.19 illustrates the dewetting process. As an example, remember from Section 8.2 that the attractive energy between two planar surfaces separated by a distance x is given by

$$W_{vdW} = -\frac{A_H}{12\pi x^2}. \tag{8.9}$$

If a film of thickness x separates a substrate from the top surface then it is likely that the thinner the film, the stronger the attractive forces acting across it and the greater the driving force for dewetting.

We discussed earlier in this chapter how spinodal decomposition could be perturbed by a surface, creating a lamellar structure. Similarly, our understanding of dewetting is not limited to creating uniform two-dimensional patterns, because we can also perturb the structure to create a series of lines. This anisotropic dewetting can be achieved by simply rubbing a substrate in a particular direction. This simple act imposes a direction on the dewetting. The dewetting still has a dominant wavelength that is governed by the compromise between the creation of surface area and thermodynamic repulsion (Figure 8.20). In other words, we can control the direction but the physics controls the wavelength. Our ability to alter the size of the structure is limited to our freedom over how far we can control the thickness of the film.

The long-range dispersion forces responsible for spinodal dewetting have also been shown to create ordered structures in free-standing polymer films capped at the top and bottom with an evaporated layer of silicon oxide. The film buckles on heating the

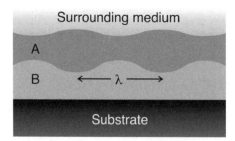

Figure 8.19 If long-range forces between a fluid B and the surrounding medium are greater than those between the surrounding medium and another fluid A, which initially lies on top of B, then the film will break up (film A will dewet film B). There will be a dominating isotropic wavelength governing the dewetting, known as the spinodal wavelength, λ. The amplitude of the fluctuations at each interface is a complicated function of parameters such as viscosities of the components

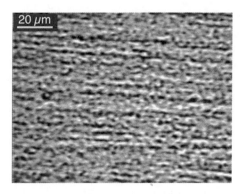

Figure 8.20   A film of PMMA dewetting an underlayer film of polystyrene. The PMMA has been prepared on a glass substrate which had been rubbed preferentially in a particular direction. The direction of rubbing imposes directionality on the dewetting. Apart from the anisotropy of the dewetting, the mechanism is expected to be the same as for Figure 8.19. Reproduced with permission from A. M. Higgins and R. A. L. Jones, *Nature* **404**, 476–478 (2000) Copyright 2000 Nature Publishing Group

trilayer structure to a temperature high enough to melt the polymer, and to leave it very mobile (but not so high as to degrade the polymer or to melt the solid capping layers). The size of the deformation is again a compromise between different physical effects. The bending of the capping layers will oppose the dispersion forces acting across the film. The patterns created by using anisotropic spinodal dewetting or in capped free-standing films are unlikely to be perfect, and so they would be unsuitable for micro- or nanofluidic devices. We therefore need to consider different means to create micro- and nanochannels. We consider two experiments that show ways of achieving reliable channels, one on the micron scale and one on the nanometre scale.

8.8   PRACTICAL NANOSCALE DEVICES EXPLOITING MACROMOLECULES AT INTERFACES

8.8.1   Molecular and macromolecular electronics

8.8.1.1   Polymer electronics

In Section 6.2 the principles of organic LEDs, FETs and photovoltaic devices were explained. Many of the challenges facing their creation concern optimising the structure of polymer films. Devices require more than simply a working medium, they must be connected together. The interconnections necessary for circuitry cannot be produced by self-assembly, and so some top-down method such as inkjet printing must be used. The combination between top-down and bottom-up technologies is not the only marriage of opposites in this area; molecular electronics also requires nanoscale solutions that extend beyond soft matter; the field is a fine example of the *soft–hard interface*. Electron beam lithography may be used to produce circuits instead of inkjet printing and transistors may be created using carbon nanotubes connected via metallic (gold) inter-connects. Although much of what is described below is in terms of soft matter, many of the methodologies discussed in this area may be replaced by a suitable hard alternative. Future work will declare which method is the more viable, but it is likely that a mixture of the two will be appropriate in many devices.

For example, Section 6.2.4.1 considered simple photovoltaic devices such as solar cells. Electron–hole pairs (excitons) created by a source of light need to be collected as a current if they are to be of any use. However, with an exciton diffusion length typically of the order of a few nanometres, there are two main challenges. To improve the efficiency of polymeric/organic photovoltaic devices, we need to be able to create an active region that contains a large amount of interface to ensure that an exciton is likely to recombine before some other decay process removes it without contributing to the current. Secondly, having created a region with a large interface-to-volume ratio, we need to ensure that the electrons and holes created at these interfaces have a clear pathway to the electrodes. Figure 8.21 shows an example of how a polymer blend structure might be designed to achieve these goals. One way of producing such a structure may be to exploit the idea of using electric fields to orient polymers in a blend in the same way as for block copolymers (Section 8.6.2).

The problem in fabricating efficient LED structures is rather like the photovoltaic problem, but in reverse. In this case the challenge is to inject holes and electrons into an active region, where they can combine to form an exciton. Here the ITO electrode is often coated with the PSS/PEDOT complex to provide a high work function coating (Section 6.2.1.2). It is known that the PSS dopant segregates preferentially to a free interface, so it is likely that, assuming the same effect occurs at the ITO interface, there is a gradual increase in work function to the bulk value, which may improve hole injection efficiency. This method may indeed work in parallel with the interface engin-eering of PEDOT/PSS hole injection layers (Section 8.5.2.1).

In FETs the structure of polymer interfaces is also a critical issue. Here the charge carriers move under forward bias in a polymeric *active layer* from the source to the drain within the first few nanometres of the active layer–dielectric interface. This imposes a requirement that the interface be of a high quality; a roughness (or interfacial width) of

Electrode

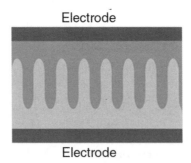

Electrode

Figure 8.21 A polymer blend photovoltaic device should benefit from phase separation on a small scale such that an exciton formed is never less than ~10 nm from an interface at which it can dissociate. The resulting holes and electrons should then have a clear path to the two electrodes so that a current may be created. The diagram shows a morphology satisfying these criteria

several nanometres can seriously inhibit carrier efficiency. It is therefore worth optimising the structure of this interface. This may involve possible solvent treatment for any of the polymers involved. Heat treatment has been shown to change the interfacial width between poly(dioctyl fluorene) (PFO) and PMMA; the correlation between the shape of the interface and performance of devices is presently the subject of research. Similarly, the FET preparation method may also affect the final interface; solvent deposition methods may broaden the interface if the solvent for the active layer even slightly interacts with the dielectric layer (or vice versa if the active layer is deposited first).

8.8.1.2 Single-molecule devices

One basic requirement of a single-molecule device is that it moves charge from an electron donor molecule to an electron acceptor, or vice versa. To carry this charge along even small distances is something of a challenge because of the very rapid exponential decay in charge transport rate constant. Any molecule that joins a donor and acceptor, a *bridge* or *spacer* molecule, should have its energy matched to the donor and acceptor. This has been shown to be possible for distances of up to 4 nm with molecular wires of oligomeric *p*-phenylene vinylene.

A single-molecule FET requires a polymer to be connected in some way to a source and a drain in order for a potential to be applied across the FET. One part of the molecule must also be in contact with an insulating dielectric material to form the gate voltage. With such a device, a voltage applied across the gate will change the electrical resistance of the molecule, because charge carriers will have their motion impeded or accelerated.

8.8.2 Nanofluidics

The prospect of both automating and miniaturising chemical operations such as analysis or even synthesis is a tremendously attractive one, and the field of *microfluidics*, in which chemical operations are carried out in channels and reactors etched by processes analogous to those used for the planar process of silicon, has experienced tremendous recent growth. The aim is often summed up in the phrase 'laboratory on a chip'. But the business of miniaturising operations that involve fluid flow raises some interesting scale issues, and an understanding of this scaling and the limitations they impose will be particularly important as the size of these devices is shrunk towards the nanoscale.

To see why scale is so important in these devices, let us review the fluid mechanics that controls the simple case of fluid flow in a pipe or channel. The flow of fluid next to a solid wall can generally be taken as zero (though there are important exceptions to this so-called no-slip boundary condition, particularly for polymer and other non-Newtonian fluids), so in order for any fluid to flow, neighbouring regions of the fluid must flow at different rates. The character of the flow depends on a single dimensionless number, the *Reynolds number*, Re. This depends on two material properties of the fluid, its density ρ and viscosity η, and two characteristics of the flow, a typical velocity v and a characteristic size l. (For fluid flow in a pipe this size would be the pipe diameter.) The Reynolds number is the dimensionless combination of these four quantities, $\mathrm{Re} = v\rho l/\eta$.

The physical significance of this number is that it expresses the relative importance of *viscosity* and *inertia* in fluid flow. At low Reynolds numbers, resistance to flow arises predominantly from viscosity, whereas at high Reynolds numbers inertia dominates. From the definition of the Reynolds number, one sees that for a given fluid, flow at smaller and smaller length scales is increasingly dominated by viscosity. Alternatively, the flow of water through a 1 μm channel will have the same character as the flow of a fluid one million times as viscous as water through a macroscopic, 1 m channel. Water, on the micro- and nanoscales, does not behave as the free-flowing liquid we are used to on the macroscale; rather its properties are as dominated by viscosity as much as treacle or molasses are on the macroscale.

At low Reynolds number, fluid flow through pipes or channels is *laminar*; there is a smooth gradient of velocity from the walls, where the fluid is stationary, to the centre. For a pipe of circular cross section, the rate of flow is given by the *Poiseuille equation*; the flow rate is proportional to the pressure gradient, inversely proportional to the viscosity, and proportional to the *fourth power* of the radius of the pipe. This very strong dependence of flow rate on the dimensions of the pipe or channel poses problems for nanofluidics and microfluidics, in that it makes the simple application of pressure a much less attractive driving force for fluid motion.

One attractive alternative to pressure-driven flows exploits the *electro-osmotic effect*. For aqueous systems containing free ions, one can use an electric field to move the fluid around. This effect relies on the fact that if an ion-containing solution is in proximity to a charged surface, near that surface there is a layer enriched in ions of the opposite charge to the surface. This is the so-called *electric double layer*. If an electric field is applied in the direction of the channel, then this will exert a force on the electric double layer, dragging the fluid along with it. The attractiveness of this method for moving fluids at the micro- and nanoscales comes from the fact that the force is applied to the fluid at the wall, leading to a velocity profile across the channel that is much flatter than it would be for Poiseuille flow. The flow can also be modulated by changing the surface charge.

Another distinctive feature of flows at low Reynolds numbers is that *turbulence* is always absent. The turbulence that is a feature of flow at high Reynolds numbers (see Figure 8.22 for a comparison between lamellar and turbulent flow) leads to an increase in resistance to flow, but it is also of immense value in chemical engineering operations as it provides a very efficient way of mixing fluids. The absence of turbulence at low Reynolds numbers makes operations that rely on mixing fluids rather problematic – in these cases mixing only takes place due to the much slower process of molecular diffusion.

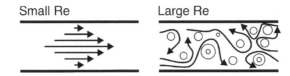

Figure 8.22 Laminar flow occurs at small values of the Reynolds number (Re). This occurs in high-viscosity fluids at low velocities in small dimensions. Turbulent flow occurs at large values of Re, when a low-viscosity fluid is travelling rapidly in a large container

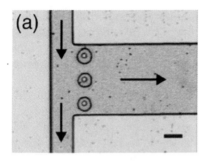 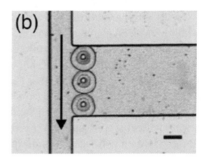

Figure 8.23 A hydrogel is attached to posts at the junction of a microchannel. (a) In poor solvent conditions the gel is collapsed, allowing fluid flow in all channels. (b) In good solvent conditions the gel expands and blocks the flow in one channel. The scale bar is 300 μm. Reproduced with permission from D. J. Beebe, J. S. Moore, J. M. Bauer, Q. Yu, R. H. Liu, C. Devadoss and B.-H. Jo, *Nature* **404**, 588–590 (2000). Copyright 2000 Nature Publishing Group

The absence of turbulent mixing in laminar flows has actually been exploited to make nanostructures. If two capillaries containing different reactants can be brought into contact, then a chemical reaction can be localised to this interface, creating a thin pattern, avoiding the need for conventional lithography. On the other hand, it is more often going to be necessary to promote mixing. One way of doing this is to pattern the walls of the channel with obstacles perpendicular to the flow direction, which even in the absence of turbulence break up the smooth laminar flow and allow mixing to take place more efficiently.

Having created submicron-sized channels, it may well be desirable to control the flow of liquids in these channels. The miniaturisation of valves that are used in macroscopic pipes is not going to be a practical solution, so new ideas are necessary. Polymeric gels are a useful starting point for the control of fluids because they can be expanded and contracted to a given size in response to their environment. Figure 8.23 shows optical microscopy images of such a system. A hydrogel composed of a weak polyacid swells at high pH, impeding the motion of a fluid from travelling through its channel. Of course, the effectiveness of such a barrier will be dependent on the hydrogel and the liquid medium, because some material will often be capable of passing through the mesh of a gel. Although this mechanism has been demonstrated using a pH-responsive gel, the temperature-dependent volume transition in the poly(*N*-isopropyl acrylamide) (PNIPAM), mentioned in Section 7.3.4, could also be used for such a purpose.

8.8.3 Filtration and sorting

Many of the tools of modern science require very pure, calibrated materials and the task of separating, filtrating and sorting materials is not a trivial task if it needs to be performed with a high degree of accuracy. This may well be exacerbated by a lack of material to begin with; if one works with very few molecules or if one *needs* to work with very few molecules, then the sizes of typical devices or vessels is inappropriate. Working with small amounts of material is a necessity in chemical micro- or nanoreactors, where the smaller length scale *increases* reaction rates because the likelihood of molecules meeting each other can be controlled more efficiently in systems where the length scales are small and the molecular motion is controlled by directed diffusion rather than turbidity. The task therefore is to find a means of moving the required molecule from A to B.

Entropy is an important thermodynamic tool to sort molecular behaviour by the use of *entropic traps*. Macromolecules trapped in materials or devices will try to find locations whereby they can maximise their *configurational entropy*. If it is allowed to do so, a polymer will leave a channel in which it is forced to be stretched in order to find a region where it can exhibit a random structure appropriate to its chemical environment. As an example, a cross-linked polymer (network) will normally contain cross-links randomly distributed throughout the material. Uncross-linked polymers that are mobile within the polymer network will generally seek out regions of material where they can experience the most conformations; these are less cross-linked regions, since cross-links will impede the polymer from having a random walk configuration. Therefore polymer networks can also be expected to act as a form of molecular sieve. Entropy may well also play a role of attracting polymers to surfaces; in a mixture of polymers of different lengths, the longer polymers would be expected to segregate to the surface. This is because the number of configurations of a polymer increases more slowly as a function of chain length than the polymer radius of gyration; therefore, per unit surface, there is a smaller entropy cost in having larger molecules present at the surface. However, in practice smaller polymers tend to segregate to surfaces because of the effect of chain ends.

Molecular sorting devices have been demonstrated using entropy through *confinement*. To move a molecule requires some form of gradient to drive the molecular motion. An electric field is an obvious candidate for such a gradient, although others can be envisaged (e.g., pressure, temperature, chemical potential, and magnetic field). Electric fields are most effective with charged polymers, but as we have seen, they should be of general use in confined spaces because the polarisability of many molecules will ensure that their motion can be controlled. A second requirement is a means of sorting the molecules so that their diffusional behaviour is different. The coupling of confinement and electric fields has been achieved with DNA. By allowing DNA to diffuse in a bath, separated from another bath by a slit of width smaller than 100 nm, an applied electric field can be used to force the DNA though the slit. Eventually the DNA will arrive in the second bath, with the arrival time being dependent on the size of the molecule. Perhaps surprisingly, the longer molecules diffused more rapidly. This is because, when inside the confining slit, the longer DNA can explore more of the slit, and so find the exit more easily; the rest of the molecule is dragged out efficiently (Figure 8.24). A related idea consists of the diffusion of DNA through a membrane pore. As the DNA passes through the pore, the pore is blocked and no current is measured at the electrodes (Figure 8.25). In such an experiment the diffusion coefficients of the DNA can be measured. The size of the DNA molecule can be correlated with the time the electrodes are blocked, creating a means of measuring DNA size, for example.

Responsive polymer brushes can also be used as a means for collecting and rejecting molecules. We have discussed how brush height varies as a function of chain length, grafting density and solvent quality. If these quantities can be varied then the brush

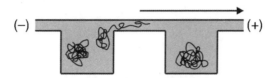

Figure 8.24   DNA will stretch in narrow pores (<100 nm); the larger the DNA molecule, the more stretched it will be and the faster it will find the exit to another reservoir (a few microns in size). The DNA is forced to diffuse by an electrical potential difference. Adapted with permission from J. Han and H. G. Craighead, *Science* **288**, 1026–1029 (2000). Copyright 2000, AAAS

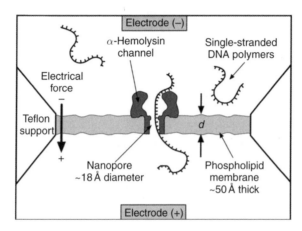

Figure 8.25   DNA can diffuse through narrow pores under the effect of an electrical potential difference. As DNA passes through the pores, it blocks a current from flowing. The time during which there is no current is a measure of DNA size and diffusion coefficient. Reproduced with permission from A. Meller, L. Nivon and D. Branton, *Phys. Rev. Lett.* **86**, 3435 (2001) Copyright 2001 by the American Physical Society

height (and consequently properties) will change accordingly. Any polyelectrolyte brush should exhibit this behaviour, but a particularly interesting example works with brushes of the thermally responsive polymer PNIPAM. PNIPAM is a particularly useful material for a responsive surface because its volume phase transition occurs in response to heat at an accessible temperature. The thermally responsive nature of the polymer is useful because the substrate can be coated with an array of heaters that will cause the brush to collapse *locally*. Experiments have been performed to show that the collapsed brush provided a suitable surface for protein adsorption. The expanded brush is considerably less amenable to protein adsorption and proteins are rejected into the bulk solution. There are many possible uses for such a device; one could, for example, envisage that the trap could be used as a means to concentrate proteins from a dilute flow before releasing them, enriched, into another bath. The method may also have benefits related to molecular weight filtration because larger proteins are more likely to adsorb onto the surface, probably because their larger size means that they have more sticking points.

BIBLIOGRAPHY

Books

For a basic text describing much of the physics contained in the present chapter, read R. A. L. Jones and R. W. Richards, *Polymers at Surfaces and Interfaces*, Cambridge University Press, Cambridge, 1999.

Texts on soft condensed matter in general will provide important background material. Two such books are particularly recommended:

R. A. L. Jones, *Soft Condensed Matter*, Oxford University Press, Oxford, 2002.

I. W. Hamley, *Introduction to Soft Matter*, Wiley, Chichester, 2000.

For a broad coverage of surface forces, only one book need be consulted:

J. N. Israelachvili, *Intermolecular and Surface Forces*, 3rd edn, Academic Press, London, 2003.

Reviews

Readers searching for more detailed information and references to many of the original research papers described in this review may want to consider the following articles.

Wetting, dewetting, pattern formation, polymer films

M. Geoghegan and G. Krausch, Wetting at polymer surfaces and interfaces. *Progress in Polymer Science* **28**, 261–302 (2003).

I. W. Hamley, Nanostructure fabrication using block copolymers. *Nanotechnology* **14**, R39–R54 (2003).

Organic electronics: single-molecule devices

C. Joachim, J. K. Gimzewski and A. Aviram, Electronics using hybrid-molecular and mono-molecular devices. *Nature* **408**, 541–48 (2000).

Soft lithography

Y. Xia and G. M. Whitesides, Soft lithography. *Angewandte Chemie International Edition* **37**, 550–75 (1998).

Microfluidics

D. J. Beebe, G. A. Mensing and G. M. Walker, Physics and applications of microfluidics in biology. *Annual Review of Biomedical Engineering* **4**, 261–86 (2002).

D. R. Meldrum and M. R. Holl, Microscale bioanalytical systems. *Science* **297**, 1197–98 (2002).

9

Bionanotechnology

The interaction between nanotechnology and cell biology is an important one, which works in two directions. Nanotechnology has provided, and will no doubt continue to provide, important new tools for biology, which help biologists to unravel the way in which the nanoscale components of living systems work together to create the remarkable examples of nanomachines and devices that biology provides. In the reverse direction, a study of the mechanisms of cell biology will give us important guidelines for making synthetic nanodevices. The process of evolution has allowed nature to find highly efficient solutions to the problems of engineering at the nanoscale, and the nanotechnologist should exploit these solutions, either by directly incorporating biological systems into nanodevices, or by building synthetic systems which mimic the operating principles of their biological analogues.

9.1 NEW TOOLS FOR INVESTIGATING BIOLOGICAL SYSTEMS

Perhaps the most useful consequence of nanotechnology for biology has been the provision of powerful new tools for the investigation of biomolecular structure, function and properties. Using these new tools, it is becoming possible to make direct measurements of structural elements of cells, and molecular recognition interactions, where in the past data could only be obtained by inference from macroscopic experiments.

9.1.1 Scanning probe microscopy for biomolecular imaging

Scanning probe microscopy (SPM) has revolutionised our understanding of the structures of solid surfaces. With the publication two decades ago of the first high-resolution scanning tunnelling microscopy (STM) images revealing the arrangements of atoms on single crystal surfaces, it was clear that powerful new capabilities were at our disposal (see Section 2.5.1). Early on, there were hopes that STM might realise similar spatial

Nanoscale Science and Technology   Edited by R. W. Kelsall, I. W. Hamley and M. Geoghegan
© 2005 John Wiley & Sons, Ltd

resolution in studies of biological molecules. DNA was a particular focus of early efforts. It was hoped that STM could be used to 'read' the sequence of a strand of DNA directly. Commonly, samples were prepared by depositing a solution containing the DNA sample onto a highly oriented pyrolytic graphite (HOPG) surface. Searching for plausible images proved difficult, but a number of groups published images that appeared to reveal the structure of DNA. There was much consternation, however, when Beebe and co-workers showed in 1000 that structures accurately matching the dimensions of DNA and appearing to exhibit the characteristic helical topology could be observed in images of clean graphite surfaces[1]. This raised important questions about the likely types of artefacts that might be associated with STM images, and contributed significantly to developing the maturity of the technique. Artefacts associated with substrate features have always been a problem in microscopy, and they remain a trap for the unwary today. Careful interrogation of the sample, adequate repetition of experiments and the use of complementary surface analytical techniques, such as X-ray photoelectron spectroscopy (XPS) and secondary ion mass spectrometry (SIMS) are key requirements if properly validated data are to be acquired. Under appropriate conditions, it is possible to acquire reliable images. For example, proteins have been successfully imaged. It is important to note that the force exerted on the sample by the STM tip may be substantial, leading to movement of the sample molecules, therefore care must be taken to immobilise them, usually by covalent attachment. An additional question concerns the imaging mechanism. Proteins are expected to be electrical insulators, and tunnelling is not expected to be very efficient. One suggestion is that the tip deforms the protein, modifying its electronic structure and generating new states at the Fermi level of the substrate. Another is that water, bound to the protein molecule, provides a conducting path for the flow of current from tip to substrate.

Atomic force microscopy (AFM) works by measurement of the force between a sharp tip and a sample, rather than measuring current as in the case of STM (see Section 2.5.1). Consequently, AFM can be used, in principle, on any material. AFM measurements on biomolecules may be performed in *contact mode*, in which the tip exerts a substantial force (on a molecular scale) on the sample. There is also a significant frictional interaction as the tip slides across the sample surface. The forces involved are usually more than adequate to displace biomolecules. A variety of approaches have been explored to solve these problems, including the use of covalent coupling schemes to tether biomolecules in place. Another approach is to crystallise the sample into a periodic array, and rely on the cohesive forces within the close-packed molecular assembly to counterbalance the disruptive influence of the tip. Although not all biomolecules may be crystallised, this approach has led to some spectacular successes, including insights into the molecular structure of membrane proteins. In studies of bacterial surface layers, or S-layers (the proteins that constitute the outermost layer of the cell wall), it was possible to examine the effects of enzymatic digestion with a spatial resolution better than 1 nm. The S-layers were deposited onto mica substrates and found to form bilayers or multilayers, The topmost layer exhibited a triangular structure when imaged at low force (100 pN); however, imaging at elevated loads (600 pN) led to removal of the top layer and exposure of a hexameric flower-shaped morphology.

[1] C. R. Clemmer and T. P. Beebe, Science **251** 640 (1991)

S-layers that had been enzymatically digested were found to be present as single layers that exhibited each type of surface with equal probability. In another study, AFM data on several membrane proteins were presented with a resolution of better than 0.7 nm. Importantly, in these studies, raw AFM data were presented that clearly exhibited substructural details of individual protein molecules. In contrast, electron micrographs with the best resolution typically represent averaged data from a large number of molecules. The AFM data enable the observation of crystal defects, or molecule-to-molecule variability in structure. Nevertheless, computational analysis of AFM images of large assemblies is still possible, leading to image averaging or more sophisticated analyses. One of the components of the photosynthetic apparatus of the bacterium *Rhodospirillum rubrum*, has been studied and shown to consist of a ring structure (the light-harvesting complex LH1) containing within it the reaction centre (RC). The RC receives energy from the LH1. Two-dimensional crystals of the complex were formed and deposited onto mica. Contact mode images revealed patterns of alternating bright and dark rows (Figure 9.1). These resulted from the existence of two distinct orientations for the RC–LH1 complex. On the cytoplasmic side of the complex (the cytoplasm is the content of the cell within the plasma membrane), the reaction centre protrudes, and is observed as a feature with bright contrast, whereas on the periplasmic side (the periplasm lies between the plasma membrane and the outer membrane of the cell) dark contrast is observed over the centre of the complex. A small number of crystal defects were observed, in which the LH1 complex adopted a different morphology; these would have been lost in electron microscopy investigations due to averaging. In some cases the reaction centre was observed to be missing, even at low loads, possibly attributable to its removal by the tip as it traversed the crystal. Imaging at loads of 200–300 pN was found to yield the best resolution. On the periplasmic side of the complex, an X-shaped structure was observed, which was attributed to the periplasmic face of the RC.

The development of *tapping mode* AFM has been one of the most useful developments for the imaging of biological specimens. Contact mode imaging may lead to the disruption of surface structure at elevated loads. For biological specimens, that may be composed of isolated molecules distributed on a solid substrate, and often interacting

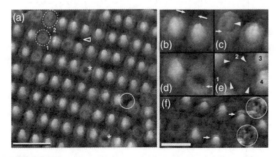

Figure 9.1 (a) High-resolution AFM image of a two-dimensional crystal of RC–LH1 complexes. The broken circle (1) and the ellipse (2) mark complexes that lack the RC–H subunit. The asterisks denote 'empty' LH1 complexes that completely lack the RC seen from the cytoplasmic side. The arrow denotes a missing RC seen from the periplasmic side. (b to e) show a variety of complexes at higher magnification. (f) shows the periplasmic side of the RC–LH1 complex imaged at higher load. The scale bars are 40 nm in (a) and 15 nm in (f). Reproduced with permission from D. Fotiadis *et al.*, J. Biol. Chem **279**, 2063 (2004)

only weakly with it, there is a real danger that the tip will move molecules around. In tapping mode, the cantilever oscillates at high frequency (100–200 kHz) and high amplitude, and only intermittently strikes the sample. This eliminates the frictional forces that contribute to damage, and reduces the rate of energy dissipation, rendering the topographical imaging of delicate materials much easier. Energy dissipation still occurs, and provides access to important additional information in the form of phase images. In *phase imaging*, the lag between the driving oscillation and the cantilever response is measured. Elastic contacts lead to a small phase lag, while contacts with viscoelastic materials, that result in a higher rate of energy dissipation, lead to a larger phase angle. Phase images reveal local variations in mechanical properties (e.g., stiffness).

Tapping mode has recently been facilitating the imaging of proteins adsorbed onto solid surfaces. In contrast to the beautiful images acquired for protein crystals, these data are typically less well resolved but do provide data for isolated molecules. For example, the von Willebrand factor (VWF), a large multimeric protein that adheres rapidly to biomaterial surfaces upon exposure to blood, has been imaged. Protein–surface interactions play a key role in regulating thrombus formation, a phenomenon of great importance when biomaterials are placed in contact with the blood because it can lead to failure of the prosthetic device. VWF adsorbed to hydrophobic monolayers of octadecyltrichlorosilane (OTS) adsorbed on glass has been compared to VWF adsorbed onto hydrophilic mica. On the OTS monolayers, VWF was found to exhibit a coiled conformation, while on mica the polypeptide chains were observed to adopt extended conformations which exhibited much larger end-to-end dimensions.

Fibronectin (FN) is another protein of considerable importance in the development of prosthetic biomaterials. FN plays an important role in cellular attachment, and is recognised by integrin receptors in cell membranes, which regulate the mechanism of attachment. FN is a dimeric protein, consisting of two polypeptide chains joined by disulphide linkages. A specific region of the molecule, containing the tripeptide sequence arginine–glycine–aspartic acid (RGD), is recognised by the integrin receptors. FN undergoes surface-specific conformational changes, and these changes in conformation lead to differences in the orientation of the cell binding domain of the molecule with respect to the solid surface on which the molecule adsorbs. The characterisation of the conformations of adsorbed proteins is very challenging, and many techniques, such as infrared spectroscopy, provide only limited information. Tapping mode AFM has been used to image FN adsorbed onto the surface of mica, single FN molecules being observed. FN was exposed to heparin-functionalised gold nanoparticles. Bound nanoparticles could be resolved as bright features situated part-way along the FN polypeptide chain, enabling the binding site to be estimated. It was concluded that there were two binding sites, based on the AFM data, attributed to the Hep I and Hep II sites previously identified using biochemical means. A difference in binding affinity for the two sites was postulated, based on the observation that twice as many functionalised nanoparticles were observed to bind to Hep I than to Hep II.

Scanning near field optical microscopy (SNOM) – see Section 2.2.4.1 – provides a route to fluorescence measurements of biological molecules with a spatial resolution of around 50 nm – well below the diffraction limit. When combined with suitable collection optics (such as an avalanche photon counting system), the detection of optical data from single molecules becomes feasible. If SNOM is combined with a Raman spectrometer, it is possible to acquire spectroscopic data with similar resolution by taking advantage of the surface-enhanced Raman effect on appropriate substrates.

Recently there has been interest in apertureless SNOM methods, in which the optical fibre is replaced with a tip. The tip is either fabricated from a noble metal, or coated with one. The tip and sample are irradiated using a laser with the tip in close proximity to the sample. A surface plasmon is excited at the tip surface, and in the region directly beneath the tip, the electrostatic field experiences a strong near-field enhancement. The field associated with a surface plasmon is spatially highly confined in the region of a noble metal asperity, leading to an intense excitation of the sample in a small defined region. Published data (including apertureless Raman microscopy) suggest a spatial resolution of 25 nm is feasible, and better resolution may well be possible.

9.1.2 Force measurement in biological systems

Force–distance measurement (also known as force spectroscopy) using the atomic force microscope has been particularly important in the investigation of biological systems. In a force–distance measurement, the AFM tip is lowered towards the sample surface. When the tip approaches very close to the surface, a mechanical instability causes it to snap into contact. Bringing the tip closer leads to a repulsive interaction, due to the quantum mechanical repulsions experienced by atoms at close proximity. The cycle is then reversed and the tip retracted from the surface. Adhesion of the tip to the surface results in hysteresis; i.e., the path followed during retraction of the tip does not exactly follow the approach path. In particular, the tip must be pulled further than the initial point of contact to separate it from the surface. Eventually the tip separates from the surface, and the adhesive load immediately prior to separation is referred to as the pull-off force or the adhesion force.

There are now a large number of illustrations of the kinds of measurement that can be made on biological systems in this way. Biotinylated bovine serum albumin (BSA) has been adsorbed to a glass microsphere attached to an AFM cantilever, and the interaction force with streptavidin distributed on a mica surface has been measured (Figure 9.2). The biotin–streptavidin lock-and-key recognition mechanism has been studied using a biotinylated agarose bean and a streptavidin-coated tip. The pull-off forces were measured and subsequently the relationship between the interaction force and enthalpy was investigated. The recognition forces between complementary strands of DNA have also been studied. The DNA strands were thiolated at their 3' and 5' ends[2] for attachment to alkylsilane monolayers attached to a silica probe and a planar surface. A slightly different approach is to attach bases to tip and surface and measure single base-pair interactionss and this approach has been adopted to try to develop a sensor. Peptide nucleic acids (PNAs) have been modified with cysteine to enable attachment to gold-coated AFM tips, and interaction forces with alkanethiol monolayers were probed

---

[2] Each of the two strands of DNA terminates in a hydroxyl (3') at one end and a phosphate (5') at the other. The strands are intertwined such that the double helix has a complementary 3' and 5' at each end.

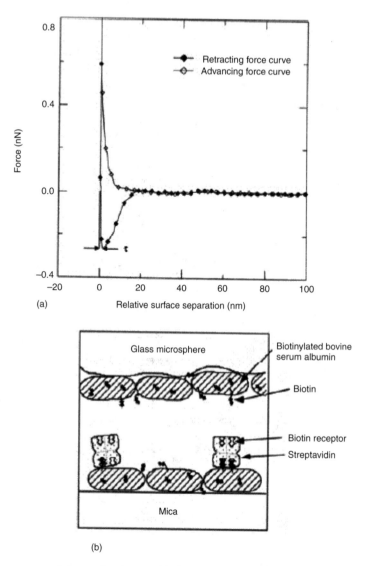

Figure 9.2 Top: measured forces between a biotinylated bovine serum albumin-coated glass microsphere attached to an AFM cantilever and a streptavidin surface in pH 7.0 phosphate buffered saline solution. Bottom: schematic diagram showing the experimental arrangement. Reproduced with permission from G. U. Lee, D. A. Kidwell and R. J. Colton, *Langmuir* **10**, 354 (1994)

before and after hybridisation with PNA or RNA. The pull-off force was reduced by hybridisation. A DNA-modified latex microparticle has been used as a probe of micron-scale patterned arrays of immobilised oligonucleotides.

These illustrations suggest the potential usefulness of techniques based on highly sensitive recognition measurements by AFM. At a fundamental level, the field of SPM-based force measurement in biology has become a very large one. Force spectroscopy has become a more or less established biophysical tool, although the interpretation of force data remains complex, and the subject of intense academic activity. Importantly for applications in other areas, the complexity of phenomena such as protein unfolding means that practitioners of biological force microscopy have had to grapple with a variety of technical issues that while less important in simpler systems, may still be significant. One illustration is the issue of rate-dependent phenomena. It has been demonstrated that the application of an external mechanical force effectively tilts the 'energy landscape' for the unfolding process, reducing the activation energy. This means it is necessary to explore unfolding processes at a range of unloading rates in order to accurately quantify the events involved. The procedure has now been widely adopted and has facilitated quantitative investigation of a range of phenomena.

The application of AFM to the characterisation of cellular structure and function presents significant experimental challenges. In particular, cells are rather soft structures. The cell membrane is a fairly fluid structure, being composed of a bilayer of lipid molecules, and the internal contents of the cell are fluid. However, AFM offers the exciting possibility of probing cellular properties and interactions with very high spatial resolution. Recent work suggests that the rewards may be considerable. Extremely elegant investigations of cellular structure have been conducted, which have demonstrated that AFM can be utilized to probe the structures of living rat liver macrophages and chicken cardiocytes. Actin is a small protein that becomes organised to form filamentous structures as cellular attachment begins to occur. The process of stress fibre formation is linked with the formation of focal adhesions that anchor cells to the substratum on which they are cultured. AFM has been used to probe the mechanical properties of actin stress fibres, which can be readily imaged. Remarkably, it proved possible to conduct investigations on migrating cells (Figure 9.3), and it was possible to detect mechanical differences between the active and stable edges of motile fibroblasts. Other cell types have also been investigated such as microbial cells, for which surprisingly well-resolved images were obtained. It is also possible to conduct force–distance measurements on cell surfaces, and to compare the adhesiveness of microbial cell surfaces. By varying the pH of the ionic fluid medium in which the pull-off forces were measured, it is possible to investigate the surface charge distribution of the cell surface.

Optical tweezers provide an alternative means to AFM for the investigation of interaction forces in biological systems. They rely on trapping a dielectric particle in a laser beam (Figure 9.4). The extremely high electric field gradient near to a tightly focused laser beam exerts a mechanical force on a dielectric particle placed close to it, drawing it towards the centre of the beam and propelling it in the direction of propagation. Although two counterpropagating beams may be used, and the name appears to imply two pincers gripping an object, in fact a single beam is sufficient and provides the basis for a typical modern instrument. The diameter of the particle is critical, and should be about 1 μm. It is usually fabricated from polymer. It is not possible to trap individual molecules using optical tweezers, so to measure the force of a biological interaction at the molecular level it is necessary to use a functionalised bead instead. Nevertheless, the method is extraordinarily powerful.

The mechanical actions of molecules involved in the contraction of muscle have been studied. When muscle contracts, mysosin pulls on actin filaments (protein fibres formed by the association of actin molecules). It was possible to investigate this action by attaching an actin fibre between two microspheres (Figure 9.4). The filament was then placed close to a third microsphere coated with myosin. The whole assembly was

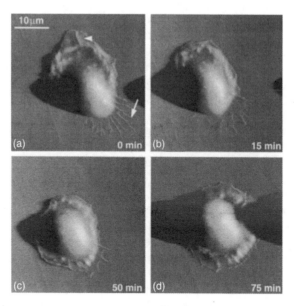

Figure 9.3 Cellular dynamics imaged in real time by AFM. Spreading is evident at the lower edge of the cell in the first frame, and by 15 min a fully developed, flat lamellipodium is evident. Simultaneously the cell's upper body is retracted and increased in height. After 50 min, all lamellipodia are retracted and the cell adopts a rounded morphology. After 75 min the cell may be observed to undergo spreading again. Reproduced with permission from C. Rotsch, F. Braet, E. Wisse and M. Radmacher, *Cell Biol. Intl.* **21**, 685 (1997)

submerged within a fluid medium containing adenosine triphosphate (ATP), which serves as a fuel supply for biological molecules. During contact between myosin and actin, an ATP molecule was digested and a tug was registered on the actin filament. The force exerted on one of the beads could be measured using optical tweezers, and the influence of the ATP concentration on the rate of the displacement process measured.

The interaction between FN and the cell cytoskeleton has been examined using laser tweezers. A short segment of FN, containing the cell binding domain, was attached to silica beads and was then allowed to interact with a cell. The bead bound to one of the integrin receptors, the $\alpha_v\beta_3$ receptor, known to recognise the part of the FN molecule containing the RGD sequence. When a bead attached to a motile lamellipodium (a part of the cell membrane that is being extended as the cell pulls itself forwards), the bead was observed to move, exerting force on the bead. The force was measured using the optical tweezers. Beads moved until the force was great enough to cleave the bond between cell and bead, at which point a sudden movement of the bead was observed, as it abruptly returned to the centre of the optical trap. The force of interaction between the functionalised bead and the cell could thus be measured. When a peptide containing the cell-binding sequence RGD was added to the culture medium, the force was observed to be reduced. The presence of talin a protein found on the interior of the cell and thought to be involved in the process of focal contact formation, linking integrin receptors to actin within the cell, is necessary to be able to measure the interaction between the cell and the bead. Remarkably, this force was estimated to be only 2 pN.

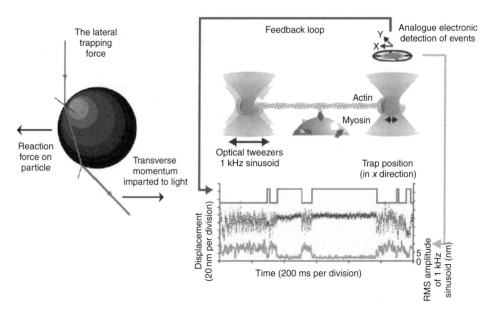

Figure 9.4 Left: schematic diagram showing the optical trapping of a dielectric particle. Right: schematic diagram showing the experimental arrangement employed by Molloy and co-workers to measure the interaction between myosin and actin. The recording shows the bead position measured in parallel with the actin filament axis versus time. To detect myosin binding with adequate time resolution, the position of one laser tweezer was oscillated at a frequency of 1 kHz and an amplitude of 35 nm RMS. The transmission of this signal to the bead in the stationary tweezer was determined from its discrete Fourier transform (grey trace) so that the position of both optical tweezers could be moved rapidly (blue trace) to apply load to the bound crossbridge. Reproduced with permission from J. E. Molloy and M. J. Padgett, *Contemp. Phys.* **43**, 241 (2002) (published by Taylor and Francis Ltd., www.tandf.co.uk/journals) and C. Viegel, J. E. Molloy, S. Schmitz and J. Kendrick-Jones, *Nature Cell Biol.* **5**, 980 (2003)

Cells respond to mechanical stresses. The process of mechanotransduction, by which cells respond to external mechanical stresses by the triggering of internal biochemical pathways, has attracted growing interest. Mechanical stress has a profound influence on the actin cytoskeleton, and it appears that physical stimuli may activate genes and signalling proteins that are also triggered by molecules that bind to specific cell surface receptors. Cells may be switched between entirely different phenotypes by alterations to the extracellular matrix structure or mechanical influences that induce changes in cell shape, independent of growth factor binding or integrin binding. These observations have stimulated a search for methods to induce cell stress in a defined fashion, and for making quantitative measurements of the mechanical behaviour of cells.

One of the best ways to induce stress in cells is to constrain their spreading by culturing them on surfaces patterned with regions that are adhesive and resistant to attachment. Much pioneering work has been carried out by the Whitesides group at Harvard. They have used self-assembled monolayers (SAMs) of alkanethiols ($HS(CH_2)_nX$) to control surface chemical structure. The SAMs may readily be patterned by microcontact printing, in which an elastomeric stamp, inked with a solution of a thiol of interest, is used to transfer it to a gold surface. The stamp is removed to leave behind a molecular pattern.

The gaps in the pattern (bare regions of gold) may be filled in by immersing the sample in a solution of a contrasting thiol in ethanol. The second thiol adsorbs at the surface, yielding a surface composed of geometrically well-defined regions with very different biological responses. If one of the thiols is terminated in an oligo(ethylene glycol) unit, it will resist the adsorption of proteins, rendering it non-adhesive to cells. However, if the other thiol is, for example, a methyl-terminated adsorbate, then proteins will adsorb and attachment will result. By defining the regions at the surface occupied by the methyl-terminated thiol, it is possible to precisely define the shapes and sizes of the areas of the surface to which cells may attach. One of the most significant results has been the observation that the area of the adhesive regions cannot be reduced arbitrarily. Eventually a point comes where the patches are so small that cell spreading is highly constrained, and the process of apoptosis, or programmed cell death, is triggered.

Several methods have been used to measure the forces exerted by cells, for example using a bed of microneedles. The microneedles are fabricated in poly(dimethyl siloxane) (PDMS). Using photolithographic methods, an array of pillars is first formed in a silicon master. PDMS is then cast onto this master and cured, leading to the formation of a PDMS master that contains an array of wells. A second batch of PDMS is then cast onto this master, and when cured and removed it consists of an array of elastomeric needles. Cells may be cultured on the needles, and as they grow and exert mechanical forces, the needles bend. If the mechanical properties of the PDMS are known, then the force exerted by the cells may be determined from the deflection of the needles.

9.1.3 Miniaturisation and analysis

The miniaturisation of analytical systems brings a variety of advantages for the investigation of biological systems. One motivation is that it is possible to carry out large numbers of experiments rapidly in parallel. There has recently been a great deal of interest in chip-based methods for the characterisation of biological systems. One approach that has been important in the emergence of genomics, the branch of science that seeks to sequence the genome of a species and then utilize the resulting information to explain and predict behaviour. The method has been enabled by the development of array-based systems of analysis. In a *microarray*, spots are deposited onto a solid support. Each spot contains a single molecule with a known identity. Because of the specificity of biological recognition each molecule will bind a specific partner, so that each spot functions as a sensor for that partner. The array is exposed to a test solution, and afterwards it is determined which spots (i.e., which molecules) have bound their partner. Usually this is achieved by the use of a fluorescent conjugate; using an optical microscope, the array is scanned and positive results scored for spots that are found to emit fluorescence. The spots in the array may range from a few tens of micrometres to 100 µm in size and the arrays may be large, containing thousands of different molecules. In DNA analysis each spot may contain a short sequence (oligonucleotide) that is complementary to a specific gene (there are approximately 40 000 in the human genome, so an array encoding the entire genome would be very extensive). The advantage for the biologist is that in a single experiment it is possible to examine whether a very large number of genes are present in the DNA sample, or even to sequence the DNA.

There are a variety of approaches to the fabrication of such arrays. In one method, oligonucleotides are synthesised in situ on a solid substrate using photolithographic methods. Photocleavable protecting groups are attached to each sequence as it grows. By selectively exposing particular spots, they may be deprotected and a particular base attached. It is essential that the chemistry is extremely effective, because even a tiny failure rate (a few percent) in the attachment of a specific base will result in a significant probability that an error will occur during the synthesis of a lengthy sequence (say 20 bases). Other approaches involve the spotting of molecules (which may be oligonucleotides or other molecules, such as proteins) using inkjet printing technology.

Although these methods offer very high throughput, there are significant drawbacks. The interaction between the solid substrate is critical and it is necessary to optimise this in order to ensure efficient binding. However, this remains problematic and hybridisation efficiencies are low in commercial systems. The sensitivities are also limited. In the case of DNA analysis, it is necessary to use polymerase chain reaction (PCR) technology to 'amplify' samples. PCR creates multiple duplicates of a piece of DNA, increasing the sample size and thus helping to counterbalance the poor sensitivity of the micro-array analysis. However, for proteins there is no equivalent. Moreover, in protein analysis, antibodies are typically bound to the solid substrate, but there are no established methods for the efficient immobilisation of antibodies such that more than a small percentage of them are able to bind to their complementary antigen. Consequently, there is a tension between the limitations of the methods, on the one hand, and their potential benefits to biologists through provision of very high throughput on the other.

The improvements in sensitivity that may potentially be realised through miniaturised systems, and the possibilities offered for high-throughput screening via large numbers of miniaturised experiments in parallel, have sparked enormous interest in the development of methods for handling small volumes of materials. Interest in microfluidic systems; i.e., miniaturised flow systems, has been growing rapidly, not just for applications in biology, but more generally across the field of analytical science. The concept of a lab-on-a-chip has now firmly taken root. The basic idea, called micro total analysis, is to undertake a complete process on a single miniaturised chip, typically fabricated from silicon or a polymer such as PDMS. The chip consists of miniaturised reactors, flow channels for manipulating samples and miniaturised valves for controlling the flow of reagents. Flow in small channels presents some interesting technological challenges. Fluid behaviour is often dominated by the interfacial free energy of the fluid–solid interface. Whereas in a macroscopic flow channel or reactor, the surface region is a small fraction of the total fluid volume, this is not true in a channel that is $\sim$ 10-100 µm wide. Reynolds numbers may also be very low, leading to laminar flow, in which separate streams of fluid converging into a single channel may proceed essentially unmixed (Section 8.8.2). Some mixing does occur, by diffusion, at the interface between separate streams, but this may be controlled such that the width of the interface region remains small compared to the width of the flow channel or cell. The approach has used to produce multiple laminar streams to expose different parts of a cell to different reagents. By controlling the flow velocity, channel geometry, channel wall interactions and diffusion coefficient of specific reagents, they could exert significant control over the composition of the laminar flow stream reaching a particular cell (Figure 9.5).

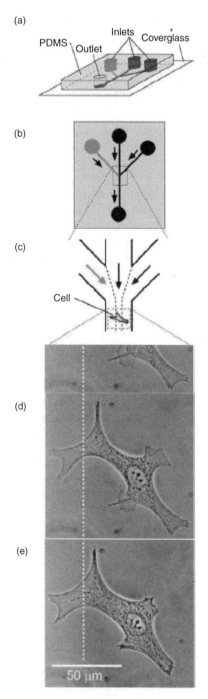

Figure 9.5 Use of microfluidic channels to generate two spatially localised populations (fluorescently labelled and unlabelled) of mitochondria within a single bovine capillary endothelial (BCE) cell. Reproduced with permission from S. Takayama, E. Ostuni, P. LeDuc, K. Naruse, D. E. Ingber and G. M. Whitesides, *Chem. and Biol.* **10**, 123 (2003)

The movement of fluids in small channels presents a variety of technological challenges. One approach is to use pressure-driven flow, by analogy with a macroscopic pumping system. Miniaturised pumps may be used. An alternative approach is to use centrifugal flow, in which the microfluidic system is spun so that fluid is forced outwards through channels. Controlling the flow of very small volumes can be problematic, and as channels become very small, larger pressure differentials are required. Electro-osmotic flow provides an alternative approach. If the walls of the flow channel are charged, then the fluid in the layer close to the walls will be charged, containing a higher than average density of counterions. If an electric potential is applied parallel to the flow channel, then this fluid will move under the influence of the resulting electric field. The result is a convective motion.

A variety of separation and analytical devices may be combined with a microfluidic network to yield the micro total analysis system. Separation is a critical operation in biological analysis, where samples may often be complex and contain multiple components. Nanotechnology is providing new tools for sample handling including separation, and these tools may ultimately be integrated into micro total analysis systems. One illustration is the use of lithographic techniques to fabricate chambers containing regular arrays of silicon pillars that provide obstacles to the movement of DNA molecules when the sample is added. The structure has a regular and controllable structure, in contrast to the gels more conventionally used in biological analysis, and is thus subjectable to a rigorous physical analysis, permitting accurate and effective separation of DNA fragments by their molecular weight.

Magnetism provides a useful approach to separation of biological components. Although biological molecules and cells are strongly influenced by electric fields, they are typically non-magnetic. Attachment to magnetic nanoparticles thus provides an elegant method of moving them in a very precisely controlled fashion through a miniaturised experimental system. Superparamagnetic particles have been used to separate DNA in a microchannel device fabricated using soft lithography. Under the influence of a magnetic field, the particles may not only be caused to move but also to organise themselves (e.g., into columns). If suitable linkers are bound to the surfaces of the particles (e.g., pairs of molecules with a strong recognition, such as biotin and streptavidin) then the arrangement may be made permanent when the magnetic field is removed.

Fluorescent labelling provides a powerful and widely used approach to the investigation of cellular structure and organisation. Biochemical probes (often antibodies) functionalised with fluorescent dyes are used to highlight structural features (e.g., the actin cytoskeleton in cells is often highlighted using FITC-conjugated phalloidin). Nanoparticles offer a different approach to analysis with ultra-high sensitivity. Quantum dots made of materials such as cadmium selenide may have high fluorescence yields and offer the possibility of conducting optical analyses at high exposures without the problems of photobleaching commonly associated with fluorescent labels. However, they are also cytotoxic so there has been interest in trying to derivatise them with molecular adsorbates in order to render them inert in the cellular environment. One approach has been to use alkanethiols, which interact reasonably strongly with CdSe nanoparticles and passivate them. Other approaches exist based on gold nanoparticles, which may also be derivatised using alkanethiols and utilized as probes for cellular structure.

9.1.4 Organisation of biomolecular structure at the nanometre scale

There has been growing interest in the manipulation of biomolecular structure on length scales approaching molecular dimensions. This is a far more challenging undertaking than the examples discussed in the preceding section, which really only explore structural effects on micron length scales. There are several motivations for seeking to manipulate molecular organisation on smaller length scales, including the possibility of fabricating highly miniaturised devices for biomolecular detection and analysis (extrapolating the gains in sensitivity and throughput realised through the adoption of miniaturised micron-scale analytical systems) and the desire to build systems for fundamental investigations of biological interactions and organisation (e.g., the fabrication of arrays of proteins molecule by molecule to challenge cells or investigate molecular recognition). There has been much more speculative discussion of bioelectronic devices too. By and large, however, it is fair to summarise the state of the art as focused on the development of new tools – working devices will require much further development.

Microcontact printing (μCP), in which an elastomeric stamp is formed by casting PDMS against a (usually) silicon master, has already been described in Section 1.4. μCP provides a rapid, straightforward and flexible method for the deposition of molecules in patterns on surfaces, and has attracted a great deal of interest as a means for organising biological molecules and cells. The earliest application of μCP was the deposition of alkanethiols onto gold. These structures were used to guide cellular attachment, or to pattern the adsorption of biological molecules. However, it is possible to stamp other molecules, including alkylsilanes onto silicon dioxide to form patterned monolayers, and polymers. The stamping of polymers is an attractive recent development, because of their stability and also because of the possibility of generating molecular relief in this way.

Although there are reports in the literature of the fabrication of structures with dimensions less than 100 nm by microcontact printing, physical limitations in the process have led researchers to explore methods based on scanning probe microscopy for the fabrication of patterns with dimensions approaching the molecular level. Dip pen nanolithography (DPN) is a method which offers much by way of analogy with microcontact printing. In DPN molecules are deposited on the surface from an atomic force microscope tip, rather than an elastomeric stamp (Figure 9.6). The tip is inked with a solution of the molecule of interest and brought into contact with the sample surface. Under normal ambient conditions, a capillary forms between an AFM tip and the surface that it contacts. In DPN this capillary functions as a liquid bridge to facilitate transfer of fluid from the tip to the surface. Control of the ambient humidity therefore has an influence on the sizes of the features formed. The first demonstrations of DPN were based on the deposition of alkanethiols onto gold surfaces. However, a variety of molecules may be deposited, including alkylsilanes, which form monolayers on silicon dioxide, and conducting polymers, by using an electrochemical AFM to polymerise 3,4-ethylenedioxythiophene during deposition. By writing simultaneously with an array of cantilevers, DPN also offers the possibility of generating multiple structures in parallel. One of the widely quoted criticisms of lithographic methods based on SPM is that, unlike conventional photolithography (a parallel process in which an entire circuit incorporating a large number of features may be fabricated in a single

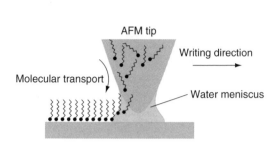

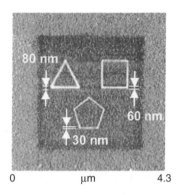

Figure 9.6   Left: schematic diagram showing the basic principle behind DPN. Right: polygons drawn by depositing mercaptohexadecanoic acid onto gold using DPN. The area around the polygons has been overwritten with a monolayer of octadecanethiol. Reproduced from S. Hong, J. Zhu and C. A. Mirkin, *Science* **286**, 523 (1999)

process), they are serial methods – features are created one after the other. Serial writing is time-consuming, and this has been an obstacle to the exploitation of electron beam lithography for electronic device manufacturing. However, the invention of the Millipede by Gerd Binnig and co-workers has provided a new paradigm for nanofabrication using SPM. The Millipede is a microelectromechanical device consisting of several thousand AFM-type cantilevers, each capable of being separately and simultaneously actuated. It thus provides the capability to implement a massive number of serial lithography functions in parallel, meaning that its capability is comparable to that of a parallel fabrication method.

A variety of biological nanofabrication processes have been demonstrated using DPN. One approach is to write thiol structures onto a surface and then attach biological molecules to these (see Figure 9.6 for an example). Patterns of mercaptohexadecanoic acid (MHA) were created on a gold substrate, then DNA was coupled to them. Gold nanoparticles functionalised with a complementary sequence of DNA were then bound to these. An alternative but equivalent approach is to deposit MHA by DPN and then use these nanopatterns as a resist for etching the underlying gold substrate. This allows gold to be removed from the sample, except where MHA is adsorbed, leading to the formation of MHA-capped gold nanostructures. After removal of the MHA by exposure to UV light, bare gold nanostructures are formed which may then be functionalised with thiol-linked DNA molecules. Gold nanoparticles derivatised with a complementary sequence may be bound to these. Thiol-linked oligonucleotides may also be deposited directly onto gold surfaces by incorporating the oligonucleotide into the ink. In this way, DNA features about 150 nm wide were written. Similar strategies have been applied to the deposition of proteins. MHA has been patterned onto gold and then bare regions between the MHA features have been filled in with an oligoethylene glycol (OEG) functionalised thiol. Proteins are extremely adhesive, and in the fabrication of any miniaturised protein structure, the inhibition of protein adsorption is a critical concern. Polyethylene glycol is a polymer that is highly resistant to the adsorption of proteins, and it has been demonstrated that OEG-terminated thiols are also highly resistant to protein adsorption and thus a powerful tool for the manipulation of biological organisation at surfaces. MHA/OEG patterns have been exposed to immunoglobulin G (IgG) leading to adsorption only on the MHA-functionalised regions of surface. It is also possible to deposit protein directly, however. An AFM tip has been inked in a solution of IgG and written directly onto bare silicon dioxide and silicon dioxide modified with aldehyde functionalities to covalently bind the protein.

DPN is a method based on the deposition of molecules, but other methods exist that enable the selective removal of material. One approach is to physically scrape adsorbates from a solid substrate, an approach known as nanoshaving. In nanoshaving, the tip of an AFM is scanned across a region of surface at elevated load while submerged beneath a solution of a thiolated molecule. As thiols are displaced from the surface, it is refunctionalised with fresh molecules adsorbing from the solution phase. The approach may be extended to the immobilisation of DNA, by shaving the thiol layer with a thiol-linked DNA strand in the solution phase. An AFM tip has been used to scratch holes in monolayers of thiol-linked DNA molecules. The thickness of the functionalised DNA monolayer was measured from the height of the step between the top of the monolayer and the adjacent region of surface. A thiolated DNA strand of a different (longer) length was then inserted into the holes. The surface was then incubated with a still longer DNA strand capable of hybridising to immobilised sequences, and again, binding could be verified by measuring the change in height in AFM images. Nanoshaving may also be used to immobilise proteins. Lysozyme has been adsorbed to charged patches introduced to methyl and oligoethylene glycol terminated surfaces, and has been covalently bound to patterned structures created by nanoshaving. Small patches have been created in an oligoethylene glycol terminated monolayer, and three different thiols have been adsorbed to them, terminated by methyl, amine and carboxylic acid groups. When the sample was exposed to a solution of lysozyme, the protein adsorbed only to the carboxylic acid functionalised region (Figure 9.7).

Photopatterning provides a convenient way to pattern self-assembled monolayers (SAMs) of alkanethiols, and has proved a useful method for patterning biological organisation at the micron scale. On exposure to UV light in the presence of air, alkanethiolates are oxidised to weakly bound alkylsulfonates which may be readily

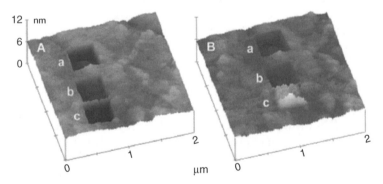

Figure 9.7 AFM topographical images showing three differently charged patches of area about 400 nm × 400 nm before (left) and after (right) exposure to lysozyme. In each case, (a) is neutral (b) is positively charged and (c) is negatively charged. Reproduced from X. Zhou, L. Wang, R. Birch, T. Rayment and C. Abell, *Langmuir* **19**, 10557 (2003)

displaced by solution-phase thiols. Exposure of the monolayer through a mask leads to spatially selective oxidation, so the subsequent immersion step results in the formation of the new chemistry in exposed regions while retaining the original chemistry in masked areas. The diffraction limit of $\lambda/2$ associated with optical techniques has generally been regarded as placing a lower limit on the sizes of features that may be created photolithographically. However, a scanning near-field optical microscope coupled to a UV laser has been used to break the diffraction limit and fabricate features as small as $\lambda/9$. This approach has been called scanning near-field photolithography (SNP). Lines were traced in fresh monolayers, leading to selective oxidation in regions as narrow as 25 nm (see Figure 8.5). Immersion of the sample in a solution of a contrasting thiol led to formation of a chemical pattern. Immersion in a solution of an etchant for the underlying gold substrate led to its selective removal, resulting in the formation of nanotrenches as narrow as 50 nm. Patterned SAMs formed in this way may be functionalised using proteins and polymer nanoparticles, opening up a variety of strategies for surface functionalisation. Finally, hydrogen-passivated silicon samples were also patterned under a layer of fluid alkene, leading to selective attachment to the surface and forming nanostructured alkylsilicon structures. The ability to pattern a wide range of materials with nanometre spatial resolution, combined with the capability to function in a fluid medium means that SNP potentially provides a route to the rapid fabrication of complex, multicomponent arrays with nanoscale features, a valuable and unique possibility.

9.2 BIOMIMETIC NANOTECHNOLOGY

Biology offers some outstandingly elegant and effective examples of the principles of self-assembly. It is also a truism that cell biology offers a compelling existence proof that a radical nanotechnology, involving sophisticated machines and mechanisms, is possible. In this section we consider some examples where mechanisms from biology can be exploited to make nanoscale devices. This is clearly an area of huge potential, and these initial efforts will surely be extensively built on in years to come.

9.2.1 DNA as a nanotechnology building block

The remarkable and very specific base-pairing mechanism that underlies the operation of DNA makes this molecule a very attractive candidate as a building block for creating complex nanostructures by self-assembly.

The key features that underly the self-assembly of DNA are as follows. A single DNA chain is a sequenced copolymer in which one of four possible bases are attached to a backbone of alternating sugar and phosphate groups. The four bases are adenine (A), guanine (G), cytosine (C) and thymine (T) and they form complementary pairs – A and T, C and G – in which strong, multiple hydrogen bonds hold the complementary pairs of bases in an edge-to-edge configuration. Two complementary strands of DNA will strongly associate to form a double helix. In nature DNA always exists as a pair of strands that are complementary to each other over their whole length; these associate to

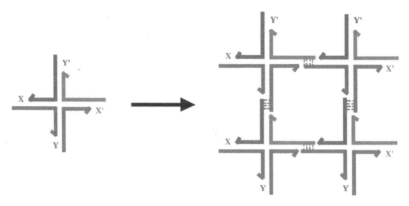

Figure 9.8 Formation of a 2D lattice from a junction with sticky ends. X and Y are sticky ends and X′ and Y′ are their complements. Four of the monomers on the left are complexed to yield the structure on the right. DNA ligase can close the gaps left in the complex, which can be extended by the addition of more monomers. From N. C. Seeman and A. M. Belcher, *Proc. Natl Acad. Sci. USA* **99**, 6451 (2002)

form a single, linear double helix. However, synthetic methodologies exist that allow one to make DNA strands with an arbitrary sequence of bases. This allows one to design more complicated topologies. Consider three strands with the following sequences. Strand 1 can be thought of as two sequences, X and Y, joined together. Strand 2 consists of X′, the complementary sequence to X, and another sequence Z. Strand 3 consists of Z′, the complementary sequence to Z, joined to Y′, the complementary sequence to Y. When these three strands are put together, they associate to form a branch point where three double helices merge. Many complex junctions can be designed in this way.

The key structural unit in building three-dimensional structures from DNA is the sticky end. Imagine a pair of complementary DNA strands, one of which has been extended by a short additional sequence. This short piece of single DNA is available for bonding with a sequence of bases on another DNA helix with the complementary sticky end. As shown in Figure 9.8, if branched DNA constructs can be prepared with properly designed sticky ends, they will self-assemble in two or even three dimensions.

This area of research provides an impressive proof-of-principle demonstration of the power of biomolecular self-assembly, but does it have any practical applications? Three possibilities stand out: firstly, as guides for the growth and interaction of nanoparticles; secondly, as templates for molecular electronics, and thirdly, to create nanomachines and motors. These are discussed in turn below.

9.2.1.1 Directed assembly using DNA

Nanoparticles will interact strongly with each other, but these interactions are typically non-specific and irreversible. It is possible, by attaching definite sequences of DNA to the surface of the nanoparticles, to program specific interactions that allow the nanoparticles to be used as building blocks to be assembled in quite precisely determined ways. Suppose we have two different types of nanoparticles, which we call A and B, and we want to arrange them in a three dimensional structure in which the A particles and B particles alternate. This may be achieved in a very elegant manner by creating two different, non-complementary sequences of DNA. One sequence is attached to particle A, the other to particle B. For example, if the nanoparticles are gold colloids, and the DNA sequences are terminated with an alkanethiol group, the DNA can be easily end-grafted at the surface of the particles to form a polymer brush. If the two types of particle are mixed, each with their different sequences of DNA grafted on, there is no interaction. The DNA sequences, not being complementary, do not interact, and the nanoparticle suspension behaves simply as a sterically stabilised colloid. However, a DNA duplex with two sticky ends – one complementary to the sequence on the A particles, and the other complementary to that on the B particles – will act as a specific linking agent which will join A particles to B particles.

9.2.1.2 DNA as a template for molecular electronics

Can one use structures self-assembled from DNA to make electrical circuits? The work described in the previous section makes it clear that circuits of quite complex topologies can be formed by self-assembly from mixtures of DNA strands with carefully designed sequences. Unfortunately, the electrical properties of DNA molecules by themselves do not seem to permit the use of the molecules as the basis of electronic devices. It seems that DNA is an insulator. The early literature on this subject is confusing, with values of resistivity being reported that vary by ten orders of magnitude, but more recent results confirm its status as a good insulator.

Although DNA molecules cannot be used directly as molecular wires, it is possible to use them as templates to grow nanowires. In the first demonstration of this, DNA with thiol terminations at each end was attached to gold electrodes, spanning the $12-16\,\mu m$ gap between them. To convert the insulating DNA strand into an electrical conductor, the first step was to exchange the sodium counterions of the negatively charged DNA chain with silver ions. These silver ions were then reduced using a basic hydroquinone solution to give a string of silver nanoparticles along the DNA backbone. The system was then exposed to an acidic solution of hydroquinone and silver ions which, when illuminated, led to the deposition of more silver on the existing nanoparticle nuclei until a continuous silver nanowire was obtained.

By combining the idea of using DNA to make nanowires with its role in directing the assembly of nanoparticles, it is possible to self-assemble a functioning molecular electronic device. A specific sequence of single-stranded DNA was complexed with a protein molecule. Self-assembly was then used to direct this protein-decorated DNA strand to a specific, complementary strand of DNA. A carbon nanotube decorated with streptavidin was then directed to this DNA strand via an antibody to the bound protein. The DNA strand was then metallised with silver; the section of DNA to which the nanotube was bound was protected against metallization by the bound protein. The result was a semiconducting nanotube well connected to two nanowires, which, when a gating voltage was applied to the substrate, behaved as a field-effect transistor.

It is not difficult to see that by combining the ability of DNA to form complex, self-assembled three-dimensional structures with these methods of selectively metallising the

DNA and controllably binding, in selected locations, semiconducting elements such as nanotubes, semiconductor nanowires or conducting polymer molecules, it may be possible to take the first steps towards creating integrated circuits based on nanoscale semiconducting elements.

9.2.1.3 DNA-based motors and nanomachines

The specificity and predictability of the base-pair interaction makes it possible to design not just sequences that produce specific structures by self-assembly, but structures that yield specified function, such as catalysis. This allows one to go beyond DNA-based structures to actual, functioning nanomachines.

Figure 9.9 shows a particularly elegant scheme that illustrates the principles. Here a catalyst molecule M has been designed which binds to the complex QL', opening up the loop. A strand L, which is complementary to L', then displaces the catalyst molecule M, which is then free to begin another cycle. During this cycle the catalyst molecule undergoes a cyclic conformational change from coil to rod and back to coil again; in principle this conformational change could be exploited to do work. The energy source for this work is the difference in base-pair binding energies between the complex QL' and the duplex LL'. The key element in the design is that each of the stages results in a lowering of the free energy of the system without the need to overcome an energy barrier.

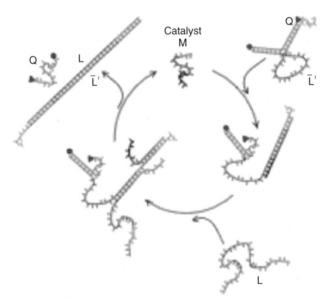

Figure 9.9 A free-running DNA nanomachine. The catalyst molecule M undergoes cyclic conformational changes when supplied with the 'fuel' – the complex QL' and strand L, the complementary strand to L'. Free Q and the LL' duplex represent the 'exhaust'. From A. J. Turberfield, J. C. Mitchell *et al.*, *Phys. Rev. Lett.* **90** (2003) 118102

9.2.2 Molecular motors

Some of the most striking nanoscale machines to be found in biology are molecular motors, nanoscale assemblies of proteins that convert chemical energy directly to mechanical energy, usually with remarkable efficiency. These motors are found in simple, single-celled prokaryotes such as the bacteria *E.coli* and in complex, multicellular eukaryotic organisms such as human beings, and they perform a wide variety of functions. These functions include propelling single-celled organisms, moving materials and structures around within eukaryotic cells, and providing the power for the muscles of multicellular animals. Many different molecular motors have evolved during the history of life, and here we consider only a few representative examples.

The study of biological molecular motors is clearly of central importance to molecular biology and biophysics. Their importance for nanotechnology is twofold. Firstly, they offer us some remarkable models to emulate. Biological motors have evolved to a considerable pitch of perfection, and the way they operate – exploiting distinctive features of the nanoworld such as the dominating presence of Brownian motion and the strength of surface forces – and the way they operate gives us valuable clues about the design rules we would need to follow in creating synthetic molecular motors. Secondly, they offer us working, off-the-shelf components that can be disassembled from the cells in which they are found and reassembled as part of hybrid structures combining synthetic and biological nanotechnology.

Before we discuss the operational mechanism of biological motors in detail, it is worth making some general comments. Biological motors are quite unlike the heat engines which we are familiar with on the macroscale. Like a petrol engine, a biological motor converts the chemical energy of a fuel into useful mechanical work. But the resemblance ends there. In a heat engine the chemical energy is first converted into heat, and then work is extracted from the flow of heat energy from a hot reservoir to a cold reservoir. In a biological motor, energy is converted from chemical energy to mechanical work directly in conditions of constant temperature. The mechanism of this conversion relies on the coupling of a cyclic chemical reaction – often the hydrolysis of the energy storage molecule adenosine triphosphate (ATP) to adenosine diphosphate (ADP) and a phosphate ion – to a conformational change that occurs in the motor protein in response to this reaction.

It is worth stressing at the outset that biological motors can reach astonishingly high efficiencies, despite the apparent difficulties of the nanoscale environment in which they operate. To someone steeped in the tradition of macroscale mechanical engineering, the prospect of making a motor that works in conditions of very high dissipation, with the constant agitation of Brownian motion, with surfaces that are prone to stick together, using components which lack the rigidity that we would assume was a prerequisite for any kind of sensible design, seems very unpromising. Biological motors succeed because they exploit these special features of the nanoworld.

Moving on to specifics, we can distinguish between linear motors, in which a molecule moves along a track, and rotary motors, which generate a spinning motion. Linear motors include various types of myosin motor, including the motors that drive our muscles, and the kinesin motors that are used for transporting organelles within cells. Rotary motors include the bacterial flagellar motor, which bacteria use to swim with, and ATP synthase, a remarkable and complex enzyme that synthesises the energy storage molecule ATP.

9.2.2.1 The operation of biological motors

The detailed way in which biological motors operate is now beginning to be understood in some detail, largely thanks to single-molecule experiments of the kind introduced in Section 8.1.2, together with detailed structural studies using X-ray diffraction and high-resolution electron microscopy. A good review for the linear motor systems myosin and kinesin is provided by Vale and Milligan.

In these linear motors, there are two key elements. There is a binding site, to which the energy storage molecule ATP binds, and there is a sticky patch which reversibly attaches the motor protein to the linear track along which it runs – actin filaments in the case of myosin, microtubules in the case of kinesin. In the case of myosin, the cycle begins with the binding of an ATP molecule to the ATP binding site. With ATP bound, the association between myosin and the actin filament is at its weakest, and the myosin head becomes detached from the track. The next stage is the hydrolysis of ATP to ADP and a phosphate group. In this state the myosin head binds to an actin filament. The phosphate group then leaves the catalytic site; this triggers a substantial conformational change in the myosin molecule. The force generated by this conformational change provides the power stroke of the motor. Dissociation of the ADP molecule soon follows, after which another ATP molecule binds to the catalytic site, the myosin becomes detached from the actin filament and the cycle begins again.

This same coupling of conformational change and chemical reaction also underlies the operation of rotary motors. The smallest rotary motor, as well as the most important, is the enzyme ATP synthase. A ubiquitous feature of all life, it uses the energy stored in a gradient of hydrogen ions to synthesise the energy-containing molecule adenosine triphosphate (ATP) from adenosine diphosphate (ADP) and a phosphate ion, and generates rotational motion in the process. Alternatively, the machine can be run backwards, using ATP as a fuel to pump hydrogen ions across a membrane. This complex assembly of proteins has two main parts. The hydrogen ion pump is called F_0; this sits in a membrane and includes the channel through which protons move. The unit that either synthesises or hydrolyses the ATP is called F_1; this consists of six subunits connected to the F_0 unit by a fixed stalk. Threaded through the middle of the six subunits of F_0 is a rotating shaft. When ATP synthase operates as a pump, conformational changes coupled to the hydrolysis of ATP in the six subunits of the F_1 component cause the shaft to rotate. The rotation of the part of the shaft that threads the F_0 component pumps hydrogen ions through the ion channel. Conversely, when the machine is synthesising ATP, it is the motion of hydrogen ions through the ion channel that causes the rotation of the shaft.

The performance of ATP synthase is astonishingly good. When it operates as a pump, each bound ATP molecule leads to a stepwise 120° rotation. The energy of hydrolysis of ATP is converted into a strain energy of about $24\,k_B T$, which is converted into mechanical energy with approaching 100% efficiency, allowing the motor to develop a torque of up to $45\,\mathrm{pN\,nm^{-1}}$. This high efficiency underlines the most important point to be made about biological motors. They are not heat engines, so their operating efficiency is not bounded by the Carnot cycle limit of classical thermodynamics. Instead they exploit Brownian motion and molecular conformational change to achieve extremely high performance; evolution has led to designs that are finely optimised for operation in the nanoscale environment.

9.2.2.2 Biological motors as components of synthetic systems

The remarkably good performance of biological motors makes the idea of using them in synthetic nanodevices extremely attractive; currently no synthetic molecular motors exist with anything like the performance of these natural machines. The drawbacks are that these systems need to operate in something approximating physiological conditions. Progress towards integrating biological molecular motors into synthetic nanodevices is currently at the proof-of-principle stage rather than the device creation stage. Nonetheless, the important principle these demonstrations establish is that biological motors can be removed from their biological contexts and operated with the same high efficiencies attained in the cell environment.

Perhaps the best-developed scheme for using biological motor proteins in synthetic environments has emerged from techniques developed to characterise protein linear motors, known as motility assays. For example, in a gliding assay, a layer of linear motor proteins is immobilised on a surface. The most commonly used motors to date have been kinesins. Microtubules are then introduced into the solution in contact with the surface. The density of motor proteins on the surface needs to be sufficiently high, and the microtubules need to be sufficiently long, so that at least three kinesin molecules attach to each microtubule. Then if fuel in the form of ATP molecules is supplied to the solution, the microtubules will move continuously across the surface, powered by the kinesin motors.

One can imagine kinesin-coated surfaces such as this serving as the basis for molecular shuttles, moving molecules around from place to place in a nanoscale chemical plant, for example. But to achieve this goal, two problems need to be overcome. Firstly, one needs to find a way of guiding the microtubules to direct their motion to where their cargos are needed, and secondly, one needs to find a way of loading and unloading the cargo onto the microtubules.

Suggested methods for guiding the motion of microtubules include selectively adsorbing the motor proteins on predetermined 'tracks'. This could be done by writing lines on the surface that would selectively adsorb the motor proteins using electron beam lithography, photolithography or a soft lithography technique. Alternatively, topography could be used; if the motor proteins were adsorbed to the bottoms of physical channels, the motion of the microtubules could be constrained to move only in the directions defined by the channels. Another approach to steering the motion of microtubules would use an externally applied field. Electric fields could be applied by means of a patterned array of electrodes on the surfaces, while magnetic fields might be practical if the microtubules were decorated with magnetic microspheres.

Binding of cargo molecules to the microtubules could be carried out by specific protein–ligand bonding pairs. One demonstration of this principle involved treating the microtubules with biotin, and attaching a streptavidin-coated bead to the microtubule using the specific interaction between streptavidin and biotin. However, it is probably fair to say that a great deal of work remains to be done before the gliding assay can be converted into a mechanism for directed nanoscale transport in useful devices.

One other spectacular demonstration of the integration of biological motors with synthetic systems involves the rotary motor part of ATP synthase, F1-ATPase (Figure 9.10). An array of nanostructured nickel posts was made by electron beam lithography, and self-assembly was used to mount a rotary motor on each post. Finally, nickel

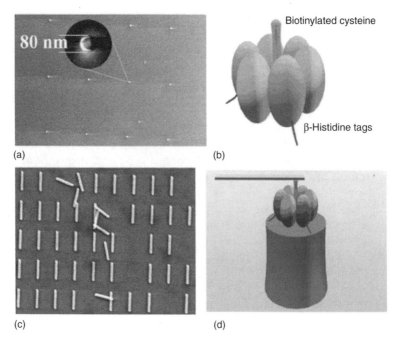

Figure 9.10 Schematic diagram of an F1-ATPase biomolecular motor-powered nanomechanical device, consisting of (A) an Ni post (height 200 nm, diameter 80 nm), (B) the F1-ATPase biomolecular motor, and (C) a nanopropeller (length 750–1400 nm, diameter 150 nm). The device (D) was assembled using sequential additions of individual components and differential attachment chemistries. From R. K. Soong, *et al.*, *Science* **290**, 1555 (2000)

nano propellers were attached to each motor using the specific binding between streptavidin and biotin. In the resulting structure, the propellers turned when exposed to a solution of ATP.

9.2.3 Artificial photosynthesis

There are very few situations in nature where coherent electron transport is important, so molecular electronics does not have many biological analogues. The vital exception to this rule is in photosynthesis, where highly optimised complexes of proteins and dye groups efficiently harvest light energy and convert it into chemical energy. Considerable efforts have been made to understand this process, and on the basis of this understanding, attempts are being made to replicate the process synthetically. In some cases only the broad outline of the design philosophy is used to make novel solar cells, whereas in other cases photosynthesis is copied more directly.

Photosynthesis takes place in structures that are enclosed in membranes (Figure 9.11). In the simplest photosynthesising systems, purple bacteria, the photosynthetic membrane is located just within the cell wall and encloses the whole cell, whereas in green plants, photosynthesis takes place in the membranes of specialised organelles called chloroplasts.

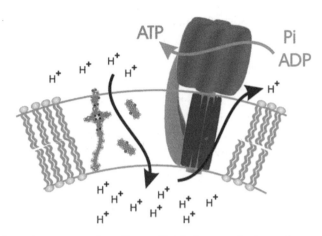

Figure 9.11 Schematic representation of an artificial photosynthetic membrane. The lipid bilayer of a liposome vesicle contains the components of a light-driven proton pump, which when illuminated leads to transport of hydrogen ions into the liposome interior. The resulting hydrogen ion gradient drives the production of ATP in ATP synthase assemblies inserted in the membrane. From D. Gust, T. A. Moore *et al.*, *Acc. Chem. Res.* **34**, 40 (2001)

At the heart of the photosynthetic system is the dye molecule chlorophyll (or bacterio-chlorophyll in purple bacteria). Light energy is converted here into an electron–hole pair, an exciton. To harvest the energy of light, this exciton needs to be separated before the electron and hole have a chance to recombine. This is accomplished by the transfer of the electron through three different dye molecules, all of which are held in a well-defined spatial relationship with each other, with the chlorophyll molecules and with the membrane in a large complex of protein molecules – the photosynthetic reaction centre. The electron is finally transferred to a mobile dye molecule, quinone. This initiates a series of reactions which have the ultimate effect of oxidising a source of hydrogen and creating a higher concentration of hydrogen ions inside the the membrane than outside. This hydrogen ion gradient is then used to drive membrane-bound ATP-synthase complexes, storing the energy in the form of ATP, which can then fuel all the other biochemical processes the cell needs to carry out.

In the simplest photosynthetic bacteria, the source of the hydrogen ions is the reduction of hydrogen sulfide. But with profound consequences for the earth, evolution devised a modified scheme involving two photosynthetic systems running in tandem, whereby hydrogen ions could be obtained from water, leaving oxygen as a by-product. This is the process used by green plants.

One final refinement that should be mentioned is that the efficiency of photosynthesis is greatly increased by the use of *light-harvesting complexes*. These complexes consist of a great many dye molecules, perhaps tens or hundreds, bound together with proteins in a precise spatial relationship. The different dye molecules absorb light in different wavelength bands, and the resulting exciton is passed extremely rapidly through the complex until it reaches the chlorophyll molecules of the reaction centre. In this way, the efficiency of photosynthesis can be maximised by ensuring that all wavelengths of light are efficiently converted into electrical energy.

Photosynthesis offers a powerful model for the conversion of solar energy into electrical and chemical energy. Attempts to emulate it synthetically fall into two categories; those systems in which the mechanisms of photosynthesis are copied rather faithfully, and systems in which only the most general operating principles are emulated.

The first category includes synthetic molecules, typically derivatives of porphyrins, sometimes bound to C_{60} fullerene, which can be inserted into a lipid bilayer. Using the spatially extended conjugation of the molecules, an exciton generated by the absorption of light can be split into an electron and a hole, and the energy used to pump hydrogen ions across the membrane. If these molecules are inserted into the membrane of a lipid vesicle or liposome, then the interior of the vesicle will be steadily made more acid on exposure to light. This requires, of course, that these asymmetric molecules are all inserted into the membrane with the same sense. The hydrogen ion gradient can then be used to power ATP synthase complexes, resulting in the storage of the light energy in ATP.

This synthetic scheme follows the biological example very closely, and results in the synthesis of the biological energy-carrying molecule ATP. At the other end of the spectrum are schemes which borrow only the most general concepts from photosynthesis. The most well developed of these are the photovoltaic and photoelectrochemical devices employing titanium dioxide sensitised by the surface adsorption of dyes, invented by Grätzel. The analogy between these important devices and photosynthesis lies purely at a conceptual level. There are two important processes that must take place in a photovoltaic device. The energy of a photon must be absorbed in the creation of a bound electron–hole pair (an exciton), and then the exciton must be split into a separate electron and hole, which are then transported to their respective electrodes. In a conventional photovoltaic, both processes take place in a bulk semiconductor. As we have seen, photosynthesis relies on excitons being formed in dye molecules highly optimised for the purpose. Grätzel cells similarly separate the processes of exciton formation and charge transport; light is absorbed by a dye that is absorbed at the surface of a nanostructured wide band gap semiconductor (typically titanium dioxide). An electron is injected into the TiO_2, and the circuit is completed by a hole-carrying electrolyte or p-type semiconductor. The advantage of Grätzel cells over conventional semiconductor photovoltaic cells is that they combine reasonable efficiencies with the potential for low-cost solution-based processing.

9.3 CONCLUSIONS

Nanotechnology has provided us with a multitude of new tools to explore biological systems on small length scales. The opening up of single-cell and single-molecule phenomena to experimental investigation represents an important step forward, both conceptually and philosophically, and promises to yield important new insights into how biological systems are assembled. The availability of miniaturised systems for use in molecular separation and analysis will not only catalyse the exploration of single-molecule phenomena but will also provide important new methodologies by which the genome revolution may ultimately be translated into practical reality through an enhanced understanding of disease and through the development of new therapeutic procedures.

Cell biology, on the other hand, offers nanotechnology some remarkable exemplars of nanoscale devices and machines. These can be exploited directly in hybrid systems, or used as inspiration for synthetic devices drawing on the same physical principles.

BIBLIOGRAPHY

General

P. Ball, Natural strategies for the molecular engineer. *Nanotechnology* **13**, R15–R28 (2003).
R. A. L. Jones, *Soft Machines*, Oxford University Press, Oxford (2004).

DNA nanotechnology

E. Braun, Y. Eichen *et al.*, DNA-templated assembly and electrode attachment of a conducting silver wire. *Nature* **391**, 775–778 (1998).
K. Keren, R. S. Berman *et al.*, DNA-templated carbon nanotube field-effect transistor. *Science* **302**, 1380–1382 (2003).
C. A. Mirkin, Programming the assembly of two- and three-dimensional architectures with DNA and nanoscale inorganic building blocks. *Inorganic Chemistry* **39**, 2258–2272 (2000).
C. M. Niemeyer, Nanoparticles, proteins, and nucleic acids: biotechnology meets materials science. *Angewandte Chemie International Edition* **40**, 4128–4158 (2001).
C. M. Niemeyer and M. Adler, Nanomechanical devices based on DNA. *Angewandte Chemie International Edition* **41**, 3779–3783 (2002).
N. C. Seeman and A. M. Belcher, Emulating biology: building nanostructures from the bottom up. *Proceedings of the National Academy of Sciences of the United States of America* **99**, 6451–6455 (2002).
A. J. Turberfield, J. C. Mitchell *et al.*, DNA fuel for free-running nanomachines. *Physical Review Letters* **90**, 118102 (2003).

Molecular motors

H. Hess and V. Vogel, Molecular shuttles based on motor proteins: active transport in synthetic environments. *Reviews in Modern Biotechnology* **82**, 67–85 (2001).
R. K. Soong, G. D. Bachand *et al.*, Powering an inorganic nanodevice with a biomolecular motor. *Science* **290**, 1555–1558 (2000).
R. D. Vale and R. A. Milligan, The way things move: looking under the hood of molecular motor proteins. *Science* **288**, 88–95 (2000).

Photosynthesis

D. Gust, T. A. Moore *et al.*, Mimicking photosynthetic solar energy transduction. *Accounts of Chemical Research* **34**, 40–48 (2001).
M. Grätzel, (2001) Photoelectrochemical cells. *Nature* **414**, 338–344 (2001).

Index

Nanoscale Science and Technology   Edited by R. W. Kelsall, I. W. Hamley and M. Geoghegan
© 2005 John Wiley & Sons, Ltd